Birkhäuser

Operator Theory: Advances and Applications

Volume 306

Founded in 1979 by Israel Gohberg

Albrecht Böttcher • Oleksiy Karlovych •
Eugene Shargorodsky • Ilya M. Spitkovsky
Editors

Achievements and Challenges in the Field of Convolution Operators

The Yuri Karlovich Anniversary Volume

Birkhäuser

Editors
Albrecht Böttcher
Fakultät für Mathematik
TU Chemnitz
Chemnitz, Germany

Oleksiy Karlovych
Faculdade de Ciências e Tecnologia,
Departamento de Matemática
Universidade Nova de Lisboa
Caparica, Portugal

Eugene Shargorodsky
Department of Mathematics
King's College London
London, UK

Ilya M. Spitkovsky
Division of Science and Mathematics
New York University Abu Dhabi
Abu Dhabi, United Arab Emirates

ISSN 0255-0156 ISSN 2296-4878 (electronic)
Operator Theory: Advances and Applications
ISBN 978-3-031-80485-4 ISBN 978-3-031-80486-1 (eBook)
https://doi.org/10.1007/978-3-031-80486-1

Mathematics Subject Classification: 01A70, 45E10, 47B35, 05C50, 15B05, 47B36, 15A18, 41A60, 65F15, 82B20

This book is published under the imprint Birkhäuser, www.birkhauser-science.com by the registered company Springer Nature Switzerland AG
The registered company address is: Gewerbestrasse 11, 6330 Cham, Switzerland

Preface

This volume celebrates Yuri Karlovich's remarkable five-decade-long mathematical career, honoring his 75th birthday on April 18, 2024. He is a renowned expert in various branches of Analysis. During his fruitful career, Yuri published more than 190 research papers, many of them with coauthors, whose list contains more than 50 colleagues from several countries. He coauthored four research monographs and supervised eight PhD theses.

The volume consists of two parts. The first part contains biographical information about Yuri, his list of publications, and personal reminiscences by one of us. To illustrate that Yuri is an outstanding team-worker, we included the reprint of a paper that was published in 1996 and was, in a sense, the culmination of a collaborative endeavor of seven enthusiasts working on the so-called "N projections theorem."

The larger second part of the volume consists of nine research papers on topics in operator theory and its applications by colleagues many of whom collaborated with Yuri or were in some other way influenced by his work. We thank all the authors who contributed to this volume for their efforts as well as the referees who in many cases generously helped to improve the manuscripts. We dedicate this volume to Yuri, with gratitude for the many things we learned from him, and we wish him many fruitful years to come.

Chemnitz, Germany — Albrecht Böttcher
Lisboa, Portugal — Oleksiy Karlovych
London, UK — Eugene Shargorodsky
Williamsburg, VA, USA — Ilya Spitkovsky
October, 2024

Contents

Part I
Yuri Karlovich and His Work

Yuri Karlovich's Path in Mathematics

Oleksiy Karlovych and Eugene Shargorodsky

Yuri Ivanovich Karlovich was born on April 18, 1949 in Leningrad, Soviet Union (now Saint Petersburg, Russia). His father, Ivan Grigorievich Karlovich, was a military officer. His mother, Olga-Matilde Davidovna Karlovich, was a housewife.

In 1966, Yuri Karlovich graduated from the secondary school No. 33 (Odessa, Ukraine) with a gold medal. During the ten years of study at school, he participated in and was a winner of several regional and republican olympiads in mathematics, physics and chemistry, and then chose mathematics for his further education under the influence of his school mathematics teacher Leonid Georgievich Zhbankov.

After graduating from school, Yuri Karlovich became a student of the Department of Mathematics and Mechanics of the Odessa State University. He completed his Master of Science degree with Distinction in 1971. From his third year at the University, he was a regular participant of the weekly Odessa City Seminar on boundary value problems and singular integral equations led by Professor Georgii Semenovich Litvinchuk. This seminar, which was active for more than 25 years, had a big influence on the scientific careers of its participants. In particular, Yuri Karlovich started to study singular integral operators with shifts under the supervision of Professor Litvinchuk.

O. Karlovych
Centro de Matemática e Aplicações, Departamento de Matemática, Faculdade de Ciências e Tecnologia, Universidade Nova de Lisboa, Caparica, Portugal
e-mail: oyk@fct.unl.pt

E. Shargorodsky (✉)
Department of Mathematics, King's College London, London, United Kingdom
e-mail: eugene.shargorodsky@kcl.ac.uk

A. Böttcher et al. (eds.), *Achievements and Challenges in the Field of Convolution Operators*, Operator Theory: Advances and Applications 306,
https://doi.org/10.1007/978-3-031-80486-1_1

From 1974 to 2009, Yuri Karlovich worked at institutions of the National Academy of Sciences of Ukraine: first, at the Department of Mathematical Modeling of Economical and Ecological Systems of the Odessa Branch of the Institute of Economics in 1974–1986, and then at the Department of Integral Equations and Boundary Value Problems of Hydroacoustics in the Hydroacoustic Branch of the Marine Hydrophysical Institute (Odessa) from 1986, taking subsequently the positions of Researcher in 1975, Senior Researcher in 1976 and Leading Researcher in 1992. In that period he was involved in the development of mathematical models and applications of mathematical methods to the study of concrete systems in economy, ecology, and hydroacoustics. This activity resulted in a cycle of papers in applied mathematics, e.g., [68, 104, 105, 117] and led to the book [116].

In 1998, Yuri was awarded, for the successful scientific work over many years, a Medal commemorating the 80th Anniversary of the National Academy of Sciences of Ukraine. He maintained scientific relations with colleagues in the Hydroacoustic Branch of the Marine Hydrophysical Institute (Odessa) until 2009.

In February 1975, Yuri defended his PhD thesis *Singular integral operators with a shift of the contour of integration in the domain and their applications* (1974, 111 pp.) at the Odessa State University under the supervision of Professor Litvinchuk (Fig. 1). The PhD degree in Physics and Mathematics (theory of functions and functional analysis) was approved by the Higher Attestation Commission in Moscow in February 1976.

In the subsequent years, Yuri jointly with Viktor Kravchenko studied Banach algebras of singular integral operators with piecewise continuous coefficients and

Fig. 1 Yuri Karlovich during his PhD defense, February 1975

a non-Carleman shift (that is, a shift generating an infinite cyclic group) on Lebesgue spaces. At that time, a big group of mathematicians were involved in this topic: A.B. Antonevich and A.V. Lebedev, A.G. Myasnikov and L.I. Sazonov, V.N. Semenyuta and A.B. Khevelev, A.P. Soldatov. During 1976–1994, Yuri Karlovich and Viktor Kravchenko wrote a series of important papers on the Fredholm symbol calculi for such Banach algebras, establishing Fredholm criteria and index formulas for the operators in these algebras (see the six papers [40–45] in Soviet Math. Dokl., [46] in Math. USSR Sbornik, [69] in Differential Equations, and [70] in Math. USSR Izvestiya). The first local principle for studying nonlocal Banach algebras with shifts and discontinuous coefficients was developed in their work. It allowed them to construct a noncommutative Fredholm symbol calculus for these algebras. These investigations are described in the important survey [106]. They made a significant contribution to the book [115].

This line of investigation of Banach algebras of singular integral operators with discrete groups of shifts was continued by Yuri and enabled him to extend the existing local principles to nonlocal Banach algebras as well as to construct Fredholm symbol calculi for Banach algebras of singular integral operators with discrete groups of shifts and various classes of discontinuous coefficients on Lebesgue spaces. Fredholm criteria were also established.

Later, in 1983, Yuri Karlovich and Ilya Spitkovsky began studying almost periodic (AP) factorization of matrix-valued functions and its applications to singular integral operators with matrix semi-almost periodic coefficients on Lebesgue spaces and Wiener-Hopf operators with semi-almost periodic matrix symbols on Lebesgue spaces and spaces of Bessel potentials. Convolution type operators on a union of intervals were also studied. These investigations were inspired by results of R.V. Duduchava and A.I. Saginashvili on the scalar analogues of such operators. Yuri and Ilya studied two-sided and one-sided invertibility of the mentioned operators with almost periodic matrix data as well as Fredholmness, index formulas and $n(d)$-normality. In contrast to the scalar AP functions of the Wiener class, APW, for the matrix APW functions, APW factorizations might not always exist, and studying the existence of AP (APW) factorization is a difficult problem related to arbitrary real-valued partial AP indices. On this topic, Yuri and Ilya wrote the papers [94, 96] in Soviet Math. Dokl. and [97] in Math. USSR Izvestiya, and also the manuscript [95].

With boundless enthusiasm Yuri continued developing a local-trajectory machinery applicable to nonlocal C^*-algebras associated with C^*-dynamical systems. It may be regarded as a nonlocal analogue of the Allan-Douglas local principle and it turned out to be a right tool for establishing isomorphism theorems for C^*-algebras (see [6, 48] and [55]). The local-trajectory method allows one to study the invertibility of elements of C^*-algebras in terms of invertibility of their local representatives. This method is based on a close relation between C^*-algebras associated with C^*-dynamical systems and the crossed products of C^*-algebras and groups of their automorphisms. As a result, faithful representations of several nonlocal C^*-algebras were obtained (see, e.g., [49]). In concrete applications, the local-trajectory method and isomorphism theorems give a powerful and convenient machinery for studying C^*-algebras of nonlocal type operators with discontinuous data.

An isomorphism theorem and a local-trajectory method applicable to C^*-algebras $\mathcal{A}$ extended by subexponential or admissible groups G of unitary elements U_g ($g \in G$) generating the automorphisms $a \mapsto U_g a U_g^*$ of $\mathcal{A}$ were elaborated by A.B. Antonevich, V.V. Brenner, and A.V. Lebedev (see the book [1]). The isomorphism theorem in the case of subexponential discrete groups G was proved making use of an estimate for the growth of the number of words of length n. Yuri extended those results to arbitrary C^*-algebras $\mathcal{A}$ with non-trivial central subalgebras $\mathcal{Z}$ and arbitrary amenable discrete groups G.

Taking advantage of the isomorphism of the C^*-algebra of singular integral operators with $n \times n$ almost periodic matrix coefficients acting on the Lebesgue space $L_n^2(\mathbb{R})$ and a C^*-algebra of operators acting on the Besicovitch space B_n^2, Yuri proved the crucial result on the equivalence of the invertibility of the Wiener-Hopf operators $W(a)$ with matrix symbols $a \in APW$ and the existence of a canonical right APW factorization for a. We should emphasize that isomorphism theorems form a bridge between algebras of singular integral operators with shifts and convolution operators with semi-almost periodic symbols because convolution operators with symbols $e^{i\lambda x}$ are translation operators.

Yuri also generalized the local method of studying the Fredholmness of singular integral operators with infinite cyclic groups of shifts, which was developed jointly with V.G. Kravchenko in [45, 70]. Applying the separation of arbitrary finite sets of different orbits, Yuri was able to study the Fredholmness of operators in Banach algebras of singular integral operators with discrete subexponential groups of shifts on the Lebesgue spaces L^p with the help of an essential modification of an earlier local principle [45] for nonlocal Banach algebras (see [50] and its further treatment and application [92, 93] developed jointly with B. Silbermann).

Yuri also studied the invertibility of functional operators and the Fredholmness of singular integral operators with a non-Carleman shift on generalized Hölder spaces [47, 79, 80, 103] and reflexive Orlicz spaces [2] (also see [51, 66, 107]). It was established that weighted shift operators on such spaces have massive spectra.

In total, Yuri with coauthors published 16 papers in Soviet Math. Dokl. and 3 papers in Doklady Mathematics.

The results described above were included in Yuri's Doctoral (Habilitation) thesis *Algebras of convolution type operators with discrete groups of shifts and oscillating coefficients*, Odessa, 1990, 380 pp. (Russian), which was defended at the Andrea Razmadze Mathematical Institute (Tbilisi, Georgia) in October 1991. The Doctor of Science degree in Physics and Mathematics (theory of functions and functional analysis) was awarded by the Higher Attestation Commission in Moscow in January 1992 (a month after the dissolution of the Soviet Union).

Along with his work at the Hydroacoustic Branch of the Marine Hydrophysical Institute, Yuri also delivered lectures in mathematics at the Odessa State University from 1990 to 1993 as an Associate Professor (1990–1992) and a Full Professor (1992–1993).

In December 1993, Yuri was invited by Albrecht Böttcher to Chemnitz Technical University (Chemnitz, Germany) as a Visiting Professor. Two and half years in Chemnitz were a very fruitful time for Yuri due to numerous mathematical discussions with Albrecht Böttcher, Bernd Silbermann and others members of their seminar. During this time, Yuri and Albrecht constructed the Fredholm theory for Banach algebras of singular integral operators with piecewise continuous (PC) coefficients on weighted Lebesgue spaces over general Carleson (Ahlfors-David) curves with general Muckenhoupt weights (see monograph [20] and also Albrecht's article included in this volume). As is known, the Cauchy singular integral operator is bounded on the weighted Lebesgue space $L^p(\Gamma, w)$ with $p \in (1, \infty)$ if and only if Γ is a Carleson curve and w is a Muckenhoupt weight.

Until 1990, the Fredholm theory for singular integral operators (SIO's) with PC coefficients on the spaces $L^p(\Gamma, w)$ was known for piecewise smooth curves Γ and power (Khvedelidze) weights w mainly due to results obtained by H. Widom, I.B. Simonenko, I. Gohberg and N. Krupnik. The local spectrum of such operators at every discontinuity point t of the coefficients is a circular arc the shape of which depends on p and the exponent of the power weight at the point t. In 1990, Ilya Spitkovsky proved that in the case of a piecewise smooth curve Γ, the local spectrum of SIO's at every discontinuity point of the coefficients on the space $L^p(\Gamma, w)$ with a general Muckenhoupt weight is a horn with a boundary consisting of two circular arcs. Subsequently, Albrecht and Yuri completely described the shapes of these local spectra for general Carleson curves Γ and general Muckenhoupt weights w. The form of the local spectrum depends on the "interference" between the geometry of the curve Γ and the properties of the weight w. The local spectra can be spiralic horns, logarithmic leaves with a median separating point and so called general leaves. These basic results of the modern Fredholm theory for Banach algebras of singular integral operators were published in the papers [17–19, 21, 22] and in the monograph [20], which was awarded the Ferran Sunyer i Balaguer 1997 prize of the Institut d'Estudis Catalans (Barcelona, Spain).

In connection with the study of the structure of Banach algebras of SIO's, the search for an appropriate N idempotents theorem was a hot topic in the 1990s. It is therefore no surprise that such a theorem was independently and almost simultaneously established by three groups: A. Böttcher, Yu. Karlovich, and I. Spitkovsky forming one of them, I. Gohberg and N. Krupnik being the second, and S. Roch and B. Silbermann constituting the third. The outcome of the effort of these seven enthusiasts was summarized in paper [25], which was an important cornerstone in the entire development and which is reprinted in this volume.

The collaboration of Yu. Karlovich, A. Böttcher and V.S. Rabinovich started in Chemnitz and originated from the idea to use Mellin pseudodifferential operators in studying SIO's on weighted Lebesgue spaces with Muckenhoupt weights over Carleson curves. This approach allowed them to clarify the nature of local spectra for SIO's and to discover new applications of the limit operators method (see the papers [26–28, 89]).

The stay in Chemnitz allowed Yuri to renew his previous investigations with Ilya Spitkovsky of matrix AP factorability and its applications to convolution type operators [98–101]. Albrecht, Yuri and Ilya founded a research group under the name *Toeplitz operators and algebras of convolution type operators*, which was supported by a NATO Collaborative Research Grant from 1995 to 1999. Their joint results on convolution type operators with semi-almost periodic matrix coefficients or symbols, related to the interesting and difficult problem of the matrix AP factorization, were published in the papers [24, 29, 31] and the monograph [30].

From December 1996 to May 1999, Yuri continued his research at the Instituto Superior Técnico (Lisbon, Portugal) as a Visiting Professor and a member of the group *Operator Factorization and Applications to Physics and Mathematics*, supported by JNICT (Portugal), 1996–1998. The collaboration of Yuri with A.F. dos Santos, M.A. Bastos, and I.M. Spitkovsky on problems of matrix factorization, its relations with the corona problem, and applications to the operator theory resulted in the cycle of papers [13–16, 92] (see also [3]).

In May 1999, Yuri moved to Mexico at the invitation by Nikolai Vasilevski for a research stay at the Department of Mathematics, CINVESTAV del I. P. N., Mexico City, supported by CONACYT (México) grant, Cátedra Patrimonial de Excelencia Nivel II, 1999–2001. During that time, Yuri studied shift-invariant algebras of SIO's with oscillating coefficients and singular integral operators with fixed singularities [89, 90], the invertibility of functional operators and the compactness of commutators [66, 67, 108] (Fig. 2).

From January 2002, Yuri is Full Professor at the Autonomous University of the State Morelos in Cuernavaca, Mexico. He has a wide range of interests in real, complex and functional analysis, as well as in operator theory. His areas of interest include studying the invertibility of functional operators with shifts and the Fredholmness of singular integral operators with discrete groups of shifts and various classes of discontinuous coefficients (piecewise slowly oscillating and piecewise quasicontinuous), including Banach algebras of these operators. Such operators and their C^*-algebras were studied, e.g., in [7–9, 23, 55] in the case of piecewise slowly oscillating coefficients and in [10–12] in the case of piecewise quasicontinuous coefficients. The Fredholm theory for Banach algebras of SIO's with shifts and piecewise slowly oscillating coefficients was constructed in the cycle of papers including, e.g., [56, 109, 110, 112–114].

Investigations of C^*-algebras of Bergman and poly-Bergman type operators with piecewise continuous coefficients (see [85–88]) and C^*-algebras of Bergman type operators with piecewise continuous and piecewise slowly oscillating coefficients in domains with piecewise-smooth boundaries having angles (see [36–39, 59]) are another important area of Yuri's interest. Studying one-dimensional Fourier and Mellin pseudodifferential operators with non-regular symbols admitting discontinuities in spatial and dual variables, algebras of such pseudodifferential operators with limited smoothness of their symbols, and applications of these operators is another interesting field of Yuri's investigations (see [52–54, 58, 61–63]).

The Fredholm theory of Wiener-Hopf integral operators with semi-almost periodic and slowly oscillating matrix symbols on weighted Lebesgue spaces on $\mathbb{R}_+$ with Muckenhoupt weights (see [71–73, 75]) and studying the Fredholmness of operators in the Banach algebras generated by the operators of multiplication by piecewise slowly oscillating functions and convolution operators with piecewise slowly oscillating symbols on weighted Lebesgue spaces with a subclass of

Fig. 2 Yuri Karlovich during the excursion to *Vallum Hardiani* at IWOTA 2004 in Newcastle

Muckenhoupt weights (see [74, 76, 77] and also [35, 78]) form other important investigation topics for Yuri.

Yuri also continued the study of the problem of AP factorability of matrix functions and its applications to Toeplitz and Wiener-Hopf operators (see the papers [4, 32–34, 102, 111]), the Haseman boundary value problem with discontinuous data [57, 60, 65, 91], C^*-algebras of two-dimensional singular integral operators with shifts [82, 84] (see also [83]), singular integral operators on curves with cusps [81], and index formulas [5, 64].

It is now for decades that Yuri has been continuing his collaboration with colleagues at the Mathematics Department of the Instituto Superior Técnico of the University of Lisbon. He also collaborates with colleagues at the NOVA School of Science and Technology of the NOVA University Lisbon.

Eight PhD theses were defended under the direction of Yuri Karlovich:

1. Vidady Aslanov: *Functional and Singular Integral Operators with a Shift on Orlicz Spaces*, Azerbaijan Oil Academy, Baku, Azerbaijan, 1992 (with G.S. Litvinchuk).
2. António José Vieira Bravo: *Factorization of semi-almost periodic matrix functions and convolution type operators with oscillating presymbols*, Instituto Superior Técnico, Lisboa, Portugal, 2003 (with M.A. Bastos).
3. Luís Filipe Serrazes Ventura de Barros Pessoa: *Two-dimensional singular integral operators with Bergman kernel and shifts*, Instituto Superior Técnico, Lisboa, Portugal, 2006 (with M.A. Bastos).
4. Cláudio António Rainha Aires Fernandes: *Algebras of singular integral operators with discontinuous coefficients and shifts*, Instituto Superior Técnico, Lisboa, Portugal, 2006 (with M.A. Bastos).
5. Juan Loreto Hernández: *Convolution type operators with oscillating symbols*, Institute of Mathematics, National Autonomous University of Mexico, Cuernavaca, Morelos, México, 2008.
6. Iván Loreto Hernández: *Algebras of nonlocal integral operators with discontinuous data*, Institute of Mathematics, National Autonomous University of Mexico, Cuernavaca, Morelos, México, 2013.
7. Enrique Espinoza Loyola: *Algebras of Bergman type operators with discontinuous data in domains with angles*, Institute of Mathematics, National Autonomous University of Mexico, Cuernavaca, Morelos, México, 2019.
8. Jennyffer Rosales Méndez: *The Haseman boundary value problem with piecewise quasicontinuous coefficients*, Autonomous University of the State of Morelos, Cuernavaca, Morelos, México, 2024.

During the six years from 2009 to 2015, Yuri was the head of an international project in Mexico that included 6 scientific groups from Poland (Institute of Mathematics, University of Szczecin, Szczecin), Russia (Institute for Information Transmission Problems (IITP) of the Russian Academy of Sciences, Moscow), and Mexico: Autonomous University of the State Morelos (Cuernavaca), National Polytechnical Institute (Mexico City), CINVESTAV del I.P.N. (Mexico City), Michoacan University of San Nicolas de Hidalgo (Morelia). This project allowed him and his colleagues to organize three International Workshops in Ixtapa (Mexico).

In addition, Yuri was an organizer of four International Workshops on "Analysis, Operator Theory and Applications" in Mexico (Cancún, April 28–May 2, 2008; Ixtapa, March 1–5, 2010; Ixtapa, January 23–27, 2012; Ixtapa, February 24–28, 2014).

Yuri Karlovich is on the editorial board of the journal "Advances in Operator Theory" and a reviewer for many mathematical journals. He is also a reviewer for

Fig. 3 Evguenia and Yuri Karlovich in Cuernavaca, Mexico

the zbMATH Open. Yuri has participated with plenary and contributed talks in more than 150 international and national conferences.

In 2019, the session *Toeplitz Operators, Convolution type Operators and Operator Algebras* of the XXX International Workshop on Operator Theory and Applications (IWOTA 2019), Lisbon, July 22–26, was dedicated to Yuri Karlovich on his 70th birthday.

Yuri married Evguenia Karlovich, nee Zhila, in 1972. She is Yuri's reliable companion who always helps him in all his plans and endeavours (Fig. 3). Yuri and Evguenia have two children, son Oleksiy and daughter Anna, and 6 grandchildren.

We wish Yuri a long and happy life! We believe that his dedication to and passion for mathematics will continue for many years to come.

References

1. Antonevich, A.B.: Linear Functional Equations. Operator Approach. Oper. Theory Adv. Appl., vol. 83. Birkhäuser, Basel (1996)
2. Aslanov, V.D., Karlovich, Yu.I.: One-sided invertibility of functional operators in reflexive Orlicz spaces. Dokl. Akad. Nauk AzSSR **45**(11–12), 3–7 (1989), in Russian
3. Ball, J.A., Karlovich, Yu.I., Rodman, L., Spitkovsky, I.M.: Sarason interpolation and Toeplitz corona theorem for almost periodic matrix functions. Integral Equ. Oper. Theory **32**(3), 243–281 (1998)

4. Bastos, M.A., Bravo, A., Karlovich, Yu.I., Spitkovsky, I.M.: Constructive factorization of some almost periodic triangular matrix functions with a quadrinomial off diagonal entry. J. Math. Anal. Appl. **376**(2), 625–640 (2011)
5. Bastos, M.A., Carvalho, C.C., Karlovich, Yu.I.: Index formula for convolution type operators with data functions in alg(SO, PC). J. Math. Anal. Appl. **443**(1), 453–477 (2016)
6. Bastos, M.A., Fernandes, C.A., Karlovich, Yu.I.: C^*-algebras of integral operators with piecewise slowly oscillating coefficients and shifts acting freely. Integral Equ. Oper. Theory **55**(1), 19–67 (2006)
7. Bastos, M.A., Fernandes, C.A., Karlovich, Yu.I.: C^*-algebras of singular integral operators with shifts having the same nonempty set of fixed points. Complex Anal. Oper. Theory **2**, 241–272 (2008)
8. Bastos, M.A., Fernandes, C.A., Karlovich, Yu.I.: A nonlocal C^*-algebra of singular integral operators with shifts having periodic points. Integral Equ. Oper. Theory **71**(4), 509–534 (2011)
9. Bastos, M.A., Fernandes, C.A., Karlovich, Yu.I.: A C^*-algebra of singular integral operators with shifts admitting distinct fixed points. J. Math. Anal. Appl. **413**(1), 502–524 (2014)
10. Bastos, M.A., Fernandes, C.A., Karlovich, Yu.I.: A C^*-algebra of singular integral operators with shifts and piecewise quasicontinuous coefficients. Oper. Theory Adv. Appl. **267**, 25–64 (2018)
11. Bastos, M.A., Fernandes, C.A., Karlovich, Yu.I.: Invertibility criteria in C^*-algebras of functional operators with shifts and PQCcoefficients. Integral Equ. Oper. Theory **91**(3), Paper No. 19, 25 pp. (2019)
12. Bastos, M.A., Fernandes, C.A., Karlovich, Yu.I.: On C*-algebras of singular integral operators with PQC coefficients and shifts with fixed points. Complex Var. Elliptic Equ. **67**(3), 581–614 (2022)
13. Bastos, M.A., Karlovich, Yu.I., dos Santos, A.F.: The invertibility of convolution type operators on a union of intervals and the corona theorem. Integral Equ. Oper. Theory **42**(1), 22–56 (2002)
14. Bastos, M.A., Karlovich, Yu.I., dos Santos, A.F., Tishin, P.M.: The corona theorem and the existence of canonical factorization of triangular AP-matrix functions. J. Math. Anal. Appl. **223**(2), 494–522 (1998)
15. Bastos, M.A., Karlovich, Yu.I., dos Santos, A.F., Tishin, P.M.: The corona theorem and the canonical factorization of triangular AP-matrix functions – Effective criteria and explicit formulas. J. Math. Anal. Appl. **223**(2), 523–550 (1998)
16. Bastos, M.A., Karlovich, Yu.I., Spitkovsky, I.M., Tishin, P.M.: On a new algorithm for almost periodic factorization. Oper. Theory Adv. Appl. **103**, 53–74 (1998)
17. Bishop, C.J., Böttcher, A., Karlovich, Yu.I., Spitkovsky, I.M.: Local spectra and index of singular integral operators with piecewise continuous coefficients on composed curves. Math. Nachr. **206**, 5–83 (1999)
18. Böttcher, A., Karlovich, Yu.I.: Toeplitz and singular integral operators on general Carleson Jordan curves. Oper. Theory Adv. Appl. **90**, 119–152 (1996)
19. Böttcher, A., Karlovich, Yu.I.: Submultiplicative functions and spectral theory of Toeplitz operators. Integral Transform. Spec. Funct. **4**(1), 181–202 (1996)
20. Böttcher, A., Karlovich, Yu.I.: Carleson Curves, Muckenhoupt Weights, and Toeplitz Operators. Progress in Mathematics, vol. 154. Birkhäuser (1997)
21. Böttcher, A., Karlovich, Yu.I.: Toeplitz operators with PC symbols on general Carleson Jordan curves with arbitrary Muckenhoupt weights. Trans. Amer. Math. Soc. **351**(8), 3143–3196 (1999)
22. Böttcher, A., Karlovich, Yu.I.: Cauchy's singular integral operator and its beautiful spectrum. Oper. Theory Adv. Appl. **129**, 109–142 (2001)
23. Böttcher, A., Karlovich, Yu.I., Silbermann, B.: Singular integral equations with PQC coefficients and freely transformed argument. Math. Nachr. **166**, 113–133 (1994)
24. Böttcher, A., Karlovich, Yu.I., Spitkovsky, I.M.: Toeplitz operators with semi-almost periodic symbols on spaces with Muckenhoupt weight. Integral Equ. Oper. Theory **18**, 261–276 (1994)

25. Böttcher, A., Gohberg, I., Karlovich, Yu., Krupnik, N., Roch, S., Silbermann, B., Spitkovsky, I.: Banach algebras generated by N idempotens and applications. Oper. Theory Adv. Appl. **90**, 19–54 (1996)
26. Böttcher, A., Karlovich, Yu.I., Rabinovich, V.S.: Emergence, persistence, and disappearance of logarithmic spirals in the spectra of singular integral operators. Integral Equ. Oper. Theory **25**, 406–444 (1996)
27. Böttcher, A., Karlovich, Yu.I., Rabinovich, V.S.: The method of limit operators for one-dimensional singular integrals with slowly oscillating data. J. Oper. Theory **43**(1), 171–198 (2000)
28. Böttcher, A., Karlovich, Yu.I., Rabinovich, V.S.: Singular integral operators with complex conjugation from the viewpoint of pseudodifferential operators. Oper. Theory Adv. Appl. **121**, 36–59 (2001)
29. Böttcher, A., Karlovich, Yu.I., Spitkovsky, I.: Toeplitz operators with semi-almost-periodic matrix symbols on Hardy spaces. Acta Appl. Math. **65**(1–3), 115–136 (2001)
30. Böttcher, A., Karlovich, Yu.I., Spitkovsky, I.M.: Convolution Operators and Factorization of Almost Periodic Matrix Functions. Operator Theory: Advances and Applications, vol. 131. Birkhäuser, Basel (2002)
31. Böttcher, A., Karlovich, Yu.I., Spitkovsky, I.M.: The C^*-algebra of singular integral operators with semi-almost periodic coefficients. J. Funct. Anal. **204**(2), 445–484 (2003)
32. Câmara, M.C., Karlovich, Yu.I., Spitkovsky, I.M.: Almost periodic factorization of some triangular matrix functions. Oper. Theory Adv. Appl. **190**, 171–190 (2009)
33. Câmara, M.C., Karlovich, Yu.I., Spitkovsky, I.M.: Constructive almost periodic factorization of some triangular matrix functions. J. Math. Anal. Appl. **367**(2), 416–433 (2010)
34. Câmara, M.C., Karlovich, Yu.I., Spitkovsky, I.M.: Kernels of asymmetric Toeplitz operators and applications to almost periodic factorization. Complex Anal. Oper. Theory **7**(2), 375–407 (2013)
35. De La Cruz Rodríguez, I., Karlovich, Yu.I., Loreto-Hernández, I.: Commutators of convolution type operators with piecewise quasicontinuous data. Commun. Math. Anal. **17**(2), 131–150 (2014)
36. Espinoza-Loyola, E., Karlovich, Yu.I.: C^*-algebras of Bergman type operators with piecewise continuous coefficients over bounded polygonal domains. Oper. Theory Adv. Appl. **258**, 145–171 (2017)
37. Espinoza-Loyola, E., Karlovich, Yu.I.: C^*-algebras of Bergman type operators with piecewise continuous coefficients over domains with Dini-smooth corners. Complex Anal. Oper. Theory **13**(1), 151–192 (2019)
38. Espinoza-Loyola, E., Karlovich, Yu.I.: C^*-algebras of poly-Bergman type operators with piecewise slowly oscillating coefficients. J. Math. Sci. (N.Y.) **266**(1), 77–102 (2022)
39. Espinoza Loyola, E., Karlovich, Yu.I., Vilchis-Torres, O.: C^*-algebras of Bergman type operators with piecewise constant coefficients over sectors. Integral Equ. Oper. Theory **83**(2), 243–269 (2015)
40. Karlovič, Ju.I., Kravčenko, V.G.: A Noether theory for a singular integral operator with a shift having periodic points. Dokl. Akad. Nauk SSSR **231**(2), 277–280 (1976), in Russian. Engl. transl. in Soviet Math. Dokl. **17**(6), 1547–1551 (1976)
41. Karlovič, Ju.I., Kravčenko, V.G.: On a singular integral operator with non-Carleman shifts on an open contour. Dokl. Akad. Nauk SSSR **236**(4), 792–795 (1977), in Russian. Engl. transl. in Soviet Math. Dokl. **18**(5), 1263–1267 (1977)
42. Karlovič, Ju.I., Kravčenko, V.G.: On systems of functional and integro-functional equations with a non-Carleman shift. Dokl. Akad. Nauk SSSR **236**(5), 1064–1067 (1977), in Russian. Engl. transl. in Soviet Math. Dokl. **18**(5), 1319–1322 (1977)
43. Karlovič, Ju.I., Kravčenko, V.G.: On an algebra of singular integral operators with non-Carleman shift. Dokl. Akad. Nauk SSSR **239**(1), 138–41 (1978), in Russian. Engl. transl. in Soviet Math. Dokl. **19**(2), 267–271 (1978)
44. Karlovič, Ju.I., Kravčenko, V.G.: On integro-functional equations with a non-Carleman shift in the space $L_2^n(0,1)$. Dokl. Akad. Nauk SSSR **246**(2), 268–271 (1979), in Russian. Engl. transl. in Soviet Math. Dokl. **20**(3), 455–458 (1979)

45. Karlovič, Ju.I., Kravčenko, V.G.: On a general method of investigating operators of singular integral type with non-Carleman shift. Dokl. Akad. Nauk SSSR **253**(1), 18–22 (1980), in Russian. Engl. transl. in Soviet Math. Dokl. **22**(1), 10–14 (1980)
46. Karlovič, Ju.I., Kravčenko, V.G.: Systems of singular integral equations with a shift. Matem. Sbornik **116**(158)(1), 87–110 (1981), in Russian. Engl. transl. in Math. URSS Sbornik **44**(1), 75–95 (1983)
47. Karlovich, Yu.I.: On the invertibility of functional operators with non-Carleman shift in Hölder spaces. Differentsial'nye Uravneniya **20**(12), 2165–2169 (1984), in Russian
48. Karlovich, Yu.I.: The local-trajectory method of studying invertibility in C^*-algebras of operators with discrete groups of shifts. Dokl. Akad. Nauk SSSR **299**(3), 546–550 (1988), in Russian. Engl. transl. in Soviet Math. Dokl. **37**(2), 407–412 (1988)
49. Karlovich, Yu.I.: C^*-algebras of operators of convolution type with discrete groups of shifts and oscillating coefficients. Dokl. Akad. Nauk SSSR **302**(3), 535–540 (1988), in Russian. Engl. transl. in Soviet Math. Dokl. **38**(2), 301–307 (1989)
50. Karlovich, Yu.I.: On algebras of singular integral operators with discrete groups of shifts in L_p-spaces. Dokl. Akad. Nauk SSSR **304**(2), 274–280 (1989), in Russian. Engl. transl. in Soviet Math. Dokl.**39**(1), 48–53 (1989)
51. Karlovich, Yu.I.: The continuous invertibility of functional operators in Banach spaces. Diss. Math. **340**, 115–136 (1995)
52. Karlovich, Yu.I.: An algebra of pseudodifferential operators with slowly oscillating symbols. Proc. Lond. Math. Soc. (3) **92**(3), 713–761 (2006)
53. Karlovich, Yu.I.: Pseudodifferential operators with compound slowly oscillating symbols. Oper. Theory Adv. Appl. **171**, 189–224 (2007)
54. Karlovich, Yu.I.: Algebras of pseudodifferential operators with discontinuous symbols. Oper. Theory Adv. Appl. **172**, 207–233 (2007)
55. Karlovich, Yu.I.: Pseudodifferential operators with compound non-regular symbols. Math. Nachr. **280**(9–10), 1128–1144 (2007)
56. Karlovich, Yu.I.: Nonlocal singular integral operators with slowly oscillating data. Oper. Theory Adv. Appl. **181**, 229–261 (2008)
57. Karlovich, Yu.I.: The Haseman boundary value problem with slowly oscillating data. In: Kilbas, A.A., Rogozin, S.V. (eds.) Analytic Methods of Analysis and Differential Equations: AMADE-2006, pp. 81–110. Cambridge Scientific Publishers, Cambridge (2008)
58. Karlovich, Yu.I.: Boundedness and compactness of pseudodifferential operators with non-regular symbols on weighted Lebesgue spaces. Integral Equ. Oper. Theory **73**(2), 217–254 (2012)
59. Karlovich, Yu.I.: C^*-algebras of Bergman type operators with continuous coefficients on polygonal domains. Oper. Matrices **9**(4), 773–802 (2015)
60. Karlovich, Yu.I.: The Haseman boundary value problem with slowly oscillating coefficients and shifts. Oper. Theory Adv. Appl. **259**, 463–500 (2017)
61. Karlovich, Yu.I.: Pseudodifferential operators with compound non-regular symbols. Oper. Theory Adv. Appl. **268**, 331–353 (2018)
62. Karlovich, Yu.I.: On Mellin pseudodifferential operators with quasicontinuous symbols. Math. Methods Appl. Sci. **44**(12), 9782–9816 (2021)
63. Karlovich, Yu.: Algebras of Mellin pseudodifferential operators with quasicontinuous symbols. J. Pseudo-Differ. Oper. Appl. **13**(2), Paper No. 18, 36 pp. (2022)
64. Karlovich, Yu.I.: Index calculation for Wiener-Hopf operators with slowly oscillating coefficients. Linear Algebra Appl. **697**, 93–130 (2024)
65. Karlovich, Yu., Guerrero-Pérez, M.: The Haseman boundary value problem with matrix coefficient of semi-almost periodic type. Banach J. Math. Anal. **17**(2), Paper No. 28, 43 pp. (2023)
66. Karlovich, A.Yu., Karlovich, Yu.I.: One-sided invertibility of binomial functional operators with a shift on rearrangement-invariant spaces. Integral Equ. Oper. Theory **42**(2), 201–228 (2002)

67. Karlovich, A.Yu., Karlovich, Yu.I.: Compactness of commutators arising in the Fredholm theory of singular integral operators with shifts. In: Factorization, Singular Operators and Related Problems (Funchal, 2002), pp. 111–129. Kluwer Acad. Publ., Dordrecht (2003)
68. Karlovich, Yu.I., Katkov, V.N.: A problem of long-term planning of economical-ecological systems of exploitation of sea basins bioresources. In: The Use of Sea Resources in the National Economy, pp. 11–22. Institute of Economics, Ukrainian Academy of Sciences, Kiev (1981), in Russian
69. Karlovich, Yu.I., Kravchenko, V.G.: Singular integral equations with non-Carleman translations on an open contour. Differentsial'nye Uravn. **17**(12), 2212–2223 (1981), in Russian. Engl. transl. in Differential Equations **17**(12), 1408–1417 (1981)
70. Karlovich, Yu.I., Kravchenko, V.G.: An algebra of singular integral operators with piecewise-continuous coefficients and a piecewise-smooth shift on a composite contour. Izv. Akad. Nauk SSSR, Ser. Matem. **47**(5), 1030–1077 (1983), in Russian. Engl. transl. in Math. USSR Izvest. **23**(2), 307–352 (1984)
71. Karlovich, Yu.I., Loreto Hernández, J.: Wiener-Hopf operators with matrix semi-almost periodic symbols on weighted Lebesgue spaces. Integral Equ. Oper. Theory **62**(1), 85–128 (2008)
72. Karlovich, Yu.I., Loreto Hernández, J.: Wiener-Hopf operators with slowly oscillating matrix symbols on weighted Lebesgue spaces. Integral Equ. Oper. Theory **64**(2), 203–237 (2009)
73. Karlovich, Yu.I., Loreto Hernández, J.: Wiener-Hopf operators with oscillating symbols on weighted Lebesgue spaces. Oper. Theory Adv. Appl. **210**, 123–145 (2010)
74. Karlovich, Yu.I., Loreto Hernández, I.: Algebras of convolution type operators with piecewise slowly oscillating data. I: Local and structural study. Integral Equ. Oper. Theory **74**(3), 377–415 (2012)
75. Karlovich, Yu.I., Loreto Hernández, J.: Convolution type operators with oscillating symbols on weighted Lebesgue spaces on a union of intervals. Mathematica **54(77)**, Special Number, 104–119 (2012)
76. Karlovich, Yu.I., Loreto Hernández, I.: On convolution type operators with piecewise slowly oscillating data. Oper. Theory Adv. Appl. **228**, 185–207 (2013)
77. Karlovich, Yu.I., Loreto Hernández, I.: Algebras of convolution type operators with piecewise slowly oscillating data. II: Local spectra and Fredholmness. Integral Equ. Oper. Theory **75**(1), 49–86 (2013)
78. Karlovich, Yu.I., Loreto-Hernández, I.: C^*-algebra of nonlocal convolution type operators. J. Math. Anal. Appl. **475**(2), 1130–1161 (2019)
79. Karlovich, Yu.I., Mardiev, R.: One-sided invertibility of functional non-Carleman translation operators in Hölder spaces. Izv. Vyssh. Uchebn. Zaved., Mat. (3), 77–80 (1987), in Russian. Engl. transl. in Soviet. Math. (Iz. VUZ.) **31**(3), 106–110 (1987)
80. Karlovich, Yu.I., Mardiev, R.: One-sided invertibility of functional operators and the $n(d)$-normality of singular integral operators with translation in Hölder spaces. Differentsial'nye Uravn. **24**(3), 488–499 (1988), in Russian. Engl. transl. in Differential Equations **24**(3), 350–359 (1988)
81. Karlovich, Yu.I., Monsiváis-González, F.J.: Fredholmness of singular integral operators with complex conjugation on star-like curves with cusps. Complex Anal. Oper. Theory **17**(5), Paper No. 69, 22 pp. (2023)
82. Karlovich, Yu.I., Mozel, V.A.: On nonlocal C^*-algebras of two-dimensional singular integral operators. Oper. Theory Adv. Appl. **220**, 115–135 (2012)
83. Karlovich, Yu.I., Mozel, V.A.: C^*-algebras of Bergman type operators with piecewise continuous coefficients and shifts. Complex Var. Elliptic Equ. **57**(7–8), 841–865 (2012)
84. Karlovich, Yu.I., Mozel, V.A.: C^*-algebras of two-dimensional singular integral operators with shifts. In: Current Trends in Analysis and its Applications, Trends Math., pp. 571–581. Birkhäuser/Springer, Cham (2015)
85. Karlovich, Yu.I., Pessoa, L.: Algebras generated by the Bergman and anti-Bergman projections and by multiplications by piecewise continuous functions. Integral Equ. Oper. Theory **52**(2), 219–270 (2005)

86. Karlovich, Yu.I., Pessoa, L.V.: C^*-algebras of Bergman type operators with piecewise continuous coefficients on bounded domains. In: Begehr, H.G.W., Nicolosi, F. (eds.) More Progresses in Analysis. Proceedings of the 5th International ISAAC Congress, Catania, Italy, July 25–30, 2005, pp. 339–348. World Scientific, Hackensack, NJ (2005)
87. Karlovich, Yu.I., Pessoa, L.V.: C^*-algebras of Bergman type operators with piecewise continuous coefficients. Integral Equ. Oper. Theory **57**(4), 521–565 (2007)
88. Karlovich, Yu.I., Pessoa, L.V.: Poly-Bergman projections and orthogonal decompositions of L^2-spaces over bounded domains. Oper. Theory Adv. Appl. **181**, 263–282 (2008)
89. Karlovich, Yu.I., Ramírez de Arellano, E.: A shift-invariant algebra of singular integral operators with oscillating coefficients. Integral Equ. Oper. Theory **39**(4), 441–474 (2001)
90. Karlovich, Yu.I., Ramírez de Arellano, E.: Singular integral operators with fixed singularities on weighted Lebesgue spaces. Integral Equ. Oper. Theory **48**(3), 331–363 (2004)
91. Karlovich, Yu.I., Rosales-Méndez, J.: The Haseman boundary value problem with quasicontinuous coefficients and shifts. Complex Anal. Oper. Theory **16**(5), Paper No. 68, 28 pp. (2022)
92. Karlovich, Yu.I., Silbermann, B.: Local method for nonlocal operators on Banach spaces. Oper. Theory Adv. Appl. **135**, 235–247 (2002)
93. Karlovich, Yu.I., Silbermann, B.: Fredholmness of singular integral operators with discrete subexponential groups of shifts on Lebesgue spaces. Math. Nachr. **272**, 55–94 (2004)
94. Karlovich, Yu.I., Spitkovskiĭ, I.M.: On the Noetherian property of certain singular integral operators with matrix coefficients of class SAP and systems of convolution equations on a finite interval connected with them. Dokl. Akad. Nauk SSSR **269**(3), 531–535 (1983), in Russian. Engl. transl. in Soviet Math. Dokl. **27**(2), 358–363 (1983)
95. Karlovich, Yu.I., Spitkovskiĭ, I.M.: Factorization of Almost Periodic Matrix-valued Functions and (Semi) Fredholmness of Certain Classes of Convolution Type Equations. Odessa. Manuscript No. 4421–85, VINITI, 1985, 138 pp. (1985), in Russian
96. Karlovich, Yu.I., Spitkovskiĭ, I.M.: On the theory of systems of convolution type equations with semi-almost-periodic symbols in spaces of Bessel potentials. Dokl. Akad. Nauk SSSR **286**(4), 799–803 (1986), in Russian. Engl. transl. in Soviet Math. Dokl. **33**(1), 180–184 (1986)
97. Karlovich, Yu.I., Spitkovskiĭ, I.M.: Factorization of almost periodic matrix-valued functions and the Noether theory for certain classes of equations of convolution type. Izv. Akad. Nauk SSSR Ser. Matem. **53**(2), 276–308 (1989), in Russian. Engl. transl. in Math. USSR Izvest. **34**(2), 281–316 (1990)
98. Karlovich, Yu.I., Spitkovsky, I.M.: (Semi)-Fredholmness of convolution operators on the spaces of Bessel potentials. Oper. Theory Adv. Appl. **71**, 122–152 (1994)
99. Karlovich, Yu.I., Spitkovsky, I.M.: Factorization of almost periodic matrix functions. J. Math. Anal. Appl. **193**(1), 209–232 (1995)
100. Karlovich, Yu.I., Spitkovsky, I.M.: Almost periodic factorization: An analogue of Chebotarev's algorithm. Contemp. Math. **189**, 327–352 (1995)
101. Karlovich, Yu.I., Spitkovsky, I.M.: Semi-Fredholm properties of certain singular integral operators. Oper. Theory Adv. Appl. **90**, 264–287 (1996)
102. Karlovich, Yu.I., Spitkovsky, I.M.: Almost periodic polynomial factorization of some triangular matrix functions. Oper. Theory Adv. Appl. **202**, 337–354 (2010)
103. Karlovich, Yu.I., Tursunkulov, B.: On singular integrals with shift in a generalized Hölder space. Izv. Vyssh. Uchebn. Zaved. Mat. (3), 71–74 (1984), in Russian. Engl. transl. in Soviet. Math. (Iz. VUZ) **28**(3), 97–101 (1984)
104. Karlovich, Yu.I., Kravchenko, V.G., Litvinchuk, G.S.: On the problem of water resource quality management. In: Problems of Sea Economics, No. 5, pp. 26–32, Institute of Economics, Ukrainian Academy of Sciences, Odessa (1976), in Russian
105. Karlovich, Yu.I., Kravchenko, V.G., Litvinchuk, G.S.: Economic-mathematical aspects of the problem of water environmental protection. In: Problems of Sea Economics, No. 6, pp. 110–119, Institute of Economics, Ukrainian Academy of Sciences, Odessa (1977), in Russian

106. Karlovich, Yu.I., Kravchenko, V.G., Litvinchuk, G.S.: The Noetherian theory of singular integral translation operators. Izv. Vyssh. Uchebn. Zaved. Mat. **27**(4), 3–27 (1983), in Russian. Engl. transl. in Soviet. Math. (Iz. VUZ) **27**(4), 1–34 (1983)
107. Karlovich, Yu.I., Kravchenko, V.G., Litvinchuk, G.S.: The invertibility of functional operators on Banach spaces. In: Funktsional'no-Differentsial'nye Uravneniya, pp. 18–58. Polytechnical Institute, Perm (1990), in Russian
108. Karlovich, A.Yu., Karlovich, Yu.I., Lebre, A.B.: Invertibility of functional operators with slowly oscillating non-Carleman shifts. Oper. Theory Adv. Appl. **142**, 147–174 (2003)
109. Karlovich, A.Yu., Karlovich, Yu.I., Lebre, A.B.: Sufficient conditions for Fredholmness of singular integral operators with shifts and slowly oscillating data. Integral Equ. Oper. Theory **70**(4), 451–483 (2011)
110. Karlovich, A.Yu., Karlovich, Yu.I., Lebre, A.B.: Necessary conditions for Fredholmness of singular integral operators with shifts and slowly oscillating data. Integral Equ. Oper. Theory **71**(1), 29–53 (2011)
111. Karlovich, Yu.I., Loreto-Hernández, J., Spitkovsky, I.M.: Factorization of some triangular matrix functions and its applications. Oper. Matrices **9**(1), 1–29 (2015)
112. Karlovich, A.Yu., Karlovich, Yu.I., Lebre, A.B.: Necessary Fredholm conditions for weighted singular integral operators with shifts and slowly oscillating data. J. Integral Equ. Appl. **29**(3), 365–399 (2017)
113. Karlovich, A.Yu., Karlovich, Yu.I., Lebre, A.B.: The index of weighted singular integral operators with shifts and slowly oscillating data. J. Math. Anal. Appl. **450**(1), 606–630 (2017)
114. Karlovich, A.Yu., Karlovich, Yu.I., Lebre, A.B.: Criteria for n(d)-normality of weighted singular integral operators with shifts and slowly oscillating data. Proc. Lond. Math. Soc. (3) **116**(4), 997–1027 (2018)
115. Kravchenko, V.G., Litvinchuk, G.S.: Introduction to the Theory of Singular Integral Operators with Shift. Kluwer, Dordrecht (1994)
116. Litvinchuk, G.S., Kravchenko, V.G., Karlovich, Yu.I., Pak, E.S., Khlebnikov, E.P., Bukreev, A.N., and Katkov, V.N.: Application of Mathematical Methods to the Prognosis and Management of the Water Quality in River Basins. Naukova Dumka, Kiev (1979), in Russian
117. Litvinchuk, G.S., Karlovich, Yu.I., Katkov, V.N.: On the problem of development and distribution of reproduction, catch and processing business of bioresources at seaside region. In: Problems and Experience of Using Reserves of Biological Resources of the Azov-Black Sea Basin, pp. 19–31. Institute of Economics, Ukrainian Academy of Sciences, Kiev (1980), in Russian

List of Publications of Yuri Karlovich

Oleksiy Karlovych

Books

1. G.S. Litvinchuk, V.G. Kravchenko, Yu.I. Karlovich, E.S. Pak, E.P. Khlebnikov, A.N. Bukreev, and V.N. Katkov: *Application of Mathematical Methods to the Prognosis and Management of the Water Quality in River Basins*. Naukova Dumka, Kiev, 1979, 154 pp. (Russian).
2. Yu.I. Karlovich and I.M. Spitkovskiĭ: *Factorization of Almost Periodic Matrix-valued Functions and (Semi) Fredholmness of Certain Classes of Convolution Type Equations*. Odessa (1985). Manuscript No. 4421–85, VINITI, 1985, 138 pp. (Russian).
3. A. Böttcher and Yu.I. Karlovich: *Carleson Curves, Muckenhoupt Weights, and Toeplitz Operators*. Progress in Mathematics, vol. 154. Birkhäuser Verlag, 1997, xv+397 pp.
4. A. Böttcher, Yu.I. Karlovich, and I.M. Spitkovsky: *Convolution Operators and Factorization of Almost Periodic Matrix Functions*. Operator Theory: Advances and Applications, vol. 131. Birkhäuser Verlag, Basel, 2002, xi+462 pp.

Edited Book and Proceedings

1. G. Burlak, Yu. Karlovich, V. Rabinovich (eds.): Special issue: Proceedings of the international workshop on analysis, mathematical physics and applications,

O. Karlovych (✉)
Centro de Matemática e Aplicações, Departmento de Matemática, Faculdade de Ciências e Tecnologia, Universidade Nova de Lisboa, Caparica, Portugal
e-mail: oyk@fct.unl.pt

A. Böttcher et al. (eds.), *Achievements and Challenges in the Field of Convolution Operators*, Operator Theory: Advances and Applications 306,
https://doi.org/10.1007/978-3-031-80486-1_2

Ixtapa, México, March 1–5, 2010. Commun. Math. Anal. (2011), Conference 3, I-II, 234 pp.
2. Yu.I. Karlovich, L. Rodino, B. Silbermann, I.M. Spitkovsky (eds.): *Operator Theory, Pseudo-Differential Equations, and Mathematical Physics. The Vladimir Rabinovich Anniversary Volume*. Operator Theory: Advances and Applications, vol. 228. Birkhäuser/Springer, Basel, 2013, xxv+406 pp.
3. G. Burlak, Yu. Karlovich, V. Rabinovich (eds.): Proceedings of the international workshop "Analysis, Operator Theory, and Mathematical Physics", Ixtapa, México, January 23–27, 2012. Commun. Math. Anal. **14** (2013), No. 2, Preface, I-II, 178 pp.
4. G. Burlak, Yu. Karlovich, V. Rabinovich (eds.): Proceedings of the international workshop"Analysis, Operator Theory, and Mathematical Physics", Ixtapa, México, February 24–28, 2014. Commun. Math. Anal. **17** (2014), No. 2, Preface, I-IV, 369 pp.

Papers

1. Yu.I. Karlovich: *One boundary value problem with a shift of the boundary into the domain*. Teoria Funk. Compl. Perem. i Kraev. Zad., Cheboksary, (1974), No. 2, 45–57 (Russian).
2. Yu.I. Karlovich: *On singular integral operators with a shift of the contour-carrier into the domain*. Trudy Tbilis. Matem. Inst. **44** (1974), 113–124 (Russian).
3. Ju.I. Karlovič: *On integral operators with a shift of the contour of integration in the domain*. Dokl. Akad. Nauk SSSR **216** (1974), No. 1, 32–35 (Russian). Engl. transl. in Soviet Math. Dokl. **15** (1974), No. 3, 740–744.
4. Ju.I. Karlovič and M.V. Šapiro: *On an algebra of operators with a Carleman shift that permutes the components of the contour*. Mat. Issled. **9** (1974), No. 3, 95–104 (Russian).
5. P.G. Baškarev, Ju.I. Karlovič, and A.P. Nečaev: *On the theory of singular integral operators with a finite group of shifts*. Dokl. Akad. Nauk SSSR **219** (1974), No. 2, 272–274 (in Russian). Engl. transl. in Soviet Math. Dokl. **15** (1974), No. 6, 1553–1556.
6. Ju.I. Karlovič and V.G. Kravčenko: *A Noether theory for a singular integral operator with a shift having periodic points*. Dokl. Akad. Nauk SSSR **231** (1976), No. 2, 277–280 (Russian). Engl. transl. in Soviet Math. Dokl. **17** (1976), No. 6, 1547–1551.
7. Yu.I. Karlovich, V.G. Kravchenko, G.S. Litvinchuk, and E.S. Pak: *On a model of regulating of the water quality in a river*. In: System modelling and computer using in economical investigations, pp. 56–67, Institute of Economics, Ukrainian Academy of Sciences, Kiev, 1976 (Russian).
8. Yu.I. Karlovich, V.G. Kravchenko, and G.S. Litvinchuk: *On the problem of water resource quality management*. In: Problems of Sea Economics, No. 5,

pp. 26–32, Institute of Economics, Ukrainian Academy of Sciences, Odessa, 1976 (Russian).

9. Ju.I. Karlovič and V.G. Kravčenko: *On a singular integral operator with non-Carleman shifts on an open contour*. Dokl. Akad. Nauk SSSR **236** (1977), No. 4, 792–795 (Russian). Engl. transl. in Soviet Math. Dokl. **18** (1977), No. 5, 1263–1267.
10. Ju.I. Karlovič and V.G. Kravčenko: *On systems of functional and integro-functional equations with a non-Carleman shift*. Dokl. Akad. Nauk SSSR **236** (1977), No. 5, 1064–1067 (Russian). Engl. transl. in Soviet Math. Dokl. **18** (1977), No. 5, 1319–1322.
11. Yu.I. Karlovich, V.G. Kravchenko, and G.S. Litvinchuk: *Economic-mathematical aspects of the problem of water environmental protection*. In: Problems of Sea Economics, No. 6, pp. 110–119, Institute of Economics, Ukrainian Academy of Sciences, Odessa, 1977 (Russian).
12. Ju.I. Karlovič and V.G. Kravčenko: *On an algebra of singular integral operators with non-Carleman shift*. Dokl. Akad. Nauk SSSR **239** (1978), No. 1, 138–41 (Russian). Engl. transl. in Soviet Math. Dokl. **19** (1978), No. 2, 267–271.
13. Yu.I. Karlovich, V.G. Kravchenko, and G.S. Litvinchuk: *Economic-mathematical problems of planning of ecological systems*. In: Problems of Sea Economics, No. 7, pp. 53–65, Institute of Economics, Ukrainian Academy of Sciences, Kiev, 1978 (Russian).
14. Ju.I. Karlovič and V.G. Kravčenko: *On integro-functional equations with a non-Carleman shift in the space $L_2^n(0,1)$*. Dokl. Akad. Nauk SSSR **246** (1979), No. 2, 268–271 (Russian). Engl. transl. in Soviet Math. Dokl. **20** (1979), No. 3, 455–458.
15. Ju.I. Karlovič and V.G. Kravčenko: *On the algebra of singular integral operators with non-Carleman shift and piecewise continuous coefficients*. Dokl. Akad. Nauk SSSR **252** (1980), No. 6, 1307–1311 (Russian). Engl. transl. in Soviet Math. Dokl. **21** (1980), No. 3, 926–930.
16. Ju.I. Karlovič and V.G. Kravčenko: *On a general method of investigating operators of singular integral type with non-Carleman shift*. Dokl. Akad. Nauk SSSR **253** (1980), No. 1, 18–22 (Russian). Engl. transl. in Soviet Math. Dokl. **22** (1980), No. 1, 10–14.
17. Yu.I. Karlovich and V.G. Kravchenko: *On some new results in the Noether theory of singular integral operators with non-Carleman shift*. Sovrem. Probl. Teor. Funk., Baku, 1980, 145–150 (Russian).
18. G.S. Litvinchuk, Yu.I. Karlovich, and V.N. Katkov: *On the problem of development and distribution of reproduction, catch and processing business of bioresources at seaside region*. In: Problems and experience of using reserves of biological resources of the Azov-Black Sea basin, pp. 19–31, Institute of Economics, Ukrainian Academy of Sciences, Kiev, 1980 (Russian).
19. Ju.I. Karlovič and V.G. Kravčenko: *Systems of singular integral equations with a shift*. Matem. Sbornik **116**(158) (1981), No. 1, 87–110 (Russian). Engl. transl. in Math. URSS Sbornik **44** (1983), No. 1, 75–95.

20. Yu.I. Karlovich and V.G. Kravchenko: *Singular integral equations with non-Carleman translations on an open contour*. Differentsial'nye Uravneniya **17** (1981), No. 12, 2212–2223 (Russian). Engl. transl. in Differential Equations **17**, No. 12, 1981, 1408–1417.
21. Yu.I. Karlovich and V.N. Katkov: *A problem of long-term planning of economical-ecological systems of exploitation of sea basins bioresources*. In: The use of sea resources in the national economy, pp. 11–22, Institute of Economics, Ukrainian Academy of Sciences, Kiev, 1981 (Russian).
22. Yu.I. Karlovich and V.N. Katkov: *A problem of middle-term planning of economical-ecological systems of exploitation of sea basins bioresources*. In: Sea economics, pp. 50–55, Institute of Economics, Ukrainian Academy of Sciences, Kiev, 1982 (Russian).
23. Yu.I. Karlovich and V.G. Kravchenko: *An algebra of singular integral operators with piecewise-continuous coefficients and a piecewise-smooth shift on a composite contour*. Izv. Akad. Nauk SSSR, Ser. Matem. **47** (1983), No. 5, 1030–1077 (Russian). Engl. transl. in Math. USSR Izvestiya **23** (1984), No. 2, 307–352.
24. Yu.I. Karlovich, V.G. Kravchenko, and G.S. Litvinchuk: *The Noetherian theory of singular integral translation operators*. Izv. Vyssh. Uchebn. Zaved. Mat. (1983), No. 4, 3–27 (Russian). Engl. transl. in Soviet. Math. (Iz. VUZ) **27** (1983), No. 4, 1–34.
25. Yu.I. Karlovich and I.M. Spitkovskiĭ: *On the Noetherian property of certain singular integral operators with matrix coefficients of class SAP and systems of convolution equations on a finite interval connected with them*. Dokl. Akad. Nauk SSSR **269** (1983), No. 3, 531–535 (Russian). Engl. transl. in Soviet Math. Dokl. **27** (1983), No. 2, 358–363.
26. Yu.I. Karlovich and B. Tursunkulov: *On singular integrals with shift in a generalized Hölder space*. Izv. Vyssh. Uchebn. Zaved., Mat. (1984), No. 3, 71–74 (Russian). Engl. transl. in Soviet. Math. (Iz. VUZ) **28** (1984), No. 3, 97–101.
27. Yu.I. Karlovich and I.M. Spitkovskiĭ: *Factorization problem for almost periodic matrix-functions and Fredholm theory of Toeplitz operators with semi-almost periodic matrix symbols*. Lect. Notes in Math. **1043** (1984), 279–282; reprinted in Lect. Notes in Math. **1573** (1994), 323–325.
28. Yu.I. Karlovich: *On the invertibility of functional operators with non-Carleman shift in Hölder spaces*. Differentsial'nye Uravneniya **20** (1984), No. 12, 2165–2169 (Russian).
29. Yu.I. Karlovich: *On a system of singular integral equations with nonunivalent shift*. In: Proceedings of a commemorative seminar on boundary value problems (Minsk, 1981), pp. 161–165, "Universitetskoe", Minsk, 1985 (Russian).
30. Yu.I. Karlovich, Yu.D. Latushkin, and R. Mardiev: *Criterion of $n(d)$-normality of singular integral operators with non-Carleman shift*. In: Funktsional'no-Differetsial'nye Uravneniya, pp. 45–50, Polytechnical Institute, Perm, 1985 (Russian).

31. Yu.I. Karlovich: *The Noether theory of one class of convolution type operators*. In: Reports of the extended sessions of a seminar of the I.N. Vekua Institute of Applied Mathematics, Vol. 1, No. 1, pp. 102–105, Tbilis. Gos. Univ., Tbilisi, 1985 (Russian).
32. Yu.I. Karlovich, G.S. Litvinchuk, and I.M. Spitkovskii: *Factorization of almost periodic matrices and Noether theory of systems of Wiener-Hopf equations with semi-almost periodic presymbols*. In: Reports of the extended sessions of a seminar of the I.N. Vekua Institute of Applied Mathematics, Vol. 1, No. 1, pp. 106–109, Tbilis. Gos. Univ., Tbilisi, 1985 (Russian).
33. Yu.I. Karlovich and I.M. Spitkovskiĭ: *On the theory of systems of convolution type equations with semi-almost-periodic symbols in spaces of Bessel potentials*. Dokl. Akad. Nauk SSSR **286** (1986), No. 4, 799–803 (Russian). Engl. transl. in Soviet Math. Dokl. **33** (1986), No. 1, 180–184.
34. Yu.I. Karlovich and R. Mardiev: *One-sided invertibility of functional non-Carleman translation operators in Hölder spaces*. Izv. Vyssh. Uchebn. Zaved., Mat. (1987), No. 3, 77–80 (Russian). Engl. transl. in Soviet. Math. (Iz. VUZ.) **31** (1987), No. 3, 106–110.
35. Yu.I. Karlovich: *The Noether theory of a class of operators of convolution type with a shift*. Dokl. Akad. Nauk SSSR **295** (1987), No. 1, 24–29 (Russian). Engl. transl. in Soviet Math. Dokl. **36** (1988), No. 1, 16–22.
36. Yu.I. Karlovich and G.S. Litvinchuk: *Algebras of singular integral operators with discrete groups of shifts*. In: Current problems in mathematical physics, Vol. II, pp. 57–64, Tbilis. Gos. Univ., Tbilisi, 1987 (Russian).
37. Yu.I. Karlovich and R. Mardiev: *One-sided invertibility of functional operators and the $n(d)$-normality of singular integral operators with translation in Hölder spaces*, Differentsial'nye Uravneniya **24** (1988), No. 3, 488–499 (Russian). Engl. transl. in Differential Equations **24** (1988), No. 3, 350–359.
38. Yu.I. Karlovich: *The local-trajectory method of studying invertibility in C^*-algebras of operators with discrete groups of shifts*. Dokl. Akad. Nauk SSSR **299** (1988), No. 3, 546–550 (Russian). Engl. transl. in Soviet Math. Dokl. **37** (1988), No. 2, 407–412.
39. Yu.I. Karlovich: *C^*-algebras of operators of convolution type with discrete groups of shifts and oscillating coefficients*. Dokl. Akad. Nauk SSSR **302** (1988), No. 3, 535–540 (Russsian). Engl. transl. in Soviet Math. Dokl. **38** (1989), No. 2, 301–307.
40. A.V. Aĭzenshtat, Yu.I. Karlovich, and G.S. Litvinchuk: *On some theorem of conformal gluing and its applications*. In: Reports of the extended sessions of a seminar of the I.N. Vekua Institute of Applied Mathematics, Vol. 3, No. 1, pp. 9–12, Tbilis. Gos. Univ., Tbilisi, 1988 (Russian).
41. Yu.I. Karlovich: *On algebras of singular integral operators with discrete groups of shifts in L_p-spaces*. Dokl. Akad. Nauk SSSR **304** (1989), No. 2, 274–280 (Russian). Engl. transl. in Soviet Math. Dokl. **39** (1989), No. 1, 48–53.
42. Yu.I. Karlovich and I.M. Spitkovskiĭ: *Factorization of almost periodic matrix-valued functions and the Noether theory for certain classes of equations of*

convolution type. Izv. Akad. Nauk SSSR Ser. Matem. **53** (1989), No. 2, 276–308 (Russian). Engl. transl. in Math. USSR Izvestiya **34** (1990), No. 2, 281–316.
43. V.D. Aslanov and Yu.I. Karlovich: *One-sided invertibility of functional operators in reflexive Orlicz spaces*. Dokl. Akad. Nauk AzSSR **45** (1989), No. 11–12, 3–7 (Russian).
44. Yu.I. Karlovich and G.S. Litvinchuk: *Some classes of semi-Noetherian operators*. Izv. Vyssh. Uchebn. Zaved., Mat. (1990), No. 2, 3–16 (Russian). Engl. transl. in Soviet. Math. (Iz. VUZ) **34** (1990), No. 2, 1–15.
45. Yu.I. Karlovich, V.G. Kravchenko, and G.S. Litvinchuk: *On the Noethericity and Mikhlin symbols of operators of the type of singular integral operators with shift*. Z. Anal. Anwendungen **9** (1990), No. 1, 15–32 (Russian).
46. Yu.I. Karlovich, V.G. Kravchenko, and G.S. Litvinchuk: *The invertibility of functional operators on Banach spaces*. In: Funktsional'no-Differentsial'nye Uravneniya, pp. 18–58, Polytechnical Institute, Perm, 1990 (Russian).
47. Yu.I. Karlovich and R. Mardiev: *On $n(d)$-normality in Hölder spaces of singular integral operators with shift*. Dokl. Akad. Nauk UzSSR (1990), No. 4, 10–12 (Russian).
48. Yu.I. Karlovich: *Riemann and Haseman vector boundary value problems with oscillating coefficients*. In: Reports of the extended sessions of a seminar of the I.N. Vekua Institute of Applied Mathematics, Vol. 5, No. 1, pp. 86–89, Tbilis. Gos. Univ., Tbilisi, 1990 (Russian).
49. Yu.I. Karlovich, A.V. Aĭzenshtat, and G.S. Litvinchuk: *On the defect numbers of the Kveselava-Vekua operator with discontinuous derivative of the shift*. Dokl. Akad. Nauk SSSR **318** (1991), No. 1, 11–15 (Russian). Engl. transl. in Soviet Math. Dokl. **43** (1991), No. 3, 633–638.
50. Yu.I. Karlovich: *On the Haseman problem*. Demonstratio Math. **26** (1993), No. 3–4, 581–595.
51. Yu.I. Karlovich and G.S. Litvinchuk: *One generalization of the factorization of measurable matrix-functions and its applications to Fredholm theory for certain classes of convolution type equations*. Demonstratio Math. **26** (1993), No. 3–4, 597–604.
52. Yu.I. Karlovich: *C^*r-algebras of nonlocal quaternionic convolution type operators*. In: Clifford Algebras and their Applications in Mathematical Physics (F. Brackx, R. Delanghe, H. Serras, eds.), 109–118, Kluwer Academic Publishers, Dordrecht, 1993.
53. Yu.I. Karlovich and I.M. Spitkovsky: *(Semi)-Fredholmness of convolution operators on the spaces of Bessel potentials*. Oper. Theory Adv. Appl. **71** (1994), 122–152.
54. A. Böttcher, Yu.I. Karlovich, and B. Silbermann: *Singular integral equations with PQC coefficients and freely transformed argument*. Math. Nachr. **166** (1994), 113–133.
55. A. Böttcher, Yu.I. Karlovich, and I.M. Spitkovsky: *Toeplitz operators with semi-almost periodic symbols on spaces with Muckenhoupt weight*. Integral Equations and Operator Theory **18** (1994), 261–276.

56. Yu.I. Karlovich: *The continuous invertibility of functional operators in Banach spaces*. Dissertationes Mathematicae **340** (1995), 115–136.
57. Yu.I. Karlovich and I.M. Spitkovsky: *Factorization of almost periodic matrix functions*. J. Math. Anal. Appl. **193** (1995), No. 1, 209–232.
58. Yu.I. Karlovich and I.M. Spitkovsky: *Almost periodic factorization: An analogue of Chebotarev's algorithm*. Contemporary Mathematics **189** (1995), 327–352.
59. A. Böttcher and Yu.I. Karlovich: *Toeplitz and singular integral operators on Carleson curves with logarithmic whirl points*. Integral Equations and Operator Theory **22** (1995), 127–161.
60. A.V. Aĭzenshtat, Yu.I. Karlovich, and G.S. Livinchuk: *The method of conformal gluing for the Haseman boundary value problem on an open contour*. Complex Variables Theory Appl. **28** (1996), No. 4, 313–346.
61. A. Böttcher, I. Gohberg, Yu. Karlovich, N. Krupnik, S. Roch, B. Silbermann, and I. Spitkovsky: *Banach algebras generated by N idempotens and applications*. Oper. Theory Adv. Appl. **90** (1996), 19–54.
62. A. Böttcher and Yu.I. Karlovich: *Toeplitz and singular integral operators on general Carleson Jordan curves*. Oper. Theory Adv. Appl. **90** (1996), 119–152.
63. Yu.I. Karlovich and I.M. Spitkovsky: *Semi-Fredholm properties of certain singular integral operators*. Oper. Theory Adv. Appl. **90** (1996), 264–287.
64. A. Böttcher, Yu.I. Karlovich, and V.S. Rabinovich: *Emergence, persistence, and disappearance of logarithmic spirals in the spectra of singular integral operators*. Integral Equations and Operator Theory **25** (1996), 406–444.
65. A. Böttcher and Yu.I. Karlovich: *Submultiplicative functions and spectral theory of Toeplitz operators*. Integral Transform. Spec. Funct. **4** (1996], No. 1, 181–202.
66. A. Böttcher and Yu.I. Karlovich: *The algebra of singular integral operators on a closed Carleson curve*. Dokl. Akad. Nauk **357** (1997), No. 1, 7–10 (Russian). Engl. transl. in Doklady Mathematics **56** (1997), No. 3, 813–816.
67. Yu.I. Karlovich and I.M. Spitkovsky: *The semi-Fredholmness of singular integral operators with matrix semi-almost periodic coefficients*. Dokl. Akad. Nauk **359** (1998), No. 1, 10–12 (Russian). Engl. transl. in Doklady Mathematics **57** (1998), No. 2, 176–178.
68. A. Böttcher and Yu.I. Karlovich: *Algebra of singular integral operators on the Lebesgue space with a Muckenhoupt weight on a closed Carleson curve*. Dokl. Akad. Nauk **359** (1998), No. 2, 151–154 (Russian). Engl. transl. in Doklady Mathematics **57** (1998), No. 2, 193–196.
69. M.A. Bastos, Yu.I. Karlovich, I.M. Spitkovsky, and P.M. Tishin: *On a new algorithm for almost periodic factorization*, Oper. Theory Adv. Appl. **103** (1998), 53–74.
70. A. Böttcher, Yu.I. Karlovich, and V.S. Rabinovich: *Mellin pseudodifferential operators with slowly varying symbols and singular integrals on Carleson curves with Muckenhoupt weights*. Manuscripta Math. **95** (1998), No. 3, 363–376.

71. M.A. Bastos, Yu.I. Karlovich, A.F. dos Santos, and P.M. Tishin: *The corona theorem and the existence of canonical factorization of triangular AP-matrix functions*. J. Math. Anal. Appl. **223** (1998), No. 2, 494–522.
72. M.A. Bastos, Yu.I. Karlovich, A.F. dos Santos, and P.M. Tishin: *The corona theorem and the canonical factorization of triangular AP-matrix functions – Effective criteria and explicit formulas*. J. Math. Anal. Appl. **223** (1998), No. 2, 523–550.
73. J.A. Ball, Yu.I. Karlovich, L. Rodman, and I.M. Spitkovsky: *Sarason interpolation and Toeplitz corona theorem for almost periodic matrix functions*. Integral Equations and Operator Theory **32** (1998), No. 3, 243–281.
74. A. Böttcher and Yu.I. Karlovich: *Toeplitz operators with PC symbols on general Carleson Jordan curves with arbitrary Muckenhoupt weights*. Trans. Amer. Math. Soc. **351** (1999), No. 8, 3143–3196.
75. C.J. Bishop, A. Böttcher, Yu.I. Karlovich, and I.M. Spitkovsky: *Local spectra and index of singular integral operators with piecewise continuous coefficients on composed curves*. Math. Nachr. **206** (1999), 5–83.
76. Yu.I. Karlovich, I.M. Spitkovsky, and R.A. Walker: *Almost periodic factorization of block triangular matrix functions revisited*. Linear Algebra Appl. **293** (1999), No. 1–3, 199–232.
77. A. Böttcher, Yu.I. Karlovich, and V.S. Rabinovich: *The method of limit operators for one-dimensional singular integrals with slowly oscillating data*. J. Operator Theory **43** (2000), No. 1, 171–198.
78. A. Böttcher, Yu.I. Karlovich, and V.S. Rabinovich: *Singular integral operators with complex conjugation from the viewpoint of pseudodifferential operators*. Oper. Theory Adv. Appl. **121** (2001), 36–59.
79. Yu.I. Karlovich and E. Ramírez de Arellano: *A shift-invariant algebra of singular integral operators with oscillating coefficients*. Integral Equations and Operator Theory **39** (2001), No. 4, 441–474.
80. A. Böttcher, Yu.I. Karlovich, and I. Spitkovsky: *Toeplitz operators with semi-almost-periodic matrix symbols on Hardy spaces*. Acta Appl. Math. **65** (2001), No. 1–3, 115–136.
81. Yu.I. Karlovich and A.B. Lebre: *Algebra of singular integral operators with a Carleman backward slowly oscillating shift*. Integral Equations and Operator Theory **41** (2001), No. 3, 288–323.
82. A. Böttcher and Yu.I. Karlovich: *Cauchy's singular integral operator and its beautiful spectrum*. Oper. Theory Adv. Appl. **129** (2001), 109–142.
83. M.A. Bastos, Yu.I. Karlovich, and A.F. dos Santos: *The invertibility of convolution type operators on a union of intervals and the corona theorem*. Integral Equations and Operator Theory **42** (2002), No. 1, 22–56.
84. A.Yu. Karlovich and Yu.I. Karlovich: *One-sided invertibility of binomial functional operators with a shift on rearrangement-invariant spaces*. Integral Equations and Operator Theory **42** (2002), No. 2, 201–228.
85. Yu.I. Karlovich and B. Silbermann: *Local method for nonlocal operators on Banach spaces*. Oper. Theory Adv. Appl. **135** (2002), 235–247.

86. A.Yu. Karlovich and Yu.I. Karlovich: *Invertibility in Banach algebras of functional operators with non-Carleman shifts*. Ukrainian Mathematics Congress–2001 (Uktainian), pp. 107–124, Natsional. Akad. Nauk Ukraïny, Inst. Mat., Kiev, 2002.
87. M.A. Bastos, Yu.I. Karlovich, and A.F. dos Santos: *Oscillatory Riemann-Hilbert problems and the corona theorem*. J. Funct. Anal. **197** (2003), No. 2, 347–397.
88. M.A. Bastos, A. Bravo, and Yu.I. Karlovich: *Singular integral operators with piecewise almost periodic coefficients and Carleman shift*. Oper. Theory Adv. Appl. **142** (2003), 29–57.
89. A.Yu. Karlovich, Yu.I. Karlovich, and A.B. Lebre: *Invertibility of functional operators with slowly oscillating non-Carleman shifts*. Oper. Theory Adv. Appl. **142** (2003), 147–174.
90. A.Yu. Karlovich and Yu.I. Karlovich: *Compactness of commutators arising in the Fredholm theory of singular integral operators with shifts*. In: Factorization, Singular Operators and Related Problems (Funchal, 2002), 111–129, Kluwer Acad. Publ., Dordrecht, 2003.
91. A. Böttcher, Yu.I. Karlovich, and I.M. Spitkovsky: *The C*-algebra of singular integral operators with semi-almost periodic coefficients*. J. Funct. Anal. **204** (2003), No. 2, 445–484.
92. Yu.I. Karlovich and E. Ramírez de Arellano: *Singular integral operators with fixed singularities on weighted Lebesgue spaces*. Integral Equations and Operator Theory **48** (2004), No. 3, 331–363.
93. Yu.I. Karlovich and B. Silbermann: *Fredholmness of singular integral operators with discrete subexponential groups of shifts on Lebesgue spaces*. Math. Nachr. **272** (2004), 55–94.
94. M.A. Bastos, Yu.I. Karlovich, and B. Silbermann: *Toeplitz operators with symbols generated by slowly oscillating and semi-almost periodic matrix functions*. Proc. London Math. Soc. (3) **89** (2004), No. 3, 697–737.
95. M.A. Bastos, A. Bravo, and Yu.I. Karlovich: *Convolution type operators with symbols generated by slowly oscillating and piecewise continuous matrix functions*. Oper. Theory Adv. Appl. **147** (2004), 151–174.
96. M.A. Bastos, A. Bravo, and Yu.I. Karlovich: *Symbol calculus and Fredholmness for a Banach algebra of convolution type operators with slowly oscillating and piecewise continuous data*. Math. Nachr. **269–270** (2004), 11–38.
97. Yu.I. Karlovich and L. Pessoa: *Algebras generated by the Bergman and anti-Bergman projections and by multiplications by piecewise continuous functions*. Integral Equations and Operator Theory **52** (2005), No. 2, 219–270.
98. Yu.I. Karlovich: *Approximation approach to canonical APW factorization*. Izv. Vuzov. Sev.-Kavk. Region, Rostov-on-Don (2005), 143–151.
99. M.A. Bastos, C.A. Fernandes, and Yu.I. Karlovich: *C^*-algebras of integral operators with piecewise slowly oscillating coefficients and shifts acting freely*. Integral Equations and Operator Theory **55** (2006), No. 1, 19–67.

100. Yu.I. Karlovich: *An algebra of pseudodifferential operators with slowly oscillating symbols*. Proc. London Math. Soc. (3) **92** (2006), No. 3, 713–761.
101. Yu.I. Karlovich: *A local-trajectory method and isomorphism theorems for nonlocal C^*-algebras*. Oper. Theory Adv. Appl. **170** (2007), 137–166.
102. Yu.I. Karlovich: *Pseudodifferential operators with compound slowly oscillating symbols*. Oper. Theory Adv. Appl. **171** (2007), 189–224.
103. Yu.I. Karlovich and L.V. Pessoa: *C^*-algebras of Bergman type operators with piecewise continuous coefficients*. Integral Equations and Operator Theory **57** (2007), No. 4, 521–565.
104. Yu.I. Karlovich: *Pseudodifferential operators with compound non-regular symbols*. Math. Nachr. **280** (2007), No. 9–10, 1128–1144.
105. Yu.I. Karlovich: *Algebras of pseudodifferential operators with discontinuous symbols*. Oper. Theory Adv. Appl. **172** (2007), 207–233.
106. M.A. Bastos, C.A. Fernandes, and Yu.I. Karlovich: *Spectral measures in C^*-algebras of singular integral operators with shifts*. J. Funct. Anal. **242** (2007), 86–126.
107. Yu.I. Karlovich: *Nonlocal singular integral operators with slowly oscillating data*. Oper. Theory Adv. Appl. **181** (2008), 229–261.
108. Yu.I. Karlovich and L.V. Pessoa: *Poly-Bergman projections and orthogonal decompositions of L^2-spaces over bounded domains*. Oper. Theory Adv. Appl. **181** (2008), 263–282.
109. Yu.I. Karlovich: *Professor Georgii Semenovich Litvinchuk (1931–2006)*. Complex Anal. Oper. Theory **2** (2008), No. 2, 203–214.
110. M.A. Bastos, C.A. Fernandes, and Yu.I. Karlovich: *C^*-algebras of singular integral operators with shifts having the same nonempty set of fixed points*. Complex Anal. Oper. Theory **2** (2008), 241–272.
111. Yu.I. Karlovich: *The Haseman boundary value problem with slowly oscillating data*. In: Analytic methods of analysis and differential equations: AMADE-2006 (A.A. Kilbas, S.V. Rogozin, eds.), pp. 81–110, Cambridge Scientific Publishers, Cambridge, 2008.
112. Yu.I. Karlovich and J. Loreto Hernández: *Wiener-Hopf operators with matrix semi-almost periodic symbols on weighted Lebesgue spaces*. Integral Equations and Operator Theory **62** (2008), No. 1, 85–128.
113. Yu.I. Karlovich and L.V. Pessoa: *C^*-algebras of Bergman type operators with piecewise continuous coefficients on bounded domains*. In: More progresses in analysis. Proceedings of the 5th international ISAAC congress, Catania, Italy, July 25–30, 2005 (H.G.W. Begehr, F. Nicolosi, eds.), pp. 339–348, World Scientific, Hackensack, NJ, 2009.
114. M.A. Bastos, C.A. Fernandes, and Yu.I. Karlovich: *A C^*-algebra of functional operators with shifts having a nonempty set of periodic points*. In: Further progress in analysis. Proceedings of the 6th international ISAAC congress, Ankara, Turkey, August 13–18, 2007 (H.G.W. Begehr, A.O. Çelebi, A.P. Gilbert, eds.), pp. 111–121, World Scientific, Hackensack, NJ, 2009.

115. M.C. Câmara, Yu.I. Karlovich, and I.M. Spitkovsky: *Almost periodic factorization of some triangular matrix functions*. Oper. Theory Adv. Appl. **190** (2009), 171–190.
116. Yu.I. Karlovich and J. Loreto Hernández: *Wiener-Hopf operators with slowly oscillating matrix symbols on weighted Lebesgue spaces*. Integral Equations and Operator Theory **64** (2009), No. 2, 203–237.
117. Yu.I. Karlovich: *An algebra of shift-invariant singular integral operators with slowly oscillating data and its application to operators with a Carleman shift*. Oper. Theory Adv. Appl. **193** (2009), 81–95.
118. G. Burlak, A. Díaz-de-Anda, Yu. Karlovich, and A.B. Klimov: *Critical behavior of nanoemitter radiation in a percolation material*. Physics Letters A **373** (2009), issue 16, 1492–1499.
119. Yu.I. Karlovich and J. Loreto Hernández: *Wiener-Hopf operators with oscillating symbols on weighted Lebesgue spaces*. Oper. Theory Adv. Appl. **210** (2010), 123–145.
120. Yu.I. Karlovich and I.M. Spitkovsky: *Almost periodic polynomial factorization of some triangular matrix functions*. Oper. Theory Adv. Appl. **202** (2010), 337–354.
121. M.C. Câmara, Yu.I. Karlovich, and I.M. Spitkovsky: *Constructive almost periodic factorization of some triangular matrix functions*. J. Math. Anal. Appl. **367** (2010), No. 2, 416–433.
122. M.A. Bastos, A. Bravo, Yu.I. Karlovich, and I.M. Spitkovsky: *Constructive factorization of some almost periodic triangular matrix functions with a quadrinomial off diagonal entry*. J. Math. Anal. Appl. **376** (2011), No. 2, 625–640.
123. G. Burlak, Yu. Karlovich, and V. Rabinovich: *Fredholm property and essential spectrum of pseudodifferential operators with operator-valued symbols*. Commun. Math. Anal. (2011), Conference 3, 50–65.
124. A.E. Choque Rivero and Yu.I. Karlovich: *The time optimal control as an interpolation problem*. Commun. Math. Anal. (2011), Conference 3, 66–76.
125. Yu.I. Karlovich, V.S. Rabinovich, and N.L. Vasilevski: *Algebras of pseudodifferential operators with discontinuous oscillating symbols*. Commun. Math. Anal. (2011), Conference 3, 108–130.
126. A.Yu. Karlovich, Yu.I. Karlovich, and A.B. Lebre: *Sufficient conditions for Fredholmness of singular integral operators with shifts and slowly oscillating data*. Integral Equations and Operator Theory **70** (2011), No. 4, 451–483.
127. A.Yu. Karlovich, Yu.I. Karlovich, and A.B. Lebre: *Necessary conditions for Fredholmness of singular integral operators with shifts and slowly oscillating data*. Integral Equations and Operator Theory **71** (2011), No. 1, 29–53.
128. M.A. Bastos, C.A. Fernandes, and Yu.I. Karlovich: *A nonlocal C^*-algebra of singular integral operators with shifts having periodic points*. Integral Equations and Operator Theory **71** (2011), No. 4, 509–534.
129. Yu.I. Karlovich and J. Loreto Hernández: *Convolution type operators with oscillating symbols on weighted Lebesgue spaces on a union of intervals*. Mathematica **54 (77)** (2012), Special Number, 104–119.

130. Yu.I. Karlovich: *Boundedness and compactness of pseudodifferential operators with non-regular symbols on weighted Lebesgue spaces*. Integral Equations and Operator Theory **73** (2012), No. 2, 217–254.
131. M.C. Câmara, C. Diogo, Yu.I. Karlovich, and I.M. Spitkovsky: *Factorizations, Riemann-Hilbert problems and the corona theorem*. J. London Math. Soc. (2) **86** (2012), No. 3, 852–878.
132. Yu.I. Karlovich and V.A. Mozel: *On nonlocal C^*-algebras of two-dimensional singular integral operators*. Oper. Theory Adv. Appl. **220** (2012), 115–135.
133. Yu.I. Karlovich and V.A. Mozel: *C^*-algebras of Bergman type operators with piecewise continuous coefficients and shifts*. Complex Var. Elliptic Equ. **57** (2012), No. 7–8, 841–865.
134. A.E. Choque Rivero, Yu.I. Karlovich, A.E. Merzon, and P.N. Zhevandrov: *On the convergence of the amplitude of the diffracted nonstationary wave in scattering by wedges*. Russ. J. Math. Phys. **19** (2012), No. 3, 373–384.
135. Yu.I. Karlovich and I. Loreto Hernández: *Algebras of convolution type operators with piecewise slowly oscillating data. I: Local and structural study*. Integral Equations and Operator Theory **74** (2012), No. 3, 377–415.
136. Yu.I. Karlovich and I. Loreto Hernández: *Algebras of convolution type operators with piecewise slowly oscillating data. II: Local spectra and Fredholmness*. Integral Equations and Operator Theory **75** (2013), No. 1, 49–86.
137. M.C. Câmara, Yu.I. Karlovich, and I.M. Spitkovsky: *Kernels of asymmetric Toeplitz operators and applications to almost periodic factorization*. Complex Anal. Oper. Theory **7** (2013), No. 2, 375–407.
138. Yu.I. Karlovich and I. Loreto Hernández: *On convolution type operators with piecewise slowly oscillating data*. Oper. Theory Adv. Appl. **228** (2013), 185–207.
139. K.V. Sklyar, G.M. Sklyar, and Yu.I. Karlovich: *On bang-bang controls for some nonlinear systems*. Commun. Math. Anal. **14** (2013), No. 2, 163–178.
140. E.V. Kopylova, Yu.I. Karlovich, A.I. Komech, and A.E. Merzon: *On the spreading rate of the soliton perturbation for relativistic nonlinear wave equations*. Commun. Math. Anal. **14** (2013), No. 2, 95–102.
141. Yu.I. Karlovich: *Boundary value problems with finite groups of Lipschitz shifts*. Zb. Pr. Inst. Mat. NAN Ukr. **10** (2013), No. 4–5, 162–184.
142. M.A. Bastos, C.A. Fernandes, and Yu.I. Karlovich: *A C^*-algebra of singular integral operators with shifts adnitting distinct fixed points*. J. Math. Anal. Appl. **413** (2014), No. 1, 502–524.
143. M.A. Bastos, A. Bravo, Yu.I. Karlovich, and I.M. Spitkovsky: *On the factorization of some block triangular almost periodic matrix functions*. Oper. Theory Adv. Appl. **242** (2014), 25–52.
144. M.A. Bastos, C.A. Fernandes, and Yu.I. Karlovich: *A C^*-algebras of singular integral operators with shifts similar to affine mappings*. Oper. Theory Adv. Appl. **242** (2014), 53–79.

145. A.Yu. Karlovich, Yu.I. Karlovich, and A.B. Lebre: *Fredholmness and index of simplest singular integral operators with two slowly oscillating shifts*. Oper. Matrices **8** (2014), No. 4, 935–955.
146. A. Choque Rivero, Yu.I. Karlovich, and V.A. Yurko: *An inverse problem for differential operators on Hedgehog-type graphs with general matching conditions*. Commun. Math. Anal. **17** (2014), No. 2, 98–107.
147. I. De La Cruz Rodríguez, Yu.I. Karlovich, and I. Loreto-Hernández: *Commutators of convolution type operators with piecewise quasicontinuous data*. Commun. Math. Anal. **17** (2014), No. 2, 131–150.
148. A.Yu. Karlovich, Yu.I. Karlovich, and A.B. Lebre: *On regularization of Mellin PDO's with slowly oscillating symbols of limited smoothness*. Commun. Math. Anal. **17** (2014), No. 2, 189–208.
149. Yu.I. Karlovich and V.A. Mozel: *C^*-algebras of two-dimensional singular integral operators with shifts*. Current trends in analysis and its applications, Trends Math., pp. 571–581, Birkhäuser/Springer, Cham, 2015.
150. Yu.I. Karlovich and I. Loreto-Hernández: *C^*-algebra of nonlocal convolution type operators with piecewise slowly oscillating data*. J. Operator Theory **73** (2015), no. 1, 211–242.
151. Yu.I. Karlovich, J. Loreto-Hernández, and I.M. Spitkovsky: *Factorization of some triangular matrix functions and its applications*. Oper. Matrices **9** (2015), No. 1, 1–29.
152. E. Espinoza-Loyola, Yu.I. Karlovich, and O. Vilchis-Torres: *C^*-algebras of Bergman type operators with piecewise constant coefficients over sectors*. Integral Equations and Operator Theory **83** (2015), No. 2, 243–269.
153. Yu.I. Karlovich: *C^*-algebras of Bergman type operators with continuous coefficients on poygonal domains*. Oper. Matrices **9** (2015), No. 4, 773–802.
154. A.Yu. Karlovich, Yu.I. Karlovich, A.B. Lebre: *On a weighted singular integral operator with shifts and slowly oscillating data*. Complex Anal. Oper. Theory **10** (2016), No. 6, 1101–1131.
155. A.Yu. Karlovich, Yu.I. Karlovich, and A.B. Lebre: *The generalized Cauchy index of some semi-almost periodic functions*. Bol. Soc. Mat. Mex. (3) **22** (2016), No. 2, 473–485.
156. M.A. Bastos, C.C. Carvalho, and Yu.I. Karlovich: *Index formula for convolution type operators with data functions in* $\mathrm{alg}(SO, PC)$. J. Math. Anal. Appl. **443** (2016), No. 1, 453–477.
157. A.Yu. Karlovich, Yu.I. Karlovich, and A.B. Lebre: *One-sided invertibility criteria for binomial functional operators with shift and slowly oscillating data*. Mediterr. J. Math. **13** (2016), No. 6, 4413–4435.
158. E. Espinoza-Loyola and Yu.I. Karlovich: *C^*-algebras of Bergman type operators with piecewise continuous coefficients over bounded polygonal domains*. Oper. Theory Adv. Appl. **258** (2017), 145–171.
159. Yu.I. Karlovich: *The Haseman boundary value problem with slowly oscillating coefficients and shifts*. Oper. Theory Adv. Appl. **259** (2017), 463–500.

160. G. Fernández-Torres and Yu. Karlovich: *Two-sided and one-sided invertibility of Wiener type functional operators with a shift and slowly oscillating data.* Banach J. Math. Anal. **11** (2017), No. 3, 554–590.
161. A.Yu. Karlovich, Yu.I. Karlovich, and A.B. Lebre: *Necessary Fredholm conditions for weighted singular integral operators with shifts and slowly oscillating data.* J. Integral Equations Appl. **29** (2017), No. 3, 365–399.
162. A.Yu. Karlovich, Yu.I. Karlovich, and A.B. Lebre: *The index of weighted singular integral operators with shifts and slowly oscillating data.* J. Math. Anal. Appl. **450** (2017), No. 1, 606–630.
163. Yu.I. Karlovich: *C^*-algebras of poly-Bergman type operators over sectors.* Integral Equations and Operator Theory **87** (2017), No. 3, 435–460.
164. Yu.I. Karlovich: *Algebras of convolution-type operators with piecewise slowly oscillating data on weighted Lebesgue spaces.* Mediterr. J. Math. **14** (2017), No. 4, Paper No. 182, 20 pp.
165. G. Fernández-Torres and Yu.I. Karlovich: *Fredholmness of nonlocal singular integral operators with slowly oscillating data.* In: Integral methods in science and engineering, Vol. 1, Theoretical Techniques, pp. 95–105, Birkhäuser/Springer, Cham, 2017.
166. I. Asekritova, Yu. Karlovich, and N. Kruglyak: *One-sided invertibility of discrete operators and their applications.* Aequationes Math. **92** (2018), No. 1, 39–73.
167. A.Yu. Karlovich, Yu.I. Karlovich, and A.B. Lebre: *Criteria for n(d)-normality of weighted singular integral operators with shifts and slowly oscillating data.* Proc. London Math. Soc. (3) **116** (2018), No. 4, 997–1027.
168. M.A. Bastos, C.A. Fernandes, and Yu.I. Karlovich: *A C^*-algebra of singular integral operators with shifts and piecewise quasicontinuous coefficients.* Oper. Theory Adv. Appl. **267** (2018), 25–64.
169. A.Yu. Karlovich, Yu.I. Karlovich, and A.B. Lebre: *Semi-Fredholmness of weighted singular integral operators with shifts and slowly oscillating data.* Oper. Theory Adv. Appl. **267** (2018), 221–246.
170. Yu.I. Karlovich: *Pseudodifferential operators with compound non-regular symbols.* Oper. Theory Adv. Appl. **268** (2018), 331–353.
171. E. Espinoza-Loyola and Yu.I. Karlovich: *C^*-algebras of Bergman type operators with piecewise continuous coefficients over domains with Dini-smooth corners.* Complex Anal. Oper. Theory **13** (2019), No. 1, 151–192.
172. Yu.I. Karlovich and I. Loreto-Hernández: *C^*-algebra of nonlocal convolution type operators.* J. Math. Anal. Appl. **475** (2019), No. 2, 1130–1161.
173. M.A. Bastos, C.A. Fernandes, and Yu.I. Karlovich: *Invertibility criteria in C^*-algebras of functional operators with shifts and PQC coefficients.* Integral Equations and Operator Theory **91** (2019), No. 3, Paper No. 19, 25 pp.
174. C.A. Fernandes, A.Yu. Karlovich, and Yu.I. Karlovich: *Noncompactness of Fourier convolution operators on Banach function spaces.* Ann. Funct. Anal. **10** (2019), No. 4, 553–561.
175. C.A. Fernandes, A.Yu. Karlovich, and Yu.I. Karlovich: *Algebra of convolution type operators with continuous data on Banach function spaces.* Function

spaces XII, pp. 157–171, Banach Center Publ. **119**, Polish Acad. Sci., Inst. Math., Warsaw, 2019.

176. E. Espinoza-Loyola and Yu.I. Karlovich: *C*-algebras of Bergman type operators with piecewise slowly oscillating coefficients over domains with Dini-smooth corners*. Integral Equations and Operator Theory **91** (2019), No. 5, Paper No. 47, 30 pp.
177. Yu.I. Karlovich: *Localization of a class of Muckenhoupt weights by using Mellin pseudo-differential operators*. Advances in microlocal and time-frequency analysis, pp. 281–310, Appl. Numer. Harmon. Anal., Birkhäuser/Springer, Cham, 2020.
178. Yu.I. Karlovich: *Algebras of singular integral operators with PQC coefficients on weighted Lebesgue spaces*. Oper. Theory Adv. Appl. **279** (2020), 205–231.
179. C.A. Fernandes, A.Yu. Karlovich, and Yu.I. Karlovich: *Calkin images of Fourier convolution operators with slowly oscillating symbols*. Oper. Theory Adv. Appl. **282** (2021), 193–218.
180. Yu.I. Karlovich: *On Mellin pseudodifferential operators with quasicontinuous symbols*. Math. Methods Appl. Sci. **44** (2021), No. 12, 9782–9816.
181. C.A. Fernandes, A.Yu. Karlovich, and Yu.I. Karlovich: *Banach algebras of Fourier multipliers equivalent at infinity to nice Fourier multipliers*. Banach J. Math. Anal. **15** (2021), No. 2, Paper No. 29, 17 p.
182. L.E. Flores-Zapotitla and Yu.I. Karlovich: *One-sided invertibility of discrete operators with bounded coefficients*. Aequationes Math. **95** (2021), No. 4, 699–735.
183. L.E. Flores-Zapotitla and Yu.I. Karlovich: *Correction to: "One-sided invertibility of discrete operators with bounded coefficients"*. Aequationes Math. **95** (2021), No. 4, 797–799.
184. C.A. Fernandes, A.Yu. Karlovich, and Yu.I. Karlovich: *Fourier convolution operators with symbols equivalent to zero at infinity on Banach function spaces*. Current trends in analysis, its applications and computation, pp. 335–343, Trends Math. Res. Perspect., Birkhäuser/Springer, Cham, 2022.
185. M.A. Bastos, C.A. Fernandes, and Yu.I. Karlovich: *On C*-algebras of singular integral operators with PQC coefficients and shifts with fixed points*. Complex Var. Elliptic Equ. **67** (2022), No. 3, 581–614.
186. Yu. Karlovich: *Algebras of Mellin pseudodifferential operators with quasicontinuous symbols*. J. Pseudo-Differ. Oper. Appl. **13** (2022), No. 2, Paper No. 18, 36 pp.
187. Yu.I. Karlovich and J. Rosales-Méndez: *The Haseman boundary value problem with quasicontinuous coefficients and shifts*. Complex Anal. Oper. Theory **16** (2022), No. 5, Paper No. 68, 28 pp.
188. E. Espinoza-Loyola and Yu.I. Karlovich: *C*-algebras of poly-Bergman type operators with piecewise slowly oscillating coefficients*. J. Math. Sci. (N.Y.) **266** (2022), No. 1, 77–102.
189. Yu.I. Karlovich and J. Rosales-Méndez: *On singular integral operators with shifts*. Matrix and Operator Equations and Applications. Mathematics Online First Collections, pp. 489–523, Springer, Cham, 2023.

190. Yu. Karlovich and M. Guerrero-Pérez: *The Haseman boundary value problem with matrix coefficient of semi-almost periodic type*. Banach J. Math. Anal. **17** (2023), No. 2, Paper No. 28, 43 pp.
191. Yu.I. Karlovich and F.J. Monsiváis-González: *Fredholmness of singular integral operators with complex conjugation on star-like curves with cusps*. Complex Anal. Oper. Theory **17** (2023), No. 5, Paper No. 69, 22 pp.
192. Yu.I. Karlovich and F.J. Monsiváis-González: *Mellin pseudodifferential operators and singular integral operators with complex conjugation*. Oper. Theory Adv. Appl. **295** (2024), 249–278.
193. Yu.I. Karlovich: *Index calculation for Wiener-Hopf operators with slowly oscillating coefficients*. Linear Algebra Appl. **697** (2024), 93–130.
194. Yu.I. Karlovich: *Algebras of convolution type operators with piecewise quasicontinuous and piecewise slowly oscillating data on weighted Lebesgue spaces*. Oper. Theory Adv. Appl. **297** (2024), 145–170.
195. Yu.I. Karlovich: *Banach algebras of convolution type operators with PQC data*. Tbilisi Analysis and PDE Seminar. Extended Abstracts of the 2020–2023 Seminar Talks, pp. 109–119. Birkhäuser, Cham, 2024.

Other Publications

1. Yu.I. Karlovich and B. Tursunkulov: *Functional and singular integral operators with non-Carleman translations in generalized Hölder spaces*. Samarkand (1983), 31 pp. Manuscript No. 89 Uz-D83, deposited in UzNIINTI, 1983 (Russian).
2. Yu.I. Karlovich, Yu.D. Latushkin and R. Mardiev: *One-sided invertibility of functional operators and the $n(d)$-normality of singular integral operators with translation*. Odessa (1984), 25 pp. Manuscript No. 8361–84, deposited in VINITI, 1984 (Russian).
3. Yu.I. Karlovich and R. Mardiev: *On one-sided invertibility of functional operators and $n(d)$-normality of singular integral operators with a shift having periodic points in Hölder spaces*. Samarkand (1988), 56 pp. Manuscript No. 822-Uz88, deposited in UzNIINTI, 1988 (Russian).
4. Yu.I. Karlovich and Z.M. Lysenko: *Modification of operator approach for discontinuous shift and piecewise-smooth contour. Noether property*. Odessa (1989), 22 pp. Manuscript No. 1942-Uk89, deposited in UkrNIINTI, 1989 (Russian).
5. Yu.I. Karlovich and Z.M. Lysenko: *Modification of operator approach for discontinuous shift and piecewise-smooth contour. Index*. Odessa (1989), 22 pp. Manuscript No. 1941-Uk89, deposited in UkrNIINTI, 1989 (Russian).

Yuri Karlovich and the Metamorphosis of Spectra of Singular Integral and Toeplitz Operators

Albrecht Böttcher

For Yura Karlovich on his 75th birthday
with sincere friendship, admiration,
and hearty thanks for being my partner
in one of the most amazing mathematical
adventures of my life

Abstract This is an expository paper describing a few selected topics of Yuri Karlovich's work. I focus the attention on our joint work devoted to the identification of the local spectra of Toeplitz and singular integral operators caused by the oscillations of Carleson curves and those of Muckenhoupt weights.

1 Origin and Motivation

Singular integral equations. The Cauchy singular integral operator is the operator

$$(S_\Gamma f)(x) = \text{p.v.}\frac{1}{\pi i}\int_\Gamma \frac{f(y)}{y-x}\,dy, \qquad x \in \Gamma. \tag{1}$$

Here Γ is a locally rectifiable and oriented curve in the complex plane $\mathbf{C}$. Specific restrictions on Γ will be made later. Admissible examples include the unit circle $\mathbf{T}$, subarcs of $\mathbf{T}$, the real line $\mathbf{R}$, the real half-line $\mathbf{R}_+ = (0, \infty)$, or the circumference of a pentagon together with its five diagonals. The integral is understood in the principal value sense:

$$\text{p.v.}\frac{1}{\pi i}\int_\Gamma \frac{f(y)}{y-x}\,dy := \frac{1}{\pi i}\lim_{\varepsilon\to 0}\int_{\Gamma\setminus\Gamma(x,\varepsilon)} \frac{f(y)}{y-x}\,dy,$$

A. Böttcher (✉)
Fakultät für Mathematik, TU Chemnitz, Chemnitz, Germany
e-mail: aboettch@mathematik.tu-chemnitz.de

A. Böttcher et al. (eds.), *Achievements and Challenges in the Field of Convolution Operators*, Operator Theory: Advances and Applications 306,
https://doi.org/10.1007/978-3-031-80486-1_3

where $\Gamma(x,\varepsilon) := \{y \in \Gamma : |y-x| < \varepsilon\}$. The questions on the existence of the principal value and on spaces we can make S_Γ act on will also be discussed later.

The simplest singular integral equations are equations of the form $cf + dS_\Gamma f = h$, where c, d, h are given functions on Γ and f is a sought function on Γ. It turns out that it is often more convenient to work with the operators P_Γ, Q_Γ defined by $P_\Gamma = (I+S_\Gamma)/2$, $Q_\Gamma = (I-S_\Gamma)/2$. The equation then becomes $aP_\Gamma f + bQ_\Gamma f = h$ with $a = c+d$, $b = c-d$, and assuming that a has no zeros on Γ, we may write the equation as $P_\Gamma f + GQ_\Gamma f + g$ with $G = b/a$ and $g = h/a$.

A pivotal topic of Yura[1] Karlovich's work is so-called singular integral equations with shift. An example of such an equation reads $W_\alpha P_\Gamma f + GQ_\Gamma f = g$ where α, the shift, is a bijective map of Γ onto itself (subject to further restrictions) and W_α stands for the composition operator $(W_\alpha f)(x) = f(\alpha(x))$.

The Haseman problem. For simplicity, suppose now that $\Gamma = \mathbf{R}_+$ is the naturally oriented positive half-line. We omit the subscript Γ in $S_\Gamma, P_\Gamma, Q_\Gamma$. If f is sufficiently nice, say $f \in L^1(\mathbf{R}_+) \cap L^p(\mathbf{R}_+)$ for some $1 < p < \infty$, then the function Φ defined by the Cauchy integral

$$\Phi(z) = \frac{1}{2\pi i}\int_{\mathbf{R}_+} \frac{f(y)}{y-z}\,dy, \quad z \in \mathbf{C} \setminus [0,\infty),$$

is analytic in $\mathbf{C} \setminus [0,\infty)$ and its limits

$$\Phi^\pm(x) := \lim_{\varepsilon\to 0^+} \Phi(x \pm i\varepsilon)$$

exist for almost all $x \in \mathbf{R}_+ = (0,\infty)$. The Sokhotski-Plemelj formulas tell us that $\Phi^+(x) = (Pf)(x)$ and $\Phi^-(x) = -(Qf)(x)$ for almost all $x \in \mathbf{R}_+$. Consequently, the equation $W_\alpha Pf + GQf = g$ may be written as

$$\Phi^+(\alpha(x)) = G(x)\Phi^-(x) + g(x), \quad x \in \mathbf{R}_+. \tag{2}$$

This is a so-called Haseman problem. One seeks a function Φ that is analytic in $\mathbf{C} \setminus [0,\infty)$ and whose boundary values $\Phi^\pm$ on $\mathbf{R}_+$ satisfy a "jump condition with shifted argument". In the case of the absence of the shift, or equivalently, in the case where $\alpha(x) = x$, Eq. (2) is known as a Riemann-Hilbert problem.

Conformal welding. In the papers [1–3], problem (2) is transformed into a problem without shift via so-called conformal welding. The idea is to find an oriented and locally rectifiable curve Γ which is homeomorphic to $[0,\infty)$ and a conformal (i.e., analytic and bijective) map $\omega : \dot{\mathbf{C}} \setminus [0,\infty) \to \dot{\mathbf{C}} \setminus \Gamma$ such that

$$\omega^+(\alpha(x)) = \omega^-(x) \ \text{ for all } \ x \in (0,\infty)$$

[1] Yura is a diminutive form of the name Yuri commonly used by friends and colleagues.

and such that $\omega(z)$ converges to the endpoint of Γ as $z \to 0$. Here $\dot{\mathbf{C}} = \mathbf{C} \cup \{\infty\}$, and $\omega^+(\xi)$ and $\omega^-(\xi)$ denote the limits of $\omega(z)$ as $z \in \mathbf{C} \setminus [0, \infty)$ approaches $\xi \in (0, \infty)$ from above and from below, respectively.

To have a concrete example, take $\alpha(x) := \beta x$ with some constant $\beta > 0$ and $\beta \neq 1$. Put

$$\delta = \frac{1}{2\pi} \log \beta, \quad \Gamma = \{re^{-i\delta \log r} : r > 0\} \cup \{0\}.$$

Notice that Γ is a logarithmic spiral. Let $\log z$ be the branch of the logarithm which is analytic in $\mathbf{C} \setminus [0, \infty)$ and assumes the value $-i\pi$ at $z = -1$, and define

$$\omega(z) = z^{1/(1+i\delta)} = \exp\left(\frac{1}{1+i\delta} \log z\right)$$

for $z \in \mathbf{C} \setminus [0, \infty)$. It can be shown by elementary computations that ω maps $\dot{\mathbf{C}} \setminus [0, \infty)$ conformally onto $\dot{\mathbf{C}} \setminus \Gamma$ and that $\omega^+(\beta x) = \omega^-(x)$ for $x \in (0, \infty)$ and $\omega(z) \to 0$ as $z \to 0$. The details of the computation are also on page 375 of [9].

Removing the shift. Let α, Γ, ω be as in the preceding example. We denote by $\tau : \dot{\mathbf{C}} \setminus \Gamma \to \dot{\mathbf{C}} \setminus \mathbf{R}_+$ the inverse of ω and define Ψ on $\dot{\mathbf{C}} \setminus \Gamma$ by $\Psi(\zeta) := \Phi(\tau(\zeta))$. Then, for $y = \omega^+(\alpha(x)) = \omega^-(x)$ on Γ,

$$\begin{aligned}
\Psi^+(y) &= \lim \Phi(\tau(\zeta)) \text{ as } \zeta \to y \text{ from the left} \\
&= \lim \Phi(z) \text{ as } z \to \alpha(x) \text{ from above} \\
&= \Phi^+(\alpha(x)), \\
\Psi^-(y) &= \lim \Phi(\tau(\zeta)) \text{ as } \zeta \to y \text{ from the right} \\
&= \lim \Phi(z) \text{ as } z \to x \text{ from below} \\
&= \Phi^-(x).
\end{aligned}$$

After extending τ to Γ by defining $\tau(\omega^-(x)) = x$ we get $G(x) = G(\tau(y))$ and $g(x) = g(\tau(y))$. Consequently, the Haseman problem (2) can be transformed into the Riemann-Hilbert problem

$$\Psi^+(y) = G(\tau(y))\Psi^-(y) + g(\tau(y)), \quad y \in \Gamma. \tag{3}$$

Instead of a singular integral equation with a shift on $\mathbf{R}_+$, we now have a singular integral equation without shift but on a logarithmic spiral Γ. To show the natural emergence of integration contours like logarithmic spirals is the purpose of this introductory section and hence I want to make a temporary stop with mathematics at this point. I will say more on the singular integral operator related to (3) later.

Fig. 1 Eight years later: with Ilya Spitkovsky and Yura Karlovich during the international conference "Factorization, singular operators, and related problems in honor of Georgii Litvinchuk" on Madeira Island in January 2002

Yura in Chemnitz. In the period from December 1993 to September 1996, Yura was a visiting professor at the TU Chemnitz. He came to Chemnitz on my invitation. I wanted to learn from him about techniques for tackling non-local operators, and we planned to write a book on this topic. The first pages had already been written when we both attended the Oberwolfach workshop "Singuläre Integral- und Pseudodifferential-Operatoren" in January 1994. This workshop motivated Yura to reconsider problem (3), and some day he told me that the well-known circular arcs appearing in the spectra of singular integral operators over Lyapunov curves become logarithmic double spirals when letting the operators act on a logarithmic spiral. Just at that time I was captured by Ilya Spitkovsky's discovery [24] of horns in the spectra of singular integral operators in the presence of general Muckenhoupt weights; see also [6]. These two happy circumstances made Yura and me put the book project aside and start the adventure I will tell about in the following. Figure 1 shows the three of us.

2 Carleson Curves and Muckenhoupt Weights

General setting. We now turn to bounded curves. A set $\Gamma \subset \mathbf{C}$ is said to be a simple curve if it is either a Jordan curve, which means that it is homeomorphic to the unit circle $\mathbf{T}$, or if it is an arc, that is, if it is homeomorphic to the line segment

$[0,1]$. A weight on a simple rectifiable curve Γ is a function $w : \Gamma \to [0,\infty]$ measurable with respect to length measure such that the pre-image $w^{-1}(\{0,\infty\})$ has measure zero. Given a simple rectifiable curve Γ and a weight w on it, we denote by $L^p(\Gamma, w)$ $(1 \le p < \infty)$ the usual Lebesgue space with the norm

$$\|f\| := \left(\int_\Gamma |f(\tau)|^p w(\tau)^p \, |d\tau|\right)^{1/p}.$$

Note that the weight enters the norm in the pth power. We let $L^\infty(\Gamma)$ stand for the space of essentially bounded measurable functions on Γ. After giving Γ an orientation, which is always the positive one in the case of a Jordan curve, the singular integral operator S_Γ is defined by (1), but we now change x, y to t, τ. The operator S_Γ is said to be well-defined on $L^p(\Gamma, w)$ if $S_\Gamma f$ exists almost everywhere on Γ and is in $L^p(\Gamma, w)$ for every $f \in L^p(\Gamma, w)$.

The following theorem is one of the great achievements of mathematical analysis of the twentieth century.

> *Let Γ be a simple curve. The singular integral operator S_Γ is well-defined and bounded on $L^p(\Gamma, w)$ if and only if $1 < p < \infty$, Γ is a Carleson curve, and w is an A_p Muckenhoupt weight.*

This theorem grew out of the work of Hardy, Littlewood, M. Riesz, Mikhlin, Babenko, Khvedelidze, Helson, Szegő, Ahlfors, Widom, Forelli, Danilyuk, Shelepov, Paatashvili, Khuskivadze, and others, the decisive steps of the proof were made by Calderón, David, Hunt, Muckenhoupt, Wheeden, and more recent developments (including new proofs and generalizations) are connected with the names of Coifman, McIntosh, Meyer, Jones, Journé, Semmes, Murai, Dynkin, Osilenker, Mattila, Melnikov, Verdera, to cite only some mathematicians. Precise references are in [9, 10]. A full proof (occupying about 80 pages) is also in Chapters 3 and 4 of [9].

Carleson curves. A Carleson curve (often also called Ahlfors-David curve) is a simple rectifiable and oriented curve Γ such that $\sup_{t,\varepsilon} |\Gamma(t,\varepsilon)|/\varepsilon < \infty$, the supremum over $t \in \Gamma$ and $\varepsilon > 0$ and with $|\Gamma(t,\varepsilon)|$ denoting the length of the curve's portion $\Gamma(t,\varepsilon) := \{\tau \in \Gamma : |\tau - t| < \varepsilon\}$. This is of course satisfied for piecewise C^1 curves. Less trivial examples are arcs of the form

$$\Gamma = \{0, 1\} \cup \{\tau \in \mathbf{C} : \tau = re^{i\theta(r)}, 0 < r < 1\} \tag{4}$$

where $\theta(r) = h(\log|\log r|)\log r$ with a real-valued function $h \in C^1(\mathbf{R})$ for which h and h' are bounded on $\mathbf{R}$; see [9, Example 1.7]. For $h(x) = 0$ we get the line segment $[0,1]$, for $h(x) = \delta$ $(\neq 0)$ we obtain logarithmic spirals, and the choice $h(x) = \delta + \gamma \sin(\eta x)$ delivers quite exotic Carleson arcs. Some pictures are in [11].

Muckenhoupt weights. Given a simple rectifiable curve Γ and $p \in (1, \infty)$, we denote by $A_p(\Gamma)$ the set of weights on Γ satisfying

$$\sup_I \left(\frac{1}{|I|} \int_I w(\tau)^p |d\tau| \right)^{1/p} \left(\frac{1}{|I|} \int_I w(\tau)^{-q} |d\tau| \right)^{1/q} < \infty,$$

where $1/p + 1/q = 1$, I ranges over all subarcs of Γ, and $|I|$ is the length of I. Such weights are called Muckenhoupt weights. Weights of the form

$$w(\tau) = \prod_{j=1}^{n} |\tau - t_j|^{\mu_j} \tag{5}$$

with distinct points $t_j \in \Gamma$ and real numbers μ_j are referred to as power weights. A classical result by B. V. Khvedelidze says that such a weight is in $A_p(\Gamma)$ if and only if $-1/p < \mu_j < 1/q$ for all j. More sophisticated weights on the arcs given by (4) result from putting

$$w\left(re^{i\theta(r)}\right) := e^{v(r)}, \quad 0 < r < 1,$$

where $v(r) = g(\log|\log r|) \log r$ with a real-valued function $g \in C^2(\mathbf{R})$ such that g, g', g'' are bounded on $\mathbf{R}$. Together with Yura we proved in [9, Theorem 2.36] that such a weight w is in $A_p(\Gamma)$ if and only if

$$-\frac{1}{p} < \liminf_{x \to +\infty} \left(g(x) + g'(x)\right) \le \limsup_{x \to +\infty} \left(g(x) + g'(x)\right) < \frac{1}{q}.$$

In the special case $g(x) = \mu + \varepsilon \sin(\eta x)$ we obtain that this holds exactly if

$$-\frac{1}{p} < \mu - |\varepsilon|\sqrt{\eta^2 + 1} \le \mu + |\varepsilon|\sqrt{\eta^2 + 1} < \frac{1}{q}. \tag{6}$$

For $g(x) = \mu$, we get the power weight $w\left(e^{i\theta(r)}\right) = e^{\mu \log r} = r^\mu$ and (6) becomes the known condition $-1/p < \mu < 1/q$. Taking $g(x) = \varepsilon \sin x$, we encounter an oscillating weight and (6) shows that this weight is an $A_p(\Gamma)$ Muckenhoupt weight if and only if $|\varepsilon|\sqrt{2} < \min(1/p, 1/q)$.

3 Toeplitz Operators

Hardy spaces. Throughout this and the next three sections we suppose that Γ is a positively oriented Carleson Jordan curve, $1 < p < \infty$, and w is a weight in $A_p(\Gamma)$. It is well known that then $S_\Gamma^2 = I$ and hence $P_\Gamma = (I + S_\Gamma)/2$ is a bounded projection on $L^p(\Gamma, w)$. The Hardy (or Smirnov) space $H^p(\Gamma, w)$ is defined as the range of P_Γ, i.e., $H^p(\Gamma, w) = P_\Gamma(L^p(\Gamma, w))$.

Toeplitz operators. Let a be a function in $L^\infty(\Gamma)$. Then multiplication by a is bounded on $L^p(\Gamma, w)$. The Toeplitz operator $T(a)$ is defined as the compression of multiplication by a to $H^p(\Gamma, w)$:

$$T(a) : H^p(\Gamma, w) \to H^p(\Gamma, w), \quad f \mapsto P_\Gamma(af).$$

The function a is in this context referred to as the symbol of $T(a)$. If $L^p(\Gamma, w)$ is the Hilbert space $L^2(\mathbf{T})$ on the unit circle (with no weight), then the matrix representation of $T(a)$ in the orthonormal basis $\{e^{in\theta}/\sqrt{2\pi}\}_{n=0}^\infty$ of the corresponding Hardy space $H^2(\mathbf{T})$ is the classical Toeplitz matrix $(a_{j-k})_{j,k=0}^\infty$ composed of the Fourier coefficients a_n ($n \in \mathbf{Z}$) of a.

Spectra. The spectrum $\sigma(A)$ of a bounded linear operator A on $L^p(\Gamma, w)$ or $H^p(\Gamma, w)$ is the set of all $\lambda \in \mathbf{C}$ for which $A - \lambda I$ is not (boundedly) invertible, and the essential spectrum $\sigma_{\text{ess}}(A)$ is the set of all $\lambda \in \mathbf{C}$ for which T is not Fredholm, that is, not invertible modulo compact operators. One can show that the essential spectrum of Toeplitz oprators is local in nature, which means that there are constructions that associate a set $\sigma_t(T(a))$ with each $t \in \Gamma$ such that

$$\sigma_{\text{ess}}(T(a)) = \bigcup_{t\in\Gamma} \sigma_t(T(a)).$$

The sets $\sigma_t(T(a))$ are called local spectra and, given p, Γ, w, they depend only on a in an arbitrarily small open neighborhood $U \subset \Gamma$ of t. If a is continuous, $a \in C(\Gamma)$, then $\sigma_t(T(a)) = \{a(t)\}$, $\sigma_{\text{ess}}(T(a)) = a(\Gamma)$, and $\sigma(T(a))$ is the union of $a(\Gamma)$ and all points in $\mathbf{C}$ that are encircled by $a(\Gamma)$ with nonzero winding number. This is a classical result in the case where Γ and w are nice. It was proved in [7] for general Carleson curves and general weights; see also [9, Theorem 6.24].

Piecewise continuous symbols. A function $a \in L^\infty(\Gamma)$ is said to be piecewise continuous, $a \in PC(\Gamma)$, if the one-sided limits $a(t \pm 0)$ exist for each $t \in \Gamma$. Here $a(t-0)$ denotes the limit of $a(\tau)$ as τ approaches t following the positive orientation of Γ, while $a(t+0)$ is the limit of $a(\tau)$ as $\tau \to t$ opposite to the positive orientation of Γ. The purpose of what follows is to find $\sigma_t(T(a))$ is $a \in PC(\Gamma)$ makes a jump at t, that is, if $a(t-0) \neq a(t+0)$. The (essential) range of a on Γ is always part of $\sigma_{\text{ess}}(T(a))$. To find out what else is contained in $\sigma_{\text{ess}}(T(a))$ we may therefore assume that a is invertible in $L^\infty(\Gamma)$.

4 Spiralic Carleson Curves

Logarithmic whirls. Fix a point $t \in \Gamma$. We then have $\tau - t = |\tau - t|e^{i\arg(\tau-t)}$ for $\tau \in \Gamma \setminus \{t\}$, and the argument $\arg(\tau - t)$ may be chosen so that it is a continuous function on $\Gamma \setminus \{t\}$. Seifullaev [23] showed that for an arbitrary Carleson curve the estimate

$$\arg(\tau - t) = O(-\log|\tau - t|) \quad (\tau \to t)$$

holds. In our paper [7], written in the first half of 1994, we studied Carleson curves subject to the stronger condition

$$\arg(\tau - t) = -\delta_t \log|\tau - t| + O(1) \quad (\tau \to t) \tag{7}$$

with some $\delta_t \in \mathbf{R}$. If $\delta_t \neq 0$, we call t a logarithmic whirl point. The curve Γ is said to be a spiralic curve if for each $t \in \Gamma$ there is a $\delta_t \in \mathbf{R}$ such that (7) is satisfied.

Theorem 1 *Suppose for each t on the Carleson Jordan curve Γ there is a $\delta_t \in \mathbf{R}$ such that (7) holds and suppose w is identically 1 on Γ. Let a be a function in $PC(\Gamma)$ such that $a(t \pm 0) \neq 0$ for all $t \in \Gamma$. Then $T(a)$ is Fredholm on $H^p(\Gamma)$ if and only if*

$$\frac{1}{p} - \frac{1}{2\pi}\left(\arg\frac{a(t-0)}{a(t+0)} - \delta_t \log\left|\frac{a(t-0)}{a(t+0)}\right|\right) \notin \mathbf{Z} \tag{8}$$

for all $t \in \Gamma$.

The idea behind the proof is that $T(a)$ is Fredholm if and only if a admits a so-called Wiener-Hopf factorization

$$a(\tau) = a_-(\tau)\tau^\kappa a_+(\tau)$$

with some integer κ, which is minus the index of the Fredholm operator $T(a)$. The factors $a_\pm$ are subject to a series of requirements. Part of these requirements are harmless. The deciding one is that $|a_+^{-1}|$ must be weight in $A_p(\Gamma)$. Choose $\gamma_t \in \mathbf{C}$ such that

$$\operatorname{Re}\gamma_t = -\frac{1}{2\pi}\arg\frac{a(t-0)}{a(t+0)}, \quad \operatorname{Im}\gamma_t = \frac{1}{2\pi}\log\left|\frac{a(t-0)}{a(t+0)}\right|. \tag{9}$$

We may suppose that the origin is in the interior of the domain bounded by Γ and define g_t by $g_t(\tau) = \tau^{-\gamma_t}$ as a branch of the function on Γ which makes the jump at the point t. Note that

$$\frac{g_t(t+0)}{g_t(t-0)} = \frac{1}{e^{-2\pi i\gamma_t}} = e^{2\pi i\gamma_t} = \frac{a(t+0)}{a(t-0)}.$$

Having recourse to a localization theorem, one can show that $T(a)$ is Fredholm if and only if $T(g_t)$ is Fredholm for each $t \in \Gamma$. A Wiener-Hopf factorization of g_t is

$$g_t(\tau) = \tau^{-\gamma_t} = (1 - t/\tau)^{\gamma_t+\kappa_t}\tau^{\kappa_t}(\tau - t)^{-\gamma_t-\kappa_t}.$$

Using condition (7) it can be proved that there are positive constants C_1, C_2 such that

$$C_1|\tau - t|^{\operatorname{Re}\gamma_t+\delta_t\operatorname{Im}\gamma_t} \le |(\tau - t)^{\gamma_t}| \le C_2|\tau - t|^{\operatorname{Re}\gamma_t+\delta_t\operatorname{Im}\gamma_t} \tag{10}$$

for $\tau \in \Gamma$. Consequently, $|(\tau - t)^{\gamma_t + \kappa_t}|$ is weight in $A_p(\Gamma)$ if and only if the power weight $|\tau - t|^{\mathrm{Re}\,\gamma_t + \delta_t \mathrm{Im}\,\gamma_t + \kappa_t}$ is in $A_p(\Gamma)$, and we know that this happens exactly if

$$-1/p < \mathrm{Re}\,\gamma_t + \delta_t \mathrm{Im}\,\gamma_t + \kappa_t < 1/q.$$

Thus, we arrive at the conclusion that $T(a)$ is Fredholm if and only if

$$1/p + \mathrm{Re}\,\gamma_t + \delta_t \mathrm{Im}\,\gamma_t \notin \mathbf{Z},$$

which, by (9), is just (8).

What does (8) mean geometrically? For $\delta, \nu \in \mathbf{R}$, consider the set

$$\mathcal{S}_{\nu,\delta} := \{\xi \in \mathbf{C} \setminus \{0\} : \arg \xi - \delta \log |\xi| \in 2\pi\nu + 2\pi\mathbf{Z}\}.$$

We have $\xi \in \mathcal{S}_{\nu,\delta}$ if and only if

$$\xi = e^{\log|\xi| + i(\delta \log|\xi| + 2\pi\nu)} = e^{2\pi i \nu} r e^{i\delta \log r}$$

with some $r > 0$, which reveals that $\mathcal{S}_{\nu,\delta}$ is a logarithmic spiral for $\delta \neq 0$ and a ray starting at the origin for $\delta = 0$. Put

$$\mathcal{S}(z_1, z_2, \nu, \delta) := \left\{ \frac{z_2\xi - z_1}{\xi - 1} : \xi \in \mathcal{S}_{\nu,\delta} \right\}.$$

The Möbius transform $\xi \mapsto (z_2\xi - z_1)/(\xi - 1)$ maps 0 to z_1 and the point at infinity to z_2. Consequently, it maps the logarithmic spiral $\mathcal{S}_{\nu,\delta}$ ($\delta \neq 0$) to a kind of logarithmic double spiral $\mathcal{S}(z_1, z_2, \nu, \delta)$ scrolling out of z_1 and coiling at z_2. The set $\mathcal{S}(z_1, z_2, \nu, 0)$, being the image of a straight ray, is a circular arc between z_1 and z_2. Here is the result established in [7].

Theorem 2 *Let $t \in \Gamma$ and suppose (7) holds. Assume further that the weight w is identically 1 on Γ. Then, for every $a \in PC(\Gamma)$, the local spectrum of $T(a)$ on $H^p(\Gamma)$ at the point t equals*

$$\sigma_t(T(a)) = \{a(t-0), a(t+0)\} \cup \mathcal{S}(a(t-0), a(t+0), 1/p, \delta_t).$$

Indeed, Theorem 1 implies that $\sigma_t(T(a)) \setminus \{a(t \pm 0)\}$ equals

$$\left\{ \lambda \in \mathbf{C} : \arg \frac{a(t-0) - \lambda}{a(t+0) - \lambda} - \delta_t \log \left| \frac{a(t-0) - \lambda}{a(t+0) - \lambda} \right| \in \frac{2\pi}{p} + 2\pi\mathbf{Z} \right\}.$$

Hence, $\lambda \in \sigma_t(T(a)) \setminus \{a(t \pm 0)\}$ if and only if

$$\xi := \frac{a(t-0) - \lambda}{a(t+0) - \lambda} \in \mathcal{S}_{1/p,\delta_t},$$

which is equivalent to saying that

$$\lambda = \frac{a(t+0)\xi - a(t-0)}{\xi - 1} \quad \text{for some } \xi \in \mathcal{S}_{1/p,\delta_t}.$$

Spiralic curves with a weight. Let Γ be an arbitrary Carleson Jordan curve and $w \in A_p(\Gamma)$. In the following section we will associate with each point $t \in \Gamma$ two numbers $-1/p < \mu_t \le \nu_t < 1/q$ measuring in some sense the "powerlikeness" of the weight w at t. If w is the pure power weight (5), then $\mu_t = \nu_t = \mu_j$ for $t = t_j$ and $\mu_t = \nu_t = 0$ if $t \notin \{t_1, \ldots, t_m\}$. Let

$$\mathcal{S}(z_1, z_2, \nu_1, \nu_2, \delta) = \bigcup_{\nu \in [\nu_1, \nu_2]} \mathcal{S}(z_1, z_2, \nu, \delta).$$

We call this set a horn for $\delta = 0$ and a spiralic horn for $\delta \neq 0$. In [7] we proved that if Γ is spiralic and $w \in A_p(\Gamma)$, then, for every $a \in PC(\Gamma)$, the local spectrum of $T(a)$ on $H^p(\Gamma, w)$ at the point t equals

$$\sigma_t(T(a)) = \{a(t-0), a(t+0)\} \cup \mathcal{S}(a(t-0), a(t+0), 1/p + \mu_t, 1/p + \nu_t, \delta_t).$$

The pictures that will follow. In the mid 1990s, we still gave our talks using hand-written transparencies. To convey a flavor of those times, I decided to take the pictures behind the mathematics presented here from just these transparencies. Figure 2 shows the essential spectrum of an operator $T(a)$.

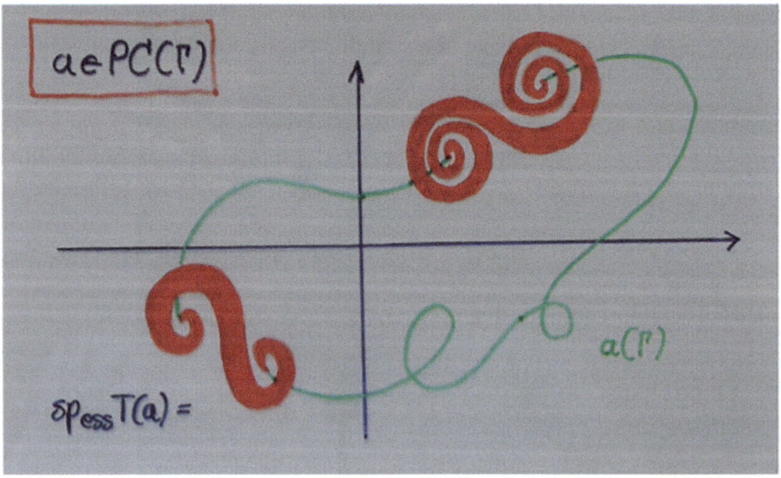

Fig. 2 The essential spectrum is the union of the range of a (green) and of spiralic horns filled in between the endpoints of the two jumps of a (red)

5 The Metamorphoses

Metamorphosis 1. The emergence of circular arcs in the business considered here was discovered by Gohberg and Krupnik [18] and Widom [25] in the 1960s. See also Krupnik's recent article [19] about the story of how the circular arcs were born. Note that all these results concerned "nice" curves Γ with power weights.

The first metamorphosis came with Ilya Spitkovsky [24] at the turn to the 1990s. He started with the observation that for "nice" curves Γ with arbitrary weight w belonging to $A_p(\Gamma)$ things may be reduced to identifying the sets

$$I_t(\Gamma, p, w) := \{\gamma \in \mathbf{R} : |\tau - t|^\gamma w(\tau) \in A_p(\Gamma)\}.$$

Ilya showed that $I_t(\Gamma, p, w)$ is always an open interval $(-1/p - \mu_t, 1/q - \nu_t)$ with certain numbers $-1/p < \mu_t \le \nu_t < 1/q$, the so-called indices of powerlikeness. (In several publications, including [6], the interval $I_t(\Gamma, p, w)$ is written as $(-\nu_t^-, 1 - \nu_t^+)$ with $\nu_t^- = 1/p + \nu_t$ and $\nu_t^+ = 1/p + \mu_t$.) In the Gohberg/Krupnik/Widom situation, $I_t(\Gamma, p, w)$ is an open interval of length 1, and the complement of $I_t(\gamma, p, w) + \mathbf{Z}$ in $\mathbf{R}$, which is comprised of the "forbidden" values, is $\{\lambda_t\} + \mathbf{Z}$ with some λ_t, i.e., the union of singletons separated at the distance 1. This caused the circular arc $\mathcal{S}(z_1, z_2, \lambda_t, 0)$. In the Spitkovsky case, the complement of $I_t(\gamma, p, w) + \mathbf{Z}$ in $\mathbf{R}$ (= the "forbidden" values) is $[1/q - \nu_t, 1/q - \mu_t] + \mathbf{Z}$, and if $\mu_t < \nu_t$, this yields an uncountable bundle of circular arcs, which constitute the horn $\mathcal{S}(z_1, z_2, 1/p + \mu_t, 1/p + \nu_t, 0)$. More details on Ilya's contribution to the topic can be found in [6].

Metamorphosis 2. The second metamorphosis of local spectra of Toeplitz operators was described in the previous section. It is the emergence of logarithmic double spirals and of spiralic horns in the case of spiralic curves. To understand this and the further metamorphoses, it is helpful to look at the plan behind them.

The blueprint. The role played by intervals $I_t(\Gamma, p, w)$ in Ilya's approach is performed by the sets

$$N_t(\Gamma, p, w) := \{\gamma \in \mathbf{C} : |(\tau - t)^\gamma| w(\tau) \in A_p(\Gamma)\} \tag{11}$$

in the general situation. Note that, as seen for example from (10), we cannot replace $|(\tau - t)^\gamma|$ by $|\tau - t|^{\mathrm{Re}\,\gamma}$. We proved that $N_t(\Gamma, p, w)$ is always an open set in the plane containing the origin and that it can be described as

$$\{\gamma = \mathrm{Re}\,\gamma + i\mathrm{Im}\,\gamma \in \mathbf{C} : -1/p < \mathrm{Re}\,\gamma + \alpha_t(\mathrm{Im}\,\gamma) \le \mathrm{Re}\,\gamma + \beta_t(\mathrm{Im}\,\gamma) < 1/q\} \tag{12}$$

with a concave function α_t and a convex function β_t. These two functions define a new set

$$Y(p, \alpha_t, \beta_t) := \{z = x + iy \in \mathbf{C} : 1/p + \alpha_t(x) \le y \le 1/p + \beta_t(x)\}.$$

We call α_t, β_t the indicator functions and $Y(p, \alpha_t, \beta_t)$ the indicator set of Γ, w, p at t. The following theorem was established in [10].

Theorem 3 *Let Γ be an arbitrary Carleson Jordan curve, $1 < p < \infty$, and $w \in A_p(\Gamma)$. Then, for $a \in PC(\Gamma)$, the local spectrum of $T(a)$ at $t \in \Gamma$ is*

$$\sigma_t(T(a)) = \left\{ \frac{a(t+0)e^{2\pi z} - a(t-0)}{e^{2\pi z} - 1} : z \in Y(p, \alpha_t, \beta_t) \right\} \cup \{a(t-0), a(t+0)\}. \tag{13}$$

The set (13) is a certain connected set joining $a(t-0)$ to $a(t+0)$. We call it a leaf. Thus, the procedure to get the leaf between two different points $z_1, z_2 \in \mathbf{C}$ is as follows: take $\zeta \in Y_t(p, \alpha_t, \beta_t)$, map it to $\xi = e^{2\pi i \zeta}$, and map ξ to

$$\lambda = M_{z_1, z_2}(\xi) := \frac{z_2 \xi - z_1}{\xi - 1}.$$

The set of all λ obtained in this way (plus the two points z_1, z_2 themselves) forms a set we denote by $\mathcal{L}(z_1, z_2, p, \alpha_t, \beta_t)$.

Metamorphoses 1 and 2 again. Let Γ be a spiralic Carleson Jordan curve with a weight $w \in A_p(\Gamma)$. Fix $t \in \Gamma$ and suppose (7) holds. For $\delta_t = 0$, we are in the context of Metamorphosis 1, for $\delta_t \neq 0$, we have Metamorphosis 2. From (10) we infer that if w is identically 1, then

$$N_t(p, \Gamma, w) = \{\gamma \in \mathbf{C} : -1/p < \operatorname{Re}\gamma + \delta_t \operatorname{Im}\gamma < 1/q\},$$

which implies that $\alpha_t(x) = \beta_t(x) = \delta_t x$ for $x \in \mathbf{R}$. In the case of an arbitrary weight w, the set $N_t(p, \Gamma, w)$ can be shown to be

$$\{\gamma \in \mathbf{C} : -1/p < \operatorname{Re}\gamma + \mu_t + \delta_t \operatorname{Im}\gamma \le \operatorname{Re}\gamma + \nu_t + \delta_t \operatorname{Im}\gamma < 1/q\}$$

with two numbers $-1/p < \mu_t \le \nu_t < 1/q$, called the indices of powerlikeness at t. In the case $\delta_t = 0$, these numbers coincide with the indices of powerlikeness introduced by Ilya. Consequently, $\alpha_t(x) = \mu_t + \delta_t x$, $\beta_t(x) = \nu_t + \delta_t x$, and

$$Y_t(p, \alpha_t, \beta_t) = \{x + iy \in \mathbf{C} : 1/p + \mu_t + \delta_t x \le y \le 1/p + \nu_t + \delta_t x\}.$$

This is a strip (degenerating to a straight line for $\mu_t = \nu_t$) with the slope δ_t. The exponential $\zeta \mapsto e^{2\pi i \zeta}$ maps this strip to the bundle of logarithmic spirals $\mathcal{S}_{\nu, \delta_t}$ $(1/p + \mu_t \le \nu \le 1/p + \nu_t)$ starting at the origin, and the Möbius transform finally maps this bundle into a spiralic horn (becoming a logarithmic double spiral in the degenerate case). For $\delta_t = 0$, further degenerations lead to horns and circular arcs.

Figures 3 and 4 illustrate Metamorphosis 1 and 2. There we took $a(t \pm 0) = \pm 1$, by virtue of which the Möbius transform is $(z+1)/(z-1)$.

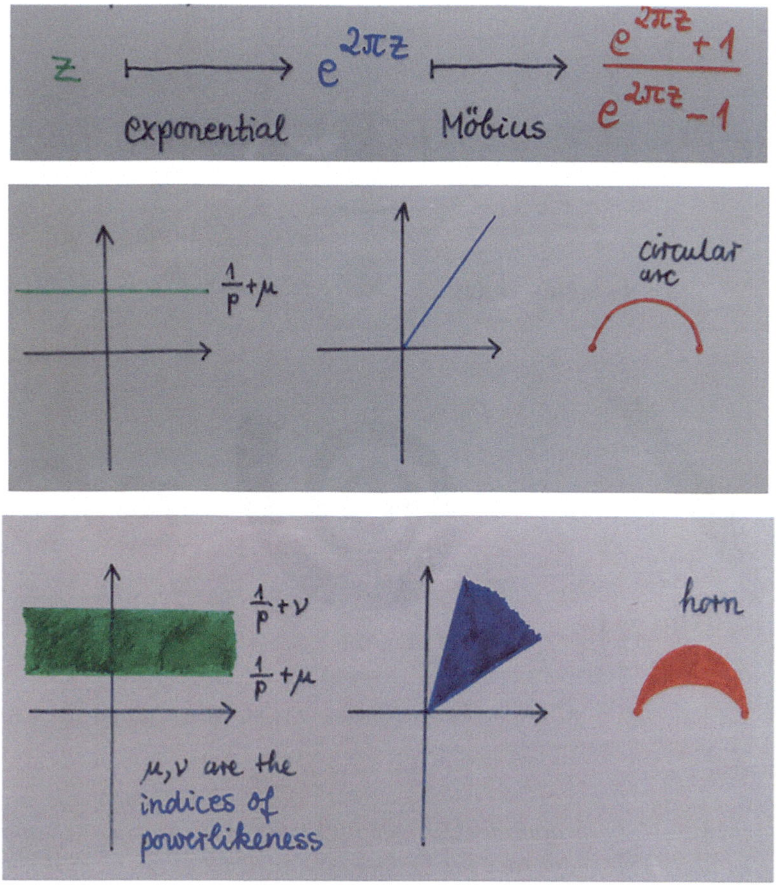

Fig. 3 Metamorphosis 1: the emergence of circular and horns on "nice" curves with arbitrary weights

Metamorphosis 3. In the first half of 1994, Yura and I had Metamorphosis 2, and it seemed to us that the further exploration of the matter, that is, the treatment of more and more general curves and weights would feed us the forthcoming years. At the annual meeting of the Deutsche Mathematiker-Vereinigung in Duisburg in 1994, I was invited to give a plenary talk, and I chose the title "Toeplitz operators with piecewise continuous symbols—a neverending story?" (in German). At that time Yura and I did not even dream of being able to settle the matter ultimately within the upcoming months. However, we made it.

Already in the early 1995 we were able to dispose of the case of general Carleson curves provided we don't have a weight on them [8]. Here is the result.

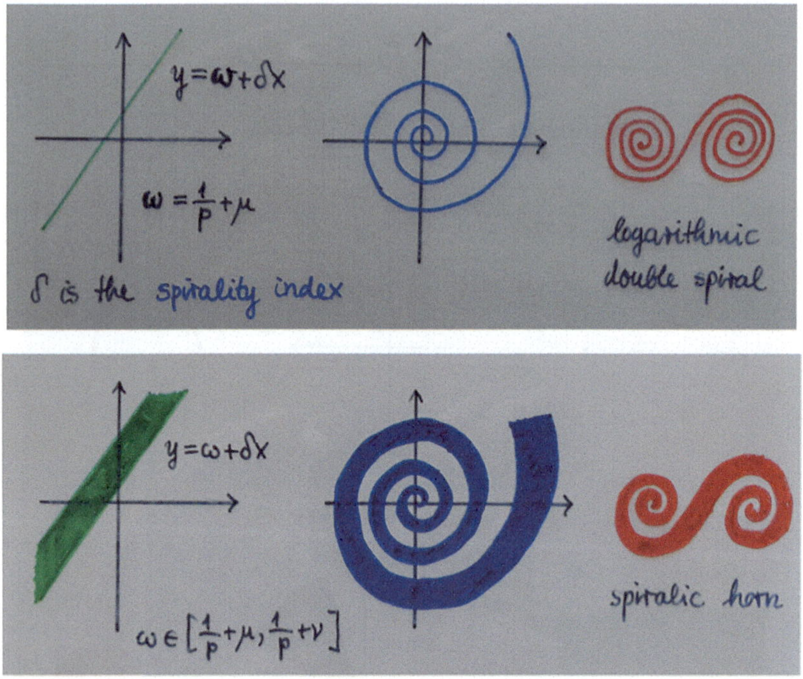

Fig. 4 Metamorphosis 2: the emergence of logarithmic double spirals and spiralic horns on spiralic curves with arbitrary weights

Theorem 4 *Let Γ be an arbitrary Carleson Jordan curve, $1 < p < \infty$, and suppose w is identically* 1. *Then for each $t \in \Gamma$, there are two numbers $\delta_t^-, \delta_t^+ \in \mathbf{R}$, the so-called spirality indices of Γ at t, such that $\delta_t^- \le \delta_t^+$ and the indicator functions in (12) are given by*

$$\alpha_t(x) = \min(\delta_t^- x, \delta_t^+ x), \quad \beta_t(x) = \max(\delta_t^- x, \delta_t^+ x) \ \textit{for} \ x \in \mathbf{R}.$$

The numbers δ_t^-, δ_t^+ can actually be shown to be independent of p; see the end of Sect. 5. If $a \in PC(\Gamma)$, then the local spectrum of $T(a)$ on $H^p(\Gamma)$ is

$$\sigma_t(T(a)) = \{a(t-0), a(t+0)\} \cup \left\{ \frac{a(t+0)e^{2\pi i\zeta} - a(t-0)}{e^{2\pi i\zeta} - 1} : \zeta \in Y_t \right\}$$

with $Y_t := \{x + iy \in \mathbf{C} : 1/p + \alpha_t(x) \le y \le 1/p + \beta_t(x)\}$. This indicator set is the union of all straight lines $\{x + iy : y = 1/p + \delta x\}$ with $\delta \in [\delta_t^-, \delta_t^+]$. Each of these straight lines is mapped via $\zeta \mapsto M_{z_1,z_2}(e^{2\pi i\zeta})$ into a logarithmic spiral between z_1 and z_2. All these logarithmic spirals pass through the point $M_{z_1,z_2}(e^{2\pi i/p})$. This is a separating point of the leaf: removing it makes the leaf split into two disjoint pieces. Thus, while oscillations of weights blow up the spirals

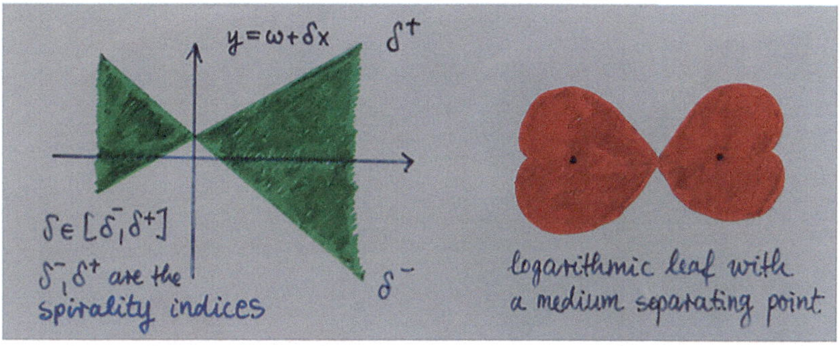

Fig. 5 Metamorphosis 3: the emergence of logarithmic leaves with a separating point in the case of arbitrary Carleson curves without weight

in the middle, oscillations of the curve blow up the spirals at their endpoints. See Fig. 5; there the blue intermediate set was omitted.

Metamorphosis 4. In the summer of 1995, we finally had an ultimate solution of the problem: we were able to characterize the local spectra for arbitrary Carleson Jordan curves Γ with arbitrary weights $w \in A_p(\Gamma)$. The solution is presented in detail in our paper [10] of 1999, the original version of which was submitted in September 1995. Due to Theorem 3, we are left with describing the indicator functions α_t, β_t. Figure 6 illustrates the following theorem.

Theorem 5 *Let Γ be a Carleson Jordan curve, $1 < p < \infty$, and $w \in A_p(\Gamma)$. Then the indicator functions α_t and β_t at a point $t \in \Gamma$ have the following properties:*

(i) $-\infty < \alpha_t(x) \le \beta_t(x) < \infty$ *for all* $x \in \mathbf{R}$*;*
(ii) $-1/p < \alpha_t(0) \le \beta_t(0) < 1/q$*;*
(iii) α_t *is a concave function and* β_t *is a convex function;*
(iv) *there exist six numbers* $\delta_t^\pm, \mu_t^\pm, \nu_t^\pm \in \mathbf{R}$ *such that*

$$\delta_t^- \le \delta_t^+, \quad -1/p < \mu_t^- \le \nu_t^- < 1/q, \quad -1/p < \mu_t^+ \le \nu_t^+ < 1/q,$$

and

$$\beta_t(x) = \nu_t^+ + \delta_t^+ x + o(1), \quad \alpha_t(x) = \mu_t^- + \delta_t^- x + o(1) \quad (x \to +\infty),$$
$$\beta_t(x) = \nu_t^- + \delta_t^- x + o(1), \quad \alpha_t(x) = \mu_t^+ + \delta_t^+ x + o(1) \quad (x \to -\infty).$$

Given $1 < p < \infty$ and any two functions α_t and β_t with these four properties, there exists a Carleson curve Γ, a weight $w \in A_p(\Gamma)$, and a point $t \in \Gamma$ such that α_t and β_t are the indicator functions at the point t.

Originally we suspected that the situation should always be as in the upper half of Fig. 6. There are indeed cases in which the indicator set is of this shape: it is completely characterized by two slopes (the spirality indices δ_t^- and δ_t^+) and the

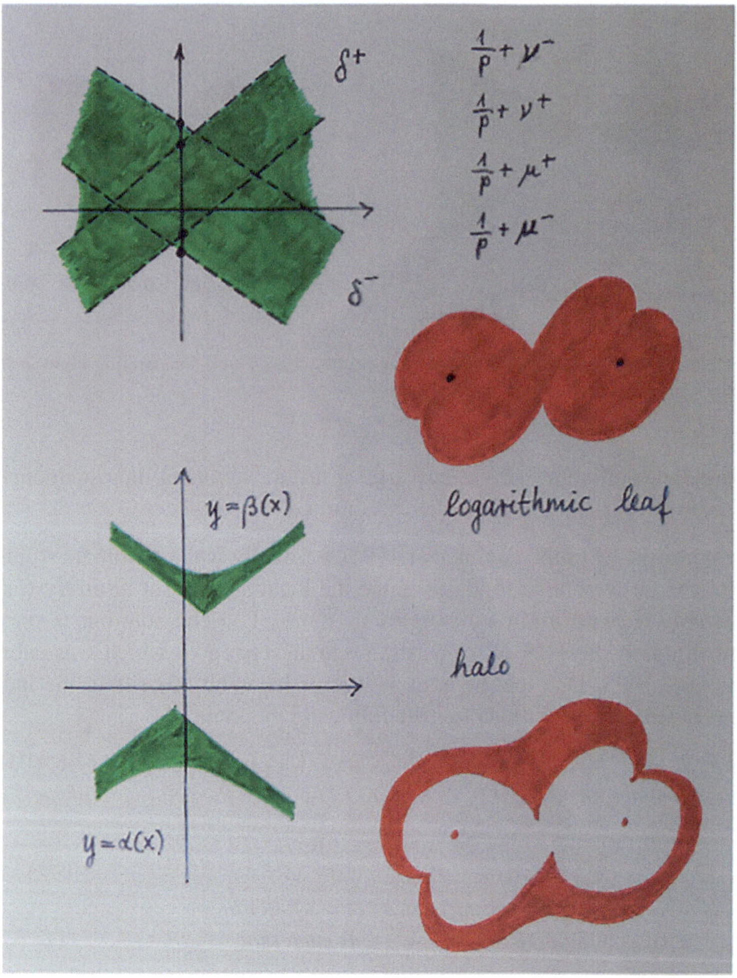

Fig. 6 Metamorphosis 4. The upper picture shows the four linear asymptotes in Theorem 5(iv), in the lower picture we see α_t, β_t. Thus, the union of the two green sets is actually *not* the indicator set as it was the case in the previous pictures. The indicator set results from shifting up the union of the two green sets by $1/p$

ordinates of the four points at which the straight boundary lines meet the imaginary axis (the numbers $1/p + \mu_t^\pm$, $1/p + \nu_t^\pm$ with the indices of powerlikeness $\mu_t^\pm$, $\nu_t^\pm$). The boundary of the resulting leaf is comprised of four pieces of logarithmic double spirals, by virtue of which we call such a leaf a logarithmic leaf.

Only after some time we understood that in general the indicator functions need not be as in the upper half of Fig. 6. Namely, some kind of interference between the oscillation of the curve and the oscillation of the weight may add two more pieces like those in the lower half of Fig. 6 to the indicator set. The upper boundary of the upper piece is given by a convex function $y = \beta_t(x)$ which has two asymptotes that coincide with the two straight upper boundary lines $y = \nu_t^{\pm} + \delta_t^{\pm} x$, while the lower boundary of the lower piece is described by a concave function $y = \alpha_t(x)$ the two asymptotes of which exist and are the two straight lower boundary lines $y = \mu_t^{\pm} + \delta_t^{\pm} x$. These two additional pieces add something to the logarithmic leaf shown in the lower half of Fig. 6. We call it a halo.[2]

The proof of Theorem 5 is very technical and I must refer the interested reader to [9, 10]. I confine myself to one key ingredient, which also gives a formula for the indicator functions. Let Γ be a Carleson Jordan curve, $1 < p < \infty$, and $w \in A_p(\Gamma)$. Fix $t \in \Gamma$. We constructed several submultiplicative functions ϱ associated with Γ, p, w, t, that is, functions $\varrho : (0, \infty) \to (0, \infty)$ satisfying $\varrho(xy) \le \varrho(x)\varrho(y)$ for all $x, y \in (0, \infty)$. For such functions one can define their so-called Boyd indices

$$\alpha := \lim_{x \to 0} \frac{\log \varrho(x)}{\log x}, \quad \beta := \lim_{x \to \infty} \frac{\log \varrho(x)}{\log x},$$

and we were able to give alternative characterizations of Muckenhoupt weights in terms of these indices. One of these constructions is as follows.

Given a weight $\psi : \Gamma \to [0, \infty]$ such that $\log \psi \in L^1(\Gamma)$, we consider the geometric means

$$(G\psi)_t(\xi R) := \exp\left(\frac{1}{\Gamma(t, \xi R)} \int_{\Gamma(t,\xi R)} \log \psi(\tau)\, |d\tau|\right),$$

where, as above, $\Gamma(t, \varepsilon)$ denotes the portion $\{\tau \in \Gamma : |\tau - t| < \varepsilon\}$, define a new function $V_t^0 \psi : (0, \infty) \to [0, \infty]$ by

$$(V_t^0 \psi)(\xi) := \begin{cases} \limsup\limits_{R \to 0} \dfrac{(G\psi)_t(\xi R)}{(G\psi)_t(R)} & \text{if } \xi \in (0, 1], \\[2ex] \limsup\limits_{R \to 0} \dfrac{(G\psi)_t(R)}{(G\psi)_t(\xi^{-1} R)} & \text{if } \xi \in [1, \infty), \end{cases}$$

and put

$$\alpha(V_t^0 \psi) := \limsup_{\xi \to 0} \frac{\log (V_t^0 \psi)(\xi)}{\log \xi}, \quad \beta(V_t^0 \psi) := \limsup_{\xi \to \infty} \frac{\log (V_t^0 \psi)(\xi)}{\log \xi}.$$

For $\tau \in \Gamma$ we have $\tau - t = |\tau - t| e^{i \arg(\tau - t)}$, and $\arg(\tau - t)$ may be chosen to be continuous of $\Gamma \setminus \{t\}$. Put $\eta_t(\tau) = e^{-\arg(\tau - t)}$ for $\tau \in \Gamma \setminus \{t\}$. It can be shown

[2] Calling it corona would be more appropriate, but the notion of corona is already used in other contexts.

that $x \log \eta_t + \log w \in L^1(\Gamma)$ for every $x \in \mathbf{R}$. Hence, we may apply the previous construction to the function $\psi := \eta_t^x w$.

Theorem 6 *The indicator functions are given by*

$$\alpha_t(x) = \alpha(V_t^0(\eta_t^x w)), \quad \beta_t(x) = \beta(V_t^0(\eta_t^x w)).$$

We also remark that the spirality indices can be expressed as $\delta_t^- = \alpha(V_t^0 \eta_t)$ and $\delta_t^+ = \beta(V_t^0 \eta_t)$, which reveals that they are indeed intrinsic characteristics of the curve Γ and independent of p and w.

6 The Spectrum of the Cauchy Singular Integral Operator

Cauchy versus Toeplitz. We now turn to the Cauchy singular integral operator. Throughout this section we let Γ stand for a Carleson Jordan curve, and we assume that $1 < p < \infty$ and $w \in A_p(\Gamma)$. Since $S_\Gamma^2 = I$, both the spectrum and the essential spectrum of S_Γ are equal to the doubleton $\{-1, 1\}$. Things become interesting if the integration curve is no longer Jordan but an arc, that is, a curve homeomorphic to line segment $[0, 1]$.

Let $E \subset \Gamma$ be an arc. We give E the orientation from Γ. Then E has a starting point $a \in \Gamma$ and a terminating point $b \in \Gamma$. The Cauchy singular integral operator S_E is defined as usual:

$$(S_E f)(t) := \text{p.v.} \frac{1}{\pi i} \int_E \frac{f(\tau)}{\tau - t} d\tau, \quad t \in E.$$

This is a well-defined and bounded operator on $L^p(E, W)$ where W denotes the restriction of w to E, i.e., $W := w|E$. It can again be shown that the essential spectrum of S_E is the union of its local spectra. Put $E^\circ := E \setminus \{a, b\}$. Thus,

$$\sigma_{\text{ess}}(S_E) = \bigcup_{t \in E} \sigma_t(S_E) = \sigma_a(S_E) \cup \sigma_b(S_E) \cup \bigcup_{t \in E^\circ} \sigma_t(S_E).$$

Let $\mathcal{L}(z_1, z_2, p, \alpha_t, \beta_t)$ be the leaf introduced in the previous section after Theorem 3.

Theorem 7 *We have $\sigma_t(S_E) = \{-1, 1\}$ for $t \in E^\circ$ and*

$$\sigma_a(S_E) = \mathcal{L}(-1, 1, p, \alpha_a, \beta_a), \quad \sigma_b(S_E) = \mathcal{L}(1, -1, p, \alpha_b, \beta_b).$$

For $t \in E^\circ$, S_E looks locally like S_Γ and hence it is not a miracle that there the local spectrum is again $\{-1, 1\}$. At the endpoints of E, we may employ the fact that in a unital Banach algebra generated by two idempotents $\boldsymbol{p}$ and $\boldsymbol{q}$ the spectra of $\boldsymbol{pqp}$ and $\boldsymbol{qpq}$ coincide. This follows from appropriate two projections theorems. Indeed, with the well-known representations

$$\boldsymbol{p} \mapsto \begin{pmatrix} 1 & 0 \\ 0 & 0 \end{pmatrix}, \quad \boldsymbol{q} \mapsto \begin{pmatrix} x & \sqrt{x(1-x)} \\ \sqrt{x(1-x)} & 1-x \end{pmatrix} \tag{14}$$

we get

$$\boldsymbol{pqp} \mapsto \begin{pmatrix} 1 & 0 \\ 0 & 0 \end{pmatrix} \begin{pmatrix} x & \sqrt{x(1-x)} \\ \sqrt{x(1-x)} & 1-x \end{pmatrix} \begin{pmatrix} 1 & 0 \\ 0 & 0 \end{pmatrix} = \begin{pmatrix} x & 0 \\ 0 & 0 \end{pmatrix}$$

and

$$\begin{aligned} \boldsymbol{qpq} &\mapsto \begin{pmatrix} x & \sqrt{x(1-x)} \\ \sqrt{x(1-x)} & 1-x \end{pmatrix} \begin{pmatrix} 1 & 0 \\ 0 & 0 \end{pmatrix} \begin{pmatrix} x & \sqrt{x(1-x)} \\ \sqrt{x(1-x)} & 1-x \end{pmatrix} \\ &= \begin{pmatrix} x^2 & x\sqrt{x(1-x)} \\ x\sqrt{x(1-x)} & x(1-x) \end{pmatrix} = U \begin{pmatrix} x & 0 \\ 0 & 0 \end{pmatrix} U^*, \\ U &:= \begin{pmatrix} \sqrt{x} & -\sqrt{1-x} \\ \sqrt{1-x} & \sqrt{x} \end{pmatrix}. \end{aligned}$$

Note that U is unitary.

Returning to Theorem 7, we first take $t = a$. Let χ_E be the function on Γ that takes the value 1 on E and is identically zero on $\Gamma \setminus E$. The operator of multiplication by χ_E will simply be denoted by χ_E as well. The point of the matter is that S_E may be identified with the operator $\chi_E S_\Gamma \chi_E = \chi_E(2P_\Gamma - I)\chi_E$ and that the operator $P_\Gamma \chi_E P_\Gamma$ may be interpreted as the Toeplitz operator $T(\chi_E)$. We have $\chi_E^2 = \chi_E$ and $P_\Gamma^2 = P_\Gamma$, and after localization we get local representatives $\boldsymbol{\chi}$ and $\boldsymbol{P}$ satisfying $\boldsymbol{\chi}^2 = \boldsymbol{\chi}$ and $\boldsymbol{P}^2 = \boldsymbol{P}$. It follows that

$$\sigma_a(S_E) - 2\sigma(\boldsymbol{\chi P \chi}) \quad 1 = 2\sigma_a(T(\chi_E)) - 1,$$

and since, by Theorem 3,

$$2\sigma_a(T(\chi_E)) - 1 = 2\mathcal{L}(0, 1, p, \alpha_a, \beta_a) - 1 = \mathcal{L}(-1, 1, p, \alpha_a, \beta_a),$$

we obtain the desired result at the starting point a. The terminating point $t = b$ becomes the starting point after reversing the orientation of the curve, which gives us

$$\sigma_b(S_E) = -\mathcal{L}(-1, 1, p, \alpha_b, \beta_b) = \mathcal{L}(1, -1, p, \alpha_b, \beta_b).$$

A rigorous proof of Theorem 7 can be based on two projections theorems. Halmos' original version was for Hilbert space operators, but in our context two projections theorems for Banach space operators are required. Such theorems were

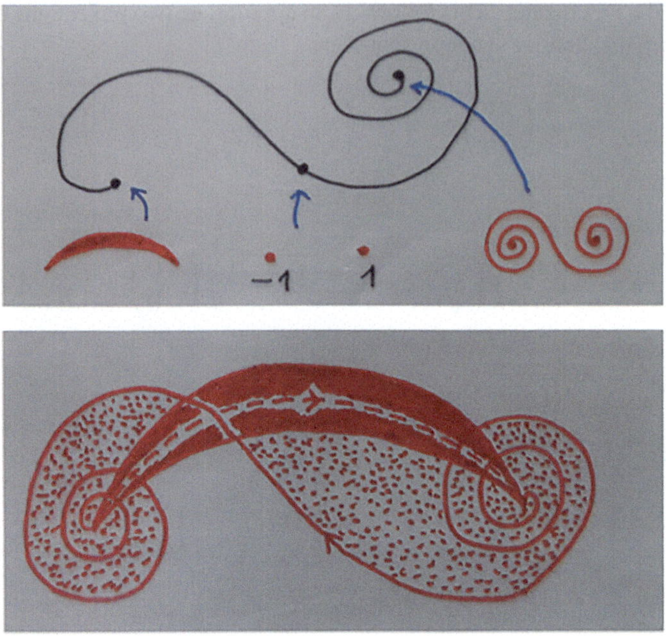

Fig. 7 Local spectra, essential spectrum, and spectrum of the operator S_E over an arc E

first established in [17, 21]. See also paper [12], which is reprinted in this volume. Theorem 1 of that paper is exactly what we need. Notice that in our concrete case the points 0 and 1 are cluster points of the spectrum of

$$pqp + (e - p)(e - q)(e - p),$$

which dispenses us from the consideration of the maps G_m appearing in part (b) of Theorem 1 of [12]. See also Chapter 8 of [9] for all technical details.[3]

Spectrum. Thus, the essential spectrum of S_E is $\sigma_{\text{ess}}(S_E) = \sigma_a(S_E) \cup \sigma_b(S_E)$, and Theorem 7 identifies the two local spectra as leaves. To get the spectrum $\sigma(S_E)$, choose any continuous curve from -1 to 1 entirely contained in $\sigma_a(S_E)$ (e.g., the dashed curve in Fig. 7) and any continuous curve from 1 to -1 in $\sigma_b(S_E)$. These two curves form a Jordan curve C, and $\sigma(S_E)$ can be shown to be the union of $\sigma_{\text{ess}}(S_E)$ and of all points in the plane that are encircled by C with nonzero winding number. See Fig. 7.

Algebras. Let $\mathcal{B}(L^p(\Gamma, w))$ be the Banach algebra of all bounded linear operators on $L^p(\Gamma, w)$. We define $\text{alg}(S, PC)$ as the smallest closed subalgebra of

[3] There are two typos in Theorem 8.7(a) of [9]: $\text{sp}_{\mathcal{B}} X$ must be $\text{sp}_{\mathcal{A}} X$ and "onto" must be "into".

Fig. 8 Examples of the set $\mathfrak{M}$ as they appeared on our transparencies in the 1990s

$\mathcal{B}(L^p(\Gamma, w))$ which contains S_Γ and all multiplication operators $M_a : f \mapsto af$ by functions $a \in PC(\Gamma)$. Let $\varphi : \mathbf{T} \to \Gamma$ be a homeomorphism and put

$$\mathfrak{M} := \bigcup_{\tau \in \mathbf{T}} \Big(\{\tau\} \times \mathcal{L}(0, 1, p, \alpha_{\varphi(\tau)}, \beta_{\varphi(\tau)}) \Big).$$

See Fig. 8. Here is a Fredholm criterion for operators in the algebra $\mathrm{alg}(S, PC)$.

Theorem 8 *For each point $m = (\tau, x) \in \mathfrak{M}$ the map*

$$\Psi_m : \{S\} \cup \{M_a : a \in PC(\Gamma)\} \to \mathbf{C}^{2\times 2}$$

sending S to $\left(\begin{smallmatrix} 1 & 0 \\ 0 & -1 \end{smallmatrix}\right)$ and M_a to

$$a(\varphi(\tau)-0) \begin{pmatrix} 1-x & -\sqrt{x(1-x)} \\ -\sqrt{x(1-x)} & x \end{pmatrix} + a(\varphi(\tau)+0) \begin{pmatrix} x & \sqrt{x(1-x)} \\ \sqrt{x(1-x)} & 1-x \end{pmatrix}$$

extends to a continuous Banach algebra homomorphism $\Psi_m : \text{alg}\,(S, PC) \to \mathbf{C}^{2\times 2}$, *and an operator* $A \in \text{alg}\,(S, PC)$ *is Fredholm on* $L^p(\Gamma, w)$ *if and only if* $\det \Psi_m(A) \neq 0$ *for all* $m \in \mathfrak{M}$. *If an operator in* $\text{alg}\,(S, PC)$ *is Fredholm, then it has a regularizer belonging to* $\text{alg}\,(S, PC)$.

The last sentence in Theorem 8 says that if $A \in \text{alg}\,(S, PC)$ is Fredholm, then there is an operator $R \in \text{alg}\,(S, PC)$ such that $AR - I$ and $RA - I$ are compact. A full proof of this theorem can be found in Chapter 8 of the book [9] or in any of the papers [5, 12]. Needless to point out that the proof of this theorem is also a struggle with inverse closedness in Banach algebras.

Extending curves and weights. So far we have assumed that the Carleson arc E is a subset of a Carleson Jordan curve Γ and that the $A_p(E)$ weight W is the restriction $W = w|E$ of a Muckenhoupt weight $w \in A_p(\Gamma)$. But what if we are merely given a Carleson arc E and a weight $W \in A_p(E)$? To put us into the situation considered above, we have to extend the Carleson arc E to a Carleson Jordan curve Γ and at the same time to continue the weight $W \in A_p(E)$ to a weight $w \in A_p(\Gamma)$. And even if we succeeded doing this, does the local spectrum of S_E at the endpoints of E depend on the specific extensions of Γ and W?

Yura and I understood that these are subtle and difficult questions, and in 1996 we turned to Ilya Spitkovsky and Chris Bishop for help. After very intense work and correspondence we were fortunately able to solve the problems, and at the end of 1996, we had our 80-pager [5] ready for publication. As expected, things were very technical and required sophisticated techniques and deep results from geometric function theory. I confine myself to citing the final result. A detailed presentation is also in Chapter 9 of the book [9].

Theorem 9 *Let* E *be a Carleson arc,* $1 < p < \infty$, *and* $W \in A_p(E)$. *Then there exists a Carleson Jordan curve* Γ *and a weight* $w \in A_p(\Gamma)$ *such that* $E \subset \Gamma$ *and* $W = w|E$. *For* $t \in E$, *the indicator set* $Y(p, \alpha_t, \beta_t)$ *of* Γ, p, w *at the point* t *is independent of the specific extensions* Γ *and* w.

Composed curves. Let K be a composed Carleson curve, that is, suppose K is a connected set which may be represented as a union of finitely many simple Carleson arcs each pair of which have at most endpoints in common. For example, K may be a pentagon together with its five diagonals or may be the star consisting of only the three diagonals of a regular hexagon. Let further $1 < p < \infty$ and $w \in A_p(K)$. The essential spectrum of S_K on $L^p(K, w)$ is again the union of the local spectra,

$$\sigma_{\text{ess}}(S_K) = \bigcup_{t \in K} \sigma_t(S_K).$$

Fix a point $t \in K$. At this point, there are n^+ outgoing arcs of K and n^- incoming arcs of K. The difference $\varepsilon := n^+ - n^-$ is called the valency of t. Pick one of the $n^+ + n^-$ arcs having t as an endpoint and orient this arc so that t becomes its starting point. We denote the arc obtained in this way by K_t and put $w_t = w|K_t$. Let $Y(p, \alpha_t, \beta_t)$ be the indicator set of K_t at t given by Theorem 9.

Theorem 10 *If* $\varepsilon = 0$, *then* $\sigma_t(S_K) = \{-1, 1\}$. *If* $\varepsilon \neq 0$, *we have*

$$\sigma_t(S_K) = \left\{\lambda \in \mathbf{C} : \left(\frac{\lambda+1}{\lambda-1}\right)^{\varepsilon} = e^{2\pi i\zeta} \text{ for some } \zeta \in Y(p, \alpha_t, \beta_t)\right\} \cup \{-1, 1\}.$$

A full proof of this theorem is in [5] and Chapter 9 of [9]. We remark that the proof does not only require the extension of Carleson arcs and weights on them to Carleson Jordan curves and weights on these curves. It rather requires the extension of so-called Carleson stars to so-called Carleson flowers with the simultaneous extension of the weight.

Résumé and taking stock of my years with Yura. In summary, the connection between a Toeplitz operator $T(a)$ on $H^p(\Gamma, w)$ with $a \in PC(\Gamma)$ and a singular integral operator S_E on $L^p(E, w|E)$ over an arc $E \subset \Gamma$ having the same orientation as Γ at the starting point t of E is given by the formula

$$\sigma_t(T(a)) = \frac{a(t-0) + a(t+0)}{2} + \frac{a(t+0) - a(t-0)}{2}\sigma_t(S_E).$$

In this sense, the problem of finding the local spectra of Toeplitz operators with piecewise continuous symbols is equivalent to the problem of describing the local spectra of the Cauchy singular integral operator over arcs. The tour taken by Yura and me was the one that first went through Toeplitz operators and ended with singular integral operators. In the next section I will briefly touch upon our joint work with Volodya Rabinovich. This work gives us the local spectra of singular integral operators without the detour through Toeplitz operators (though not for arbitrary weights in $A_p(E)$ but only for weights in a proper subset $A_p^0(E)$, which, however, is so large that it delivers all possible indicator sets).

The years with Yura were an amazing period for me, and I think also for Yura. During his stay in Chemnitz and in the following years we enjoyed several joint trips, joint participations in conferences, and joint visits to friends, including Ilya Spitkovsky, Anatoly Aizenshtat, Serezha Grudsky, and Volodya Rabinovich. Figures 9 and 10 show some photos.

I think the most painful event for Yura during his visit in Chemnitz was my proposal to give up writing a book on his beloved non-local operators and to write instead a book on Toeplitz operators over curves and with weights that could not been tackled by the existing machinery. However, Yura soon fell in love with the new project and what followed remains something like a miracle to me until today. As said, originally we thought that diving deeper and deeper into the matter would give us work for several years. But astonishingly, we had everything for general curves after only a few months and then settled the problem for general weights after only a few months more. It was a true kind of rush we experienced, and the body of results and thus the book increased week by week. At some time, Yura suggested

Fig. 9 Left: In the Negev desert in 1995, during the weekend break of the workshop "One-dimensional linear singular integral operators" in Tel Aviv and Jerusalem, organized by Israel Gohberg and myself. Right: In Xochicalco, Mexico, in 2003, during a visit to Yura in Cuernavaca

to include not only the spectral theory of Toeplitz and singular integral operators into the book but also a full proof of the boundedness criterion for them. This was an enormous amount of work and it was only due to Yura's unceasing energy that we brought this venture to a successful end. In 1996, we had the manuscript of our book [9] accomplished. To our great delight, the book won the Ferran Sunyer i Balaguer Prize of the year 1997, and Yura and I enjoyed meeting each other for the first time after his visit to Chemnitz anew in Barcelona. See Fig. 11.

During our joint work, Yura perpetually surprised me with unconventional ideas. I want to mention at least one of his strokes of genius. Representations (14) are in general use in the two projections business. The complex variable x runs through a certain set that is specified in advance. In the context of singular integral and Toeplitz operators, this set is a leaf between 0 and 1. The interesting values of x are $x \notin \{0, 1\}$. Yura observed that for such x representation (14) may be replaced by the representation

$$p \mapsto \begin{pmatrix} 1 & 0 \\ 0 & 0 \end{pmatrix}, \quad q \mapsto \begin{pmatrix} x & x-1 \\ -x & 1-x \end{pmatrix}. \tag{15}$$

The enigma's resolution is that

$$D_x \begin{pmatrix} 1 & 0 \\ 0 & 0 \end{pmatrix} D_x^{-1} = \begin{pmatrix} 1 & 0 \\ 0 & 0 \end{pmatrix}, \quad D_x \begin{pmatrix} x & \sqrt{x(1-x)} \\ \sqrt{x(1-x)} & 1-x \end{pmatrix} D_x^{-1} = \begin{pmatrix} x & x-1 \\ -x & 1-x \end{pmatrix}$$

Fig. 10 Left: In Williamsburg, Virginia, attending the Old Dominion Operator Theory and Analysis Conference in 1995. Right: In Key West, Florida, during a trip along the East Coast in 1997, which included also a visit to Anatoly Aizenshtat in Fort Lauterdale

with the diagonal matrix

$$D_x := \operatorname{diag}(\sqrt[4]{(1-x)/x}, -\sqrt[4]{x/(1-x)}).$$

Note that D_x is only needed to establish the equivalence of the representations; D_x does not appear in the representations themselves. In contrast to (14), the new representation (15) is "root-free", and this turned out to be a deciding advantage when passing from two projections to N projections. See [5, 12] or Chapter 8 of [9].

Finally, recall that Toeplitz operators have their origin in the infinite Toeplitz matrix

$$(a_{j-k})_{j,k=0}^{\infty} = \begin{pmatrix} a_0 & a_{-1} & a_{-2} & a_{-3} & \dots \\ a_1 & a_0 & a_{-1} & a_{-2} & \ddots \\ a_2 & a_1 & a_0 & a_{-1} & \ddots \\ a_3 & a_2 & a_1 & a_0 & \ddots \\ \vdots & \ddots & \ddots & \ddots & \ddots \end{pmatrix}.$$

In our context, this matrix corresponds to a Toeplitz operator on $H^2(\mathbf{T}) \cong \ell^2(\mathbf{Z}_+)$. When leaving the $p = 2$ and $\Gamma = \mathbf{T}$ case, one is entering new worlds. Besides the operators on $H^p(\Gamma, w)$ considered here, one is led to Wiener-Hopf integral operators on $L^p(\mathbf{R}_+, w)$ or discrete convolution operators on $\ell^p(\mathbf{Z}_+, w)$. In the

Fig. 11 In Barcelona in 1997 after the receipt of the Ferran Sunyer i Balaguer Prize for our monograph [9]. We both love the cute typo in the certificate

last two contexts, Ilya Spitkovsky's horns persist—no new leaves are emerging. Section 6 of [6] tells more about this part of the story.

7 Mellin Convolutions

Let me finish with what we started: the Cauchy singular integral operator over a logarithmic spiral.

Volodya Rabinovich. Some time in the mid 1990s, Yura and I came together with Volodya Rabinovich, and he surprised us with the observation that the indicator set $Y(p, \alpha_t, \beta_t)$ is always the union of straight lines:

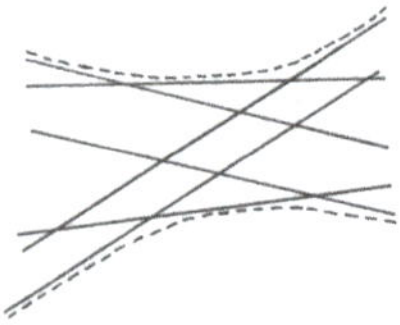

When pursuing this idea, Volodya and we developed two more approaches to the spectral theory of the Cauchy singular integral operator. The first of them is based on Fourier(-Mellin) techniques [14], the second makes use of the theory of so-called limit operators [15], and both eventually reduce things to pseudodifferential operators on the real line with slowly oscillating symbols. For example, one can associate a family of limit operators with a given operator A so that the local spectrum of A is the union of the local spectra of the limit operators. Under certain assumptions, the limit operators are sufficiently simple, which means that their indicator sets are straight lines. Clearly, this gives the characterization of the indicator set observed by Volodya. See also [11] for some illustrations.

From logarithmic spirals to Mellin convolutions. The following approach is based on our paper [13] with Volodya and follows Section 10.6 of [9]. Let $\delta \in \mathbf{R}$ and

$$\Gamma_\delta := \{x^{1-i\delta} : 0 < x < \infty\} = \{xe^{-i\delta \log x} : 0 < x < \infty\}. \tag{16}$$

Thus, Γ_δ is a logarithmic spiral. We give Γ_δ the orientation from the origin to infinity. The logarithmic spiral Γ_δ is an unbounded curve and hence the results stated above are not directly applicable. However, the Cauchy singular integral S_{Γ_δ},

$$(S_{\Gamma_\delta} f)(t) := \lim_{\varepsilon \to 0} \frac{1}{\pi i} \int_{\Gamma_\delta \setminus \Gamma_\delta(t,\varepsilon)} \frac{f(\tau)}{\tau - t}\, d\tau, \quad t \in \Gamma_\delta,$$

can be shown to be well-defined and bounded on $L^p(\Gamma_\delta, w)$ if $1 < p < \infty$ and $w(\tau) := |\tau|^\lambda$ with $-1/p < \lambda < 1/q$.[4] It can of course also be proved that the local spectrum $\sigma_0(S_{\Gamma_\delta})$ at the origin is the logarithmic double spiral

$$\mathcal{S}(-1, 1, 1/p + \lambda, \delta)$$

introduced above and that $\sigma_t(S_{\Gamma_\delta}) = \{-1, 1\}$ for $t \in \Gamma_\delta \setminus \{0\}$. Here is a completely different approach to the matter.

We put $d\mu(x) := dx/x$ and let $L^p(\mathbf{R}_+, d\mu)$ stand for the L^p space over $\mathbf{R}_+ = (0, \infty)$ with the weight $x^{-1/p}$:

$$\|f\|_{L^p(\mathbf{R}_+, d\mu)} := \left(\int_0^\infty |f(x)|^p \frac{dx}{x} \right)^{1/p}.$$

It is straightforward to check that the map

$$C_\delta : L^p(\Gamma_\delta, w) \to L^p(\mathbf{R}_+, d\mu), \quad (C_\delta h)(x) := |1 - i\delta|^{1/p} x^{1/p+\lambda} h(x^{1-i\delta})$$

[4] We now have to denote the exponent by λ since μ is in common use for the measure $d\mu(x) = dx/x$.

is an isometric isomorphism. We have

$$(C_\delta S_{\Gamma_\delta} C_\delta^{-1} f)(x) = |1-i\delta|^{1/p} x^{1/p+\lambda} (S_{\Gamma_\delta} C_\delta^{-1} f)(x^{1-i\delta})$$

$$= \frac{|1-i\delta|^{1/p} x^{1/p+\lambda}}{\pi i} \int_{\Gamma_\delta} \frac{(C_\delta^{-1} f)(\tau)}{\tau - x^{1-i\delta}}\, d\tau.$$

Making the change of variables $\tau = y^{1-i\delta}$, $d\tau = (1-i\delta) y^{1-i\delta}\, dy/y$ and taking into account that $(C_\delta^{-1} f)(y^{1-i\delta}) = |1-i\delta|^{-1/p} y^{-1/p-\lambda} f(y)$, we get

$$(C_\delta S_{\Gamma_\delta} C_\delta^{-1} f)(x) = \frac{|1-i\delta|^{1/p} x^{1/p+\lambda}}{\pi i} \int_0^\infty (1-i\delta) \frac{(C_\delta^{-1} f)(y^{1-i\delta})}{y^{1-i\delta} - x^{1-i\delta}} y^{1-i\delta} \frac{dy}{y}$$

$$= \frac{1-i\delta}{\pi i} \int_0^\infty \frac{(x/y)^{1/p+\lambda}}{1-(x/y)^{1-i\delta}} f(y) \frac{dy}{y}. \tag{17}$$

The integral in (17) is a so-called Mellin convolution. The half-line $\mathbf{R}_+$ is a locally compact abelian group with the group operation $x * y := xy$ and the invariant measure dx/x. In this context, a convolution operator K is formally given by

$$(Kf)(x) = \int_{\mathbf{R}_+} k(x * y^{-1}) f(y)\, d\mu(y) = \int_0^\infty k(x/y) f(y) \frac{dy}{y}, \quad x \in \mathbf{R}_+.$$

The characters $\chi : \mathbf{R}_+ \to \mathbf{T}$ all act by the rule $\chi_\xi(x) := x^{i\xi}$ with $\xi \in \mathbf{R}$. Consequently, we may identify the dual group $\mathbf{R}_+^*$ as the additive group $\mathbf{R}$ with Lebesgue measure. The Fourier transform corresponding to this pairing is usually denoted by M and called the Mellin transform. Formally,

$$(Mk)(\xi) := \int_{\mathbf{R}_+} k(x)\chi_\xi(x)\, d\mu(x) = \int_0^\infty k(x) x^{i\xi} \frac{dx}{x},$$

and the Mellin convolution K can be written as $K = \mathcal{M}^0(a) := M^{-1} a M$ with $a(\xi) := (Mk)(\xi)$. When considering $\mathcal{M}^0(a)$ on L^p, one is faced with a multiplier problem. However, all we need in the case at hand is that if $a \in L^\infty(\mathbf{R}) \cap PC(\mathbf{R})$ has bounded total variation, then $\mathcal{M}^0(a)$ is bounded on $L^p(\mathbf{R}_+, d\mu)$, and $\mathcal{M}^0(a)$ is invertible on $L^p(\mathbf{R}_+, d\mu)$ if and only if a is invertible in $L^\infty(\mathbf{R})$.

Computing the Mellin transform. Thus, we need the function $(Mk)(\xi)$ for $k(x) = x^{1/p+\lambda}/(1-x^{1-i\delta})$. A more general result states that if $k(x) = x^\eta/(1-x^\varrho)$ with $0 < \eta < 1$ and a complex number ϱ such that $\operatorname{Re}\varrho \ge 1$, then

$$a_{\eta,\varrho}(\xi) := \frac{\varrho}{\pi i}(Mk)(\xi) = \coth\left(\pi \frac{i\eta - \xi}{\varrho}\right), \quad \xi \in \mathbf{R},$$

the function $a_{\eta,\varrho}$ is continuous and has bounded total variation on $\mathbf{R}$, the finite limits $a(\pm\infty)$ exist, and $\mathcal{M}^0(a_{\eta,\varrho})$ acts on $L^p(\mathbf{R}_+, d\mu)$ by the rule

$$(\mathcal{M}^0(a_{\eta,\varrho})f)(x) = \lim_{\varepsilon\to 0}\frac{\varrho}{\pi i}\int_{|\log(x/y)|>\varepsilon}\frac{(x/y)^\eta}{1-(x/y)^\varrho}\,f(y)\,\frac{dy}{y}, \qquad x \in \mathbf{R}_+.$$

For $\varrho = 1$, a proof is in [16, pp. 24–25, 51–52] and [22, pp. 12–13]. For $\operatorname{Re}\varrho \geq 1$, see [13].

Computing the spectrum. By (17) and the preceding paragraph, the operator S_{Γ_δ} is similar to $\mathcal{M}^0(a_{1/p+\lambda,1-i\delta})$. Let $\overline{\mathbf{R}} := \mathbf{R} \cup \{\pm\infty\}$. Since

$$\coth z = \frac{e^z + e^{-z}}{e^z - e^{-z}} = \frac{e^{2z}+1}{e^{2z}-1} = M_{-1,1}(e^{2z}),$$

we obtain

$$a_{1/p+\lambda,1-i\delta}(\overline{\mathbf{R}}) = \left\{M_{-1,1}\left(\exp\left(2\pi\frac{i(1/p+\lambda)-\xi}{1-i\delta}\right)\right) : \xi \in \overline{\mathbf{R}}\right\},$$

and by decomposing $(i(1/p+\lambda)-\xi)/(1-i\delta)$ into real and imaginary parts we get

$$a_{1/p+\lambda,1-i\delta}(\overline{\mathbf{R}}) = \{M_{-1,1}(e^{2\pi(x+iy)}) : y = 1/p + \lambda + \delta x, x \in \mathbf{R}\} \cup \{-1, 1\}.$$

This is the logarithmic double spiral

$$\mathcal{L}(-1, 1, p, \alpha, \beta) = \mathcal{S}(-1, 1, 1/p + \lambda, \delta),$$

and we have finally arrived at the following result.

Theorem 11 *Let $\delta \in \mathbf{R} \setminus \{0\}$, $p \in (1, \infty)$, $\lambda \in (-1/p, 1/q)$, denote by Γ_δ the logarithmic spiral (16) with the orientation from 0 to infinity, and let $w(\tau) = |\tau|^\lambda$ for $\tau \in \Gamma_\delta$. Then the spectrum of S_{Γ_δ} on $L^p(\Gamma_\delta, w)$ is a logarithmic double spiral joining -1 to 1. This logarithmic double spiral is given by*

$$\left\{\frac{e^{2\pi z}+1}{e^{2\pi z}-1} : \operatorname{Im} z = \frac{1}{p} + \lambda + \delta \operatorname{Re} z\right\} \cup \{-1, 1\}.$$

We could have included the case $\delta = 0$ into the theorem: for $\delta = 0$, we get a circular arc between -1 and 1 if $1/p + \lambda \neq 1/2$, and the line segment $[-1, 1]$ if $1/p + \lambda = 1/2$.

The gain from pseudodifferential operators. Here is a generalization of the previous approach. Let

$$\Gamma := \{t + xe^{i\theta(x)} : 0 < x < s\}, \quad w(t + xe^{i\theta(x)}) := e^{v(x)} \tag{18}$$

where $s \in (0, \infty]$ and θ and v are real-valued functions in $C^\infty(0, s)$ satisfying

$$\sup_{x\in(0,s)} |(xD_x)^j|\theta(x)| < \infty, \quad \sup_{x\in(0,s)} |(xD_x)^j|v(x)(x)| < \infty \quad \text{for all } j \geq 1,$$

$$\lim_{x\to 0}(xD_x)^2\theta(x) = 0, \quad \lim_{x\to 0}(xD_x)^2 v(x) = 0,$$

$$-1/p < \liminf_{x\to 0} xv'(x) \leq \limsup_{x\to 0} xv'(x) < 1/q.$$

We add the point t to Γ, and if s is a finite number, we also add the point $t + se^{i\theta(s)}$. We orient Γ so that t is the starting point. The requirements made for θ and w ensure that Γ is a Carleson curve and that $w \in A_p(\Gamma)$. Using results from [4, 20] we showed in [14] that point $z \in \mathbf{C}$ belongs to the local spectrum $\sigma_t(S_\Gamma)$ if and only if

$$\lim_{\varepsilon\to 0} \inf_{(x,\xi)\in(0,\varepsilon)\times\mathbf{R}} \left| M_{-1,1}\left(\exp\left(2\pi \frac{i(1/p + xv'(x)) - \xi}{1 + ix\theta'(x)}\right)\right) - z\right| = 0.$$

This gives the following description of the local spectrum.[5]

Theorem 12 *Under the above assumptions, let $\mathcal{P}$ be the set of the partial limits of the map*

$$(0, s) \to (0, 1) \times \mathbf{R}, \quad x \mapsto \left(\frac{1}{p} + xv'(x), -x\theta'(x)\right)$$

as $x \to 0$. Then

$$\sigma_t(S_\Gamma) = \bigcup_{(\omega,\delta)\in\operatorname{conv}\mathcal{P}} \left\{\frac{e^{2\pi z} + 1}{e^{2\pi z} - 1} : \operatorname{Im} z = \omega + \delta \operatorname{Re} z\right\} \cup \{-1, 1\},$$

where $\operatorname{conv}\mathcal{P}$ *is the convex hull of $\mathcal{P}$. Actually it suffices to take the union only over solely $\mathcal{P}$.*

Clearly, if $\theta(x) = -\delta \log x$ and $v(x) = \lambda \log x$, then $\mathcal{P}$ is just the singleton $\{(1/p + \lambda, \delta)\}$, and Theorem 12 becomes Theorem 11.

The gain from limit operators and localization. In [15] we used the method of limit operators to localize the problem completely. The abstract of this paper is as follows. “One of the great challenges of the spectral theory of singular integral operators is a theory unifying the three ‘forces’ which determine the local spectra: the oscillation of the Carleson curve, the oscillation of the Muckenhoupt weight, and the oscillation of the coefficients. In this paper we demonstrate how by employing

[5] The minus sign in $-x\theta'(x)$ appearing in the following two theorems is unfortunately missing in Theorems 10.25 and 10.27 of [9].

the method of limit operators one can describe the spectra in case all data of the operator (the curve, the weight, and the coefficients) are slowly oscillating."

A function $f \in C^\infty(0,s) \cap L^\infty(0,s)$ is said to be slowly oscillating at the origin if

$$\lim_{x \to 0} |(xD_x)^j f(x)| = 0 \text{ for all } j \geq 1.$$

For example, if $g \in C^\infty(\mathbf{R})$ and g as well as all its derivatives are bounded, then $f(x) := g(\log(-\log x))$ $(0 < x < 1)$ is slowly oscillating at the origin. The following theorem was established in [15]. Note that it is for the case $p = 2$.

Theorem 13 *Let $p = 2$ and let Γ and w be given by (18). Suppose c and d are slowly oscillating at the origin. Put*

$$c_\Gamma(t + xe^{i\theta(x)}) := c(x), \quad d_\Gamma(t + xe^{i\theta(x)}) := d(x),$$

and consider the operator $A := c_\Gamma I + d_\Gamma S_\Gamma$ on $L^2(\Gamma, w)$. Denote by $\mathcal{P}_A$ the set of all partial limits of the map

$$(0,s) \to \mathbf{C}^2 \times (0,1) \times \mathbf{R}, \quad x \mapsto (c(x), d(x), 1/2 + xv'(x), -x\theta'(x))$$

as $x \to 0$. Then

$$\sigma_t(A) = \bigcup_{(\alpha,\beta,\omega,\delta)\in\mathcal{P}_A} \alpha + \beta\mathcal{S}(-1,1,\omega,\delta) = \bigcup_{(\alpha,\beta,\omega,\delta)\in\mathcal{P}_A} \mathcal{S}(\alpha-\beta, \alpha+\beta, \omega, \delta)$$

where, as above, $\mathcal{S}(z_1, z_2, \omega, \delta)$ is the logarithmic double spiral

$$\mathcal{S}(z_1, z_2, \omega, \delta) = \left\{ \frac{z_2 e^{2\pi z} - z_1}{e^{2\pi z} - 1} : \operatorname{Im} z = \omega + \delta \operatorname{Re} z \right\} \cup \{z_1, z_2\}.$$

Credits. The photos and illustrations are courtesy of the author.

References

1. Aizenshtat, A.V., Karlovich, Y.I., Litvinchuk, G.S.: On a theorem on conformal gluing and its application (in Russian). Dokl. Rasshir. Zased. Semin. Inst. Prikl. Matem. **3**, 9–12 (1988)
2. Aizenshtat, A.V., Karlovich, Y.I., Litvinchuk, G.S.: On the defect numbers of the Kveselava-Vekua operator with discontinuous derivative of the shift. Soviet Math. Dokl. **43**, 633–638 (1991)
3. Aizenshtat, A.V., Karlovich, Y.I., Litvinchuk, G.S.: The method of conformal gluing for the Haseman boundary value problem on an open contour. Complex Var. **28**, 313–346 (1996)
4. Beals, R.: L^p and Hölder estimates for pseudodifferential operators: sufficient conditions. Ann. Inst. Fourier **29**, 239–260 (1979)

5. Bishop, C.J., Böttcher, A., Karlovich, Y.I., Spitkovsky, I.: Local spectra and index of singular integral operators with piecewise continuous coefficients on composed curves. Math. Nachr. **206**, 5–83 (1999)
6. Böttcher, A.: Ilya Spitkovsky's pioneering work on massive local spectra of Toeplitz operators. Oper. Theory Adv. Appl. **297**, 19–39 (2024)
7. Böttcher, A., Karlovich, Y.I.: Toeplitz and singular integral operators on Carleson curves with logarithmic whirl points. Integral Equ. Oper. Theory **22**, 127–161 (1995)
8. Böttcher, A., Karlovich, Y.I.: Toeplitz and singular integral operators on general Carleson Jordan curves. Oper. Theory Adv. Appl. **90**, 119–152 (1996)
9. Böttcher, A., Karlovich, Y.I.: Carleson Curves, Muckenhoupt Weights, and Toeplitz Operators. Birkhäuser Verlag, Basel (1997)
10. Böttcher, A., Karlovich, Y.I.: Toeplitz operators with PC symbols on general Carleson Jordan curves with arbitrary Muckenhoupt weights. Trans. Am. Math. Soc. **351**, 3143–3196 (1999)
11. Böttcher, A., Karlovich, Y.I.: Cauchy's singular integral operator and its beautiful spectrum. Oper. Theory Adv. Appl. **129**, 109–142 (2001)
12. Böttcher, A., Gohberg, I., Karlovich, Y.I., Krupnik, N., Roch, S., Silbermann, B., Spitkovsky, I.: Banach algebras generated by N idempotents and applications. Oper. Theory Adv. Appl. **90**, 19–54 (1996); reprinted in this volume
13. Böttcher, A., Karlovich, Y.I., Rabinovich, V.S.: Emergence, persistence, and disappearance of logarithmic spirals in the spectra of singular integral operators. Integral Equ. Oper. Theory **25**, 406–444 (1996)
14. Böttcher, A., Karlovich, Y.I., Rabinovich, V.S.: Mellin psesdodifferential operators with slowly varying symbols and singular integrals on Carleson curves with Muckenhoupt weights. Manuscripta Math. **95**, 363–376 (1998)
15. Böttcher, A., Karlovich, Y.I., Rabinovich, V.S.: The method of limit operators for one-dimensional singular integrals with slowly oscillating data. J. Oper. Theory **43**, 171–198 (2000)
16. Duduchava, R.V.: Integral Equations with Fixed Singularities. Teubner, Leipzig (1979)
17. Fink, T., Roch, S., Silbermann, B.: Two projections theorems and symbol calculus for operators with massive local spectra. Math. Nachr. **162**, 167–185 (1993)
18. Gohberg, I., Krupnik, N.: The spectrum of singular integral operators in L_p spaces with weight. Soviet Math. Dokl. **10**, 406–410 (1969)
19. Krupnik, N.: The influence of some of B. V. Khvedelidze's results on the development of the Fredholm theory for SIOs with PC coefficients in $L_p^n(\Gamma, \varrho)$. Mem. Differ. Equ. Math. Phys. **66**, 103–117 (2015)
20. Rabinovich, V.S.: Mellin pseudodifferential operators techniques in the theory of singular integral operators on some Carleson curves. Oper. Theory Adv. Appl. **102**, 201–218 (1998)
21. Roch, S., Silbermann, B.: Algebras generated by idempotents and the symbol calculus for singular integral operators. Integral Equ. Oper. Theory **11**, 385–419 (1988)
22. Roch, S., Silbermann, B.: Algebras of Convolution Operators and their Image in the Calkin Algebra. Report R-Math-05/90, Karl-Weierstrass-Institute for Mathematics, Berlin (1990)
23. Seifullaev, R.K.: The Riemann boundary value problem on non-smooth open curves. Math. USSR Sb. **40**, 135–148 (1981)
24. Spitkovsky, I.M.: Singular integral operators with PC symbols on the spaces with general weights. J. Funct. Anal. **105**, 129–143 (1992)
25. Widom, H.: Singular integral equations in L^p. Trans. Am. Math. Soc. **97**, 131–160 (1960)

Banach Algebras Generated by N Idempotents and Applications

A. Böttcher, I. Gohberg, Yu. Karlovich, N. Krupnik, S. Roch, B. Silbermann, and I. Spitkovsky

This is a 1-1 reprint of a paper that originally appeared in Singular Integral Operators and Related Topics (Tel Aviv, 1995), Operator Theory: Advances and Applications, vol. 90, pp. 15–54, Birkhäuser, Basel (1996).

Abstract It is well known that for Banach algebras generated by two idempotents and the identity all irreducible representations are of order not greater than two. These representations have been described completely and have found important applications to symbol theory. It is also well known that without additional restrictions on the idempotents these results do not admit a natural generalization to algebras generated by more than two idempotents and the identity. In this paper we describe all irreducible representations of Banach algebras generated by N idempotents which satisfy some additional relations. These representations are of order not greater than N and allow us to construct a symbol theory with applications to singular integral operators.

A. Böttcher (✉) · Yu. Karlovich · B. Silbermann
Fakultät für Mathematik, TU Chemnitz-Zwickau, Chemnitz, Germany

I. Gohberg
School of Mathematical Sciences, Tel Aviv University, Ramat Aviv, Israel

N. Krupnik
Department of Mathematics, Bar-Ilan University, Ramat-Gan, Israel

S. Roch
Mathematisches Institut, Universität Leipzig, Leipzig, Germany

I. Spitkovsky
Department of Mathematics, The College of William and Mary, Williamsburg, VA, USA

A. Böttcher et al. (eds.), *Achievements and Challenges in the Field of Convolution Operators*, Operator Theory: Advances and Applications 306,
https://doi.org/10.1007/978-3-031-80486-1_4

Preface and Acknowledgment

The study of the structure of Banach algebras generated by two idempotents and the identity has a long history of more than 30 years and has found interesting applications to Banach algebras of singular integral operators on simple contours.

Further advances in the theory of Banach algebras of singular integral operators on non-simple contours required developing a structure theory for Banach algebras generated by N idempotents which satisfy certain relations. The authors of this paper, working in different groups, developed several approaches to this problem.

At the request of the other authors, Steffen Roch, a member of one of the groups, unified these approaches and styles, closed the gaps, and brought the paper to the form in which it is presented. All authors express their sincere gratitude to Steffen Roch for the outstanding task he performed.

1 Introduction

In the last 15–20 years, notable advance in understanding the structure of Banach algebras generated by singular integral operators has been made. Many new insights are essentially based on two observations which are characteristic for a large variety of concrete algebras.

The first one is that the Calkin image of operator algebras often contains a non-trivial center, which offers the opportunity of applying local techniques such as Allan's local principle (see below). This principle associates with each of these algebras a whole family of smaller, so-called *local*, algebras which are labeled by the points of a compact space, namely the maximal ideals of the center. Now the second observation enters the scene: in many cases, these local algebras are generated by two (concrete) idempotent cosets, and so they are subject to so-called two projections theorems (see [26, 34, 35, 37, 38, 46, 48] for the C^*-case, [46] for the W^*-case, and [18, 21, 22, 40], and [49], for the general Banach algebra case). Two projection theorems describe abstract algebras generated by two idempotents either completely (the C^*-case) or yield at least necessary and sufficient invertibility criteria for the elements of the algebra (the Banach algebra case) by associating with each element of the algebra a certain 2×2 or 1×1 matrix function. The correspondence between the elements of the algebra and the matrix function is either an isometric isomorphism (C^*-case) or a spectrum-preserving homomorphism (Banach algebra case).

Since Douglas' pioneering paper [14], the idea of combining local principles with two projections theorems has been successfully employed, e.g., for algebras generated by one-dimensional singular integral operators with piecewise continuous coefficients, for algebras of Wiener-Hopf and multiplication operators, for algebras of Toeplitz and Hankel operators with piecewise continuous or piecewise quasicontinuous generating functions, for algebras of Fourier integral operators, and for algebras of operators with Carleman shifts (see [2–5, 8, 9], [6, Chapter 4], [7, 15, 31, 36, 37, 40, 42], [44]). In all these situations, effective symbol calculi for Fredholmness are available.

Moreover, during the last few years it has become clear that the same approach also applies to certain algebras of approximating sequences for operator equations, the symbol now telling us something about the stability of the sequence. For this topic see [24, 39, 41, 43] and the monograph [25].

In the present paper we consider Banach algebras which are generated by more than two idempotents. Algebras of this type appear as local algebras of concrete operator or sequence algebras. We recall that, in general, there is no matrix-valued symbol calculus even for algebras generated by only three idempotents. However, under certain additional conditions, we establish an N projections theorem which yields exactly the two projections theorem (without additional conditions) in case $N = 2$. We also illustrate the application of our N projections theorem to the construction of a symbol calculus for algebras generated by singular integral operators with piecewise continuous coefficients.

For a first discussion of the N projections problem (but without deriving effective invertibility criteria) see [47].

The paper is organized as follows. In Sect. 2, we remind the reader of some known results on algebras generated by two idempotents and on so-called local principles. Section 3 is devoted to algebras generated by three idempotents. We there point out that such algebras do not possess a matrix symbol in general, but that a matrix symbol exists under certain additional hypotheses. Section 4 contains the main theorem (Theorem 9 in Sect. 4.4) and its proof. In Sect. 5, we illustrate how the main theorem may be applied to singular integral operators on composed curves. In Sect. 6, we record several special cases, modifications, and extensions of the main theorem.

2 Algebras Generated by Two Idempotents

The following theorem is one of the main results of [40], with a completion by [21].

Theorem 1 *Let $\mathcal{A}$ be a Banach algebra with identity e, and let p and q be idempotents in $\mathcal{A}$ (i.e. $p^2 = p$ and $q^2 = q$). Let further $\mathcal{B}$ stand for the smallest closed subalgebra of $\mathcal{A}$ which contains p, q and e. Then*

(a) for each

$$x \in \sigma_{\mathcal{B}}(pqp + (e-p)(e-q)(e-p)) \setminus \{0, 1\},$$

the mapping

$$F_x : \{e, p, q\} \to \mathbb{C}^{2\times 2},$$

given by

$$F_x(e) = \begin{pmatrix} 1 & 0 \\ 0 & 1 \end{pmatrix}, \quad F_x(p) = \begin{pmatrix} 1 & 0 \\ 0 & 0 \end{pmatrix}, \quad F_x(q) = \begin{pmatrix} x & \sqrt{x(1-x)} \\ \sqrt{x(1-x)} & 1-x \end{pmatrix},$$

where $\sqrt{x(1-x)}$ *denotes any number with* $\left(\sqrt{x(1-x)}\right)^2 = x(1-x)$ *and* $\sigma_{\mathcal{B}}(a)$ *refers to the spectrum of a in* $\mathcal{B}$*, extends to a continuous algebra homomorphism from* $\mathcal{B}$ *into* $\mathbb{C}^{2\times 2}$*;*

(b) for each

$$m \in \sigma_{\mathcal{B}}(p + 2q) \cap \{0, 1, 2, 3\},$$

the mapping

$$G_m : \{e, p, q\} \to \mathbb{C}^{1\times 1},$$

given by

$$G_0(e) = 1, \quad G_0(p) = G_0(q) = 0, \quad G_1(e) = G_1(p) = 1, \quad G_1(q) = 0,$$
$$G_2(e) = G_2(q) = 1, \quad G_2(p) = 0, \quad G_3(e) = G_3(p) = G_3(q) = 1,$$

extends to a continuous algebra homomorphism from $\mathcal{B}$ *into* $\mathbb{C}^{1\times 1}$*;*

(c) an element $a \in \mathcal{B}$ *is invertible in* $\mathcal{B}$ *if and only if the matrices* $F_x(a)$ *are invertible for all*

$$x \in \sigma_{\mathcal{B}}(pqp + (e-p)(e-q)(e-p)) \setminus \{0, 1\},$$

and the numbers $G_m(a)$ *are non-zero for all* $m \in \sigma_{\mathcal{B}}(p + 2q) \cap \{0, 1, 2, 3\}$*;*

(d) an element $a \in \mathcal{B}$ *is invertible in* $\mathcal{A}$ *if and only if the matrices* $F_x(a)$ *are invertible for all*

$$x \in \sigma_{\mathcal{A}}(pqp + (e-p)(e-q)(e-p)) \setminus \{0, 1\},$$

and the numbers $G_m(a)$ *are non-zero for all* $m \in \sigma_{\mathcal{A}}(p + 2q) \cap \{0, 1, 2, 3\}$.

For a proof see [18, 21, 22, 40] and compare also [49].

The known proofs of the two projections theorem make use of at least one of the following basic properties of the abstract two projections algebra $\mathcal{B} = \mathrm{alg}(e, p, q)$.

(a) The algebra $\mathcal{B}$ possesses a non-trivial center. In particular, the element

$$pqp + (e-p)(e-q)(e-p)$$

commutes with each other element of $\mathcal{B}$ (recall that the center of an algebra consists of all elements which commute with each other element of the algebra).

(b) The algebra $\mathcal{B}$ is an algebra with a polynomial identity. More precisely, it satisfies the standard polynomial F_4 where

$$F_{2n}(a_1, \ldots, a_{2n}) = \sum_{\sigma \in S_{2n}} (\operatorname{sign} \sigma) a_{\sigma(1)} \ldots a_{\sigma(2n)}$$

and S_{2n} refers to the group of all permutations of the set $\{1, 2, \ldots, 2n\}$, which means that

$$F_4(b_1, b_2, b_3, b_4) = 0 \quad \text{for all} \quad b_1, \ldots, b_4 \in \mathcal{B}.$$

The first property renders the algebra $\mathcal{B}$ accessible to the local principle by Allan and Douglas (see [1] and [13]), which reads as follows.

Theorem 2 *Let $\mathcal{A}$ be a Banach algebra with identity e, and let C be a subalgebra of the center of $\mathcal{A}$ which contains e. For each maximal ideal x of the (commutative) Banach algebra C, let I_x denote the smallest closed two-sided ideal of $\mathcal{A}$ which contains x. Then an element a of $\mathcal{A}$ is invertible if and only if the cosets $a + I_x$ are invertible in the quotient algebra $\mathcal{A}/I_x$ for all maximal ideals of C.*

(In case $I_x = \mathcal{A}$, the coset $a + I_x$ is invertible by definition for all a.)

Property (b) shows that the two projections algebra is also subject to another local principle, which is due to one of the authors (see [29]):

Theorem 3 *Let $\mathcal{A}$ be a Banach algebra with identity which satisfies the standard polynomial F_{2n}. Then*

(a) for each two-sided maximal ideal M of $\mathcal{A}$, the quotient algebra $\mathcal{A}/M$ is isomorphic to the matrix algebra $\mathbb{C}^{l \times l}$ with a certain $l = l(M)$ less than or equal to n;

(b) an element $a \in \mathcal{A}$ is invertible if and only if the matrices $f_M(a)$ are invertible for all two-sided maximal ideals M where $f_M = \varphi_M \pi_M$, π_M is the canonical homomorphism from $\mathcal{A}$ onto $\mathcal{A}/M$, and φ_M is the isomorphism from $\mathcal{A}/M$ onto $\mathbb{C}^{l \times l}$ given by (a).

Let us remark that this theorem remains true if $\mathcal{A}$ only satisfies a certain power F_{2n}^m of F_{2n} (see [17]).

3 Algebras Generated by Three Idempotents

Let $\mathcal{A}$ be a Banach algebra with identity. We say that $\mathcal{A}$ possesses a matrix symbol of order n if there is a family $(f_t)_{t \in T}$ of continuous algebra homomorphisms f_t from $\mathcal{A}$ into the algebra $\mathbb{C}^{l(t) \times l(t)}$ with $l(t) \leq n$ such that an element $a \in \mathcal{A}$ is invertible in $\mathcal{A}$ if and only if the matrices $f_t(a)$ are invertible for all $t \in T$. By Theorem 3, each F_{2n}^m-algebra has a matrix symbol of order n and, in particular, each algebra generated by two idempotents has a matrix symbol of order 2.

The following result is taken from [30]. It shows that the (abstract) algebra generated by three idempotents cannot possess a matrix symbol of a certain fixed order.

Theorem 4 *If $n \geq 3$ then the algebra $\mathbb{C}^{n\times n}$ is generated by three idempotents.*

Moreover, one has the following characterization of algebras generated by three idempotents. Recall that a Banach algebra is called separable if it possesses a countable dense subset.

Theorem 5 *(a) Every Banach algebra generated by three idempotents is separable. (b) Every separable Banach algebra is isomorphic to a subalgebra of an algebra generated by three idempotents.*

Proof The first assertion is evident. For the second one, we first prove that every separable Banach algebra is isomorphic to a subalgebra of a finitely generated Banach algebra.

Let $\mathcal{A}$ be a separable Banach algebra with a dense subset $\{a_1, a_2, \ldots\}$ and suppose without loss of generality that $a_n \neq 0$ for all n. For $n = 1, 2, \ldots$ and $k = 1, 2, \ldots, 2^n$ set $c_{2^n-2+k} := a_k/\|a_k\|$. Let further $\ell^2(\mathcal{A})$ stand for the Banach space of all sequences $(x_n)_{n=1}^{\infty}$ of elements of $\mathcal{A}$ such that

$$\|(x_n)\|^2 := \sum_{n=1}^{\infty} \|x_n\|^2 < \infty,$$

and writhe $L(\ell^2(\mathcal{A}))$ for the Banach algebra of all bounded linear operators on $\ell^2(\mathcal{A})$. On $\ell^2(\mathcal{A})$ we consider the following operators:

$$A : (x_n) \mapsto (y_n), \qquad y_n = c_n x_n,$$

$$V_1 : (x_n) \mapsto (y_n), \qquad y_n = \begin{cases} 0 & \text{if } n = 1 \\ x_{n-1} & \text{if } n > 1, \end{cases}$$

$$V_{-1} : (x_n) \mapsto (y_n), \quad y_n = x_{n+1},$$

$$W_1 : (x_n) \mapsto (y_n), \qquad y_n = \begin{cases} x_k & \text{if } n = 2^k - 1 \\ 0 & \text{if } n \neq 2^k - 1, \end{cases}$$

$$W_{-1} : (x_n) \mapsto (y_n), \quad y_n = x_{2^n-1}.$$

Obviously, $A, V_1, V_{-1}, W_1, W_{-1} \in L(\ell^2(\mathcal{A}))$, and so it makes sense to consider the smallest closed subalgebra $\mathcal{B}$ of $L(\ell^2(\mathcal{A}))$ which contains the operators $A, V_1, V_{-1}, W_1, W_{-1}$ and the identity operator I. The algebra $\mathcal{B}$ is finitely generated, and we claim that $\mathcal{A}$ is isomorphic to a subalgebra of $\mathcal{B}$. Since

$$W_{-1}AW_1 : (x_n) \mapsto (y_n), \quad y_n = c_1 x_n,$$

$$W_{-1}V_{-1}AV_1W_1 : (x_n) \mapsto (y_n), \quad y_n = c_2 x_n,$$

we conclude that the diagonal matrix $\mathrm{diag}(c_k, c_k, \dots)$ lies in $\mathcal{B}$ for $k = 1, 2$. In order to arrive at this conclusion for $k > 2$, set

$$r_k = 2^{\{\log_2 k\}} - 3 + k \quad \text{and} \quad s_k = 2^{\{\log_2 k\}} - 2$$

where $\{z\}$ refers to the smallest integer which is greater than or equal to z. Then

$$V_{-1}^{s_k} W_{-1} V_1^{s_k} \cdot V_{-1}^{r_k} A V_1^{r_k} \cdot V_{-1}^{s_k} W_1 V_1^{s_k} : (x_n) \mapsto (y_n), \qquad y_n = c_k x_n$$

for all $k > 2$. Hence, $\mathrm{diag}(c_k, c_k, \dots)$ is in $\mathcal{B}$ for all k and consequently, $\mathrm{diag}(a_k, a_k, \dots)$ belongs to $\mathcal{B}$ for all k. Now it is easy to check that the mapping

$$T : \mathcal{A} \to \mathcal{B}, \quad a \mapsto \mathrm{diag}(a, a, \dots)$$

is the desired isomorphism from $\mathcal{A}$ onto a subalgebra of $\mathcal{B}$.

To finish the proof it remains to remark that, for *each* finitely generated Banach algebra $\mathcal{B}$, the algebra $\mathcal{B}^{r \times r}$ of all $r \times r$ matrices with entries in $\mathcal{B}$ is generated by three idempotents if only r is large enough (see [32]). Thus, each finitely generated Banach algebra is isomorphic to a subalgebra of a Banach algebra generated by three idempotents (the isomorphism simply being given by

$$\mathcal{B} \to \mathcal{B}^{r \times r}, \quad b \mapsto \mathrm{diag}(b, b, \dots, b)\,),$$

and this result in combination with what has already been proved gives our claim.

□

Theorem 5 indicates that the variety of all Banach algebras generated by three idempotents is extremely large and that these algebras can show a rather involved structure. This observation suggests the study of Banach algebras generated by three (or more) idempotents with additional relations between their generators. For example, let $L^2(J)$ denote the Hilbert space of the squared integrable functions on some (finite or infinite) interval J. On $L^2(\mathbb{R})$, we introduce the operator $S_{\mathbb{R}}$ of singular integration,

$$(S_{\mathbb{R}} f)(t) = \frac{1}{\pi i} \int_{-\infty}^{\infty} \frac{f(s)}{s - t} ds, \quad t \in \mathbb{R},$$

and the operators $\chi_{\mathbb{R}^+} I$ and $\chi_{[0,1]} I$ of multiplication by the characteristic functions of the intervals $\mathbb{R}^+$ and $[0, 1]$, respectively. Let $\mathcal{A}$ denote the smallest closed sub algebra of $L(L^2(\mathbb{R}))$ which contains the operators $S_{\mathbb{R}}$, $\chi_{\mathbb{R}^+} I$ and $\chi_{[0,1]} I$. Since $S_{\mathbb{R}}^2 = I$ and $S_{\mathbb{R}}^* = S_{\mathbb{R}}$ (see [20]), we conclude that $P_{\mathbb{R}} := (I + S_{\mathbb{R}})/2$ is a projection and hence, the algebra $\mathcal{A}$ is generated by three projections and the identity operator. Let further $\mathcal{B}$ refer to the smallest closed subalgebra of $\mathcal{A}$ which contains all operators $\chi_{[0,1]}(\chi_{\mathbb{R}^+} S_{\mathbb{R}} \chi_{\mathbb{R}^+})^k \chi_{[0,1]} I$ with $k = 0, 1, \dots$. Clearly, one can think of $\mathcal{B}$ as a subalgebra of $L(L^2([0, 1]))$.

Theorem 6 *The algebra $\mathcal{B}$ (which is a subalgebra of an algebra generated by three idempotents) contains all compact operators on $L^2([0, 1])$.*

For a proof see, e.g., Theorem 8.7 in [16].

Taking into account that the ideal of all compact operators on $L^2([0, 1])$ contains a copy of $\mathbb{C}^{l\times l}$ for all l or having recourse to Corollary 22.1 in [29], we arrive at the conclusion that the algebra $\mathcal{B}$ (and hence the algebra $\mathcal{A}$) cannot possess a matrix symbol of any fixed order. Thus, even if the three idempotents are projections, and even if two of them commute, a matrix symbol need not exist. This highlights that the additional conditions we look for in order to guarantee the existence of matrix symbols have to be rather strong.

Here are two (positive) examples of algebras generated by three idempotents which possess a matrix symbol. Observe the strong relations between the generating elements.

Theorem 7 *Let $\mathcal{A}$ be a Banach algebra with identity e, and let p, q and j be elements in $\mathcal{A}$ such that*

$$p^2 = p, \quad q^2 = q, \quad j^2 = e \quad \text{and} \quad jpj = e - p, \quad jqj = e - q.$$

Then the smallest closed subalgebra $\mathcal{B}$ of $\mathcal{A}$ which contains e, p, q and j is F_4, and it possesses a matrix symbol of order 2.

For a proof (and also for the explicit derivation of the matrix symbol under an additional condition) see [40]. Let us emphasize that the algebra $\mathcal{B}$ in Theorem 7 is indeed generated by three idempotents since p and q are idempotent and $(e + j)/2$ is idempotent, too.

Theorem 8 *Let $\mathcal{A}$ be a Banach algebra with identity e, and let p, q and j be elements in $\mathcal{A}$ such that*

$$p^2 = p, \quad q^2 = q, \quad j^2 = e \quad \text{and} \quad jpj = p, \quad jqj = e - q.$$

Then the smallest closed subalgebra $\mathcal{B}$ of $\mathcal{A}$ which contains e, p, q and j possesses a matrix symbol of order 4.

For a proof, and for an explicit matrix symbol, see [31].

4 An N Projections Theorem

4.1 Choice of the Additional Conditions

We are going to describe a class of Banach algebras which are generated by a large number of idempotents and possess a matrix symbol. Our choice of the additional conditions between the generating elements of the algebras is motivated by the situation considered in Sect. 5 (and, in a sense, by the approach of the papers [19] and [23]).

Let $\mathcal{A}$ be a Banach algebra with identity element I, and let $\{p_1, \ldots, p_{2N}\}$ be a *partition of unity into projections*, i.e. suppose $p_i \neq 0$ for all i,

$$p_i \cdot p_j = \delta_{ij} p_i \quad \text{for all} \quad i, j, \tag{1}$$

where δ_{ij} is the Kronecker delta, and

$$\sum_{i=1}^{2N} p_i = I. \tag{2}$$

Let further $P \in \mathcal{A}$, put $Q = I - P$, and suppose that

$$P(p_{2i-1} + p_{2i})P = (p_{2i-1} + p_{2i})P \tag{3}$$

and

$$Q(p_{2i} + p_{2i+1})Q = (p_{2i} + p_{2i+1})Q \tag{4}$$

for all $i = 1, \ldots, N$, where $p_{2N+1} := p_1$. In what follows we use the convention $p_k := p_r$ with $r \in \{1, \ldots, 2N\}$ whenever $k - r$ is divisible by $2N$. It is clear that then (3) and (4) hold for all integers i.

The algebra $\mathcal{B}$ we are interested in is the smallest closed subalgebra of $\mathcal{A}$ which contains the set $\{p_i\}_{i=1}^{2N}$ as well as the element P. Observe that $\mathcal{B}$ contains the identity I (due to (2)) and that P and Q are complementary idempotents. Indeed, adding the identities (3) for $i = 1, \ldots, N$ yields

$$P \cdot \sum_{i-1}^{2N} p_i \cdot P = \sum_{l=1}^{2N} p_i \cdot P,$$

that is, $P^2 = P$, whence $Q^2 = Q$. Thus, $\mathcal{B}$ is actually an algebra generated by $2N + 1$ idempotents (or by $2N$ idempotents and the identity).

We will show that the algebra $\mathcal{B}$ possesses a matrix symbol of order $2N$.

4.2 Algebraic Structure of $\mathcal{B}$

We start with examining the smallest (not necessarily closed) subalgebra $\mathcal{B}^0$ which contains the partition of unity into projections $\{p_i\}_{i=1}^{2N}$ and the idempotent P. Set

$$X := \sum_{i=1}^{N} (p_{2i-1} P p_{2i-1} + p_{2i} Q p_{2i}).$$

Proposition 1 *The element X is in the center of $\mathcal{B}^0$.*

Proof Evidently, X commutes with each of the idempotents p_i. It remains to show that $PX = XP$. Let us first prove that

$$X = \sum_{i=1}^{N} ((p_{2i} + p_{2i+1}) Q p_{2i} Q + (p_{2i-1} + p_{2i}) P p_{2i-1} P). \tag{5}$$

Since the p_j form a partition of unity into projections, it is sufficient to prove that

$$p_j X = p_j \sum_{i=1}^{N} ((p_{2i} + p_{2i+1}) Q p_{2i} Q + (p_{2i-1} + p_{2i}) P p_{2i-1} P)$$

for $j = 1, \ldots, 2N$ or, equivalently, that

$$p_{2i} Q p_{2i} = p_{2i} Q p_{2i} Q + p_{2i} P p_{2i-1} P \tag{6}$$

and

$$p_{2i-1} P p_{2i-1} = p_{2i-1} Q p_{2i-2} Q + p_{2i-1} P p_{2i-1} P \tag{7}$$

for all $i = 1, \ldots, N$. For (6) we observe that

$$\begin{aligned} p_{2i} Q p_{2i} Q + p_{2i} P p_{2i-1} P &= p_{2i} Q - p_{2i} P p_{2i} Q + p_{2i} P p_{2i-1} P \\ &= p_{2i} Q - p_{2i} P p_{2i} + p_{2i} P p_{2i} P + p_{2i} P p_{2i-1} P \\ &= p_{2i} Q - p_{2i} P p_{2i} + p_{2i} P (p_{2i-1} + p_{2i}) P \\ &= p_{2i} Q - p_{2i} P p_{2i} + p_{2i} P \\ &= p_{2i} - p_{2i} P p_{2i} \\ &= p_{2i} (P + Q) p_{2i} - p_{2i} P p_{2i} \\ &= p_{2i} Q p_{2i}, \end{aligned}$$

and (7) follows analogously. Thus (5) holds. Further, axioms (3) and (4) say that

$$Q(p_{2i-1} + p_{2i}) P = P(p_{2i} + p_{2i+1}) Q = 0 \tag{8}$$

for $i = 1, \ldots, N$, and the axioms (3), (4) together with the identities (5), (8) yield

$$PX = P \cdot \sum_{i=1}^{N} ((p_{2i} + p_{2i+1}) Q p_{2i} Q + (p_{2i-1} + p_{2i}) P p_{2i-1} P)$$

$$= \sum_{i=1}^{N} (p_{2i-1} + p_{2i}) P p_{2i-1} P$$

and

$$XP = \sum_{i=1}^{N} ((p_{2i} + p_{2i+1}) Q p_{2i} Q + (p_{2i-1} + p_{2i}) P p_{2i-1} P) \cdot P$$
$$= \sum_{i=1}^{N} (p_{2i-1} + p_{2i}) P p_{2i-1} P$$

and, hence, $PX = XP$. □

Proposition 2 *Considered as module over its center, the algebra $\mathcal{B}^0$ is generated by the $(2N)^2$ elements $(p_i)_{i=1}^{2N}$ and $(p_i P p_j)_{i,j=1}^{2N}$ with $i \neq j$. To be more precise, given $A \in \mathcal{B}^0$, there are polynomials R_{ij} in X such that*

$$A = \sum_{i=1}^{2N} R_{ii}(X) p_i + \sum_{\substack{i,j=1 \\ i \neq j}}^{2N} R_{ij}(X) p_i P p_j. \tag{9}$$

Proof Let $\mathcal{B}^1$ denote the set of all elements in $\mathcal{B}^0$ which can be written as in (9). First we show that the generating elements of $\mathcal{B}^0$ belong to $\mathcal{B}^1$. This is evident for the idempotents p_i. Since further

$$p_i P p_i = p_i P p_i \cdot p_i = \begin{cases} X \cdot p_i & \text{if } i \text{ is odd} \\ (I - X) \cdot p_i & \text{if } i \text{ is even,} \end{cases} \tag{10}$$

the assertion for P can be seen as follows:

$$P = \sum_{i,j=1}^{2N} p_i P p_j = \sum_{i=1}^{N} p_{2i} P p_{2i} + \sum_{i=1}^{N} p_{2i-1} P p_{2i-1} + \sum_{\substack{i,j=1 \\ i \neq j}}^{2N} p_i P p_j$$
$$= \sum_{i=1}^{N} (I - X) p_{2i} + \sum_{i=1}^{N} X p_{2i-1} + \sum_{\substack{i,j=1 \\ i \neq j}}^{2N} p_i P p_j.$$

In the second step we are going to show that $\mathcal{B}^1$ is actually an algebra. Since the generating elements of $\mathcal{B}^0$ belong to $\mathcal{B}^1$, this automatically yields that $\mathcal{B}^0 = \mathcal{B}^1$.

The set $\mathcal{B}^1$ is evidently closed under addition. In order to get its closedness under multiplication we have to show that the product of each two of the elements $(p_i)_{i=1}^{2N}$

and $(p_i P p_j)_{i,j=l}^{2N}$ with $i \neq j$ is in $\mathcal{B}^1$ again. This is obvious if one of these elements is p_i, and so we have only to deal with the products $p_i P p_j \cdot p_k P p_l$ with $i \neq j$ and $k \neq l$. This product is 0 (which is in $\mathcal{B}^1$) if $j \neq k$ and equal to $p_i P p_j P p_l$ in case $j = k$. If j is even (say, $j = 2n$) then

$$\begin{aligned} p_i P p_{2n} P p_l &= p_i P(p_{2n-1} + p_{2n}) P p_l - p_i P p_{2n-1} P p_l \\ &= p_i (p_{2n-1} + p_{2n}) P p_l - p_i P p_{2n-1} P p_l \end{aligned} \tag{11}$$

by axiom (3), whereas in case j is odd $(j = 2n - 1)$,

$$\begin{aligned} p_i P p_{2n-1} P p_l &= p_i P(p_{2n-2} + p_{2n-1}) P p_l - p_i P p_{2n-2} P p_l \\ &= p_i P(p_{2n-2} + p_{2n-1}) p_l - p_i P p_{2n-2} P p_l \end{aligned} \tag{12}$$

by (8). The first items in (11) and (12) are in $\mathcal{B}^1$. Indeed, they are either 0 or equal to $p_i P p_l$ (in dependence on j). If $i \neq l$ then $p_i P p_l \in \mathcal{B}^1$ by definition, whereas the inclusion $p_i P p_i \in \mathcal{B}^1$ follows from (10).

Thus, identities (11) and (12) reduce the question whether $p_i P p_j P p_l \in \mathcal{B}^1$ to the problem whether $p_i P p_{j-1} P p_l \in \mathcal{B}^1$. Repeated application of this argument finally yields an element of the form $p_i P p_i P p_l$. This element is in $\mathcal{B}^1$ since

$$p_i P p_i P p_l = p_i P p_i \cdot p_i P p_l = \begin{cases} X \cdot p_i P p_l & \text{if } i \text{ is odd} \\ (I - X) \cdot p_i P p_l & \text{if } i \text{ is even} \end{cases}$$

and by (10). □

Let us have a closer look at the products $p_i P p_j \cdot p_j P p_l$ in case $i \neq j$ and $j \neq l$.

Proposition 3 *(a) If $l > j > i$ or $j > i > l$ or $i > l > j$ then*

$$p_i P p_j P p_l = (-1)^{j-1} (X - I) p_i P p_l.$$

(c) If $l > i > j$ or $j > l > i$ or $i > j > l$ then

$$p_i P p_j P p_l = (-1)^{j-1} X p_i P p_l.$$

(c) If $i = l$ and $i \neq j$ then

$$p_i P p_j P p_i = (-1)^{j-i} X (X - I) p_i.$$

Proof Let $j \neq i, l$. Then

$$p_i P p_j P p_l = p_i P(p_{j-i} + p_j) P p_l - p_i P p_{j-1} P p_l. \tag{13}$$

If, moreover, $j - i \neq i, l$, then we conclude from (3) and (8) that $p_i P(p_{j-i} + p_j)Pp_l = 0$ and, hence,

$$p_i P p_j P p_l = -p_i P p_{j-1} P p_l. \tag{14}$$

Suppose now the conditions of assertion (a) to be satisfied. Then there is a smallest positive integer k such that (all computations modulo $2N$)

$$j \neq i, l, \quad j - 1 \neq i, l, \ldots, j - (k - l) \neq i, l$$

but $j - k = i$. Consequently, repeated application of (14) gives

$$p_i P p_j P p_l = (-1)^{k-1} p_i P p_{j-(k-1)} P p_l$$

whence by virtue of (13),

$$\begin{aligned} p_i P p_j P p_l &= (-1)^{k-1}(p_i P(p_{j-k} + p_{j-(k-1)})Pp_l - p_i P p_{j-k} P p_l) \\ &= (-1)^{k-1}(p_i P(p_i + p_{i+1})Pp_l - p_i P p_i P p_l). \end{aligned}$$

Observe that our assumptions imply that $l \neq i$ and $l \neq i + 1$ (otherwise $j - (k - 1)$ would be equal to l). Thus

$$\begin{aligned} p_i P p_j P p_l &= \begin{cases} (-1)^{k-1}(p_i(p_i + p_{i+1})Pp_l - p_i P p_i P p_l) & \text{if } i \text{ is odd} \\ (-1)^{k-1}(p_i P(p_i + p_{i+1})p_l - p_i P p_i P p_l) & \text{if } i \text{ is even} \end{cases} \\ &= \begin{cases} (-1)^{k-1}(p_i P p_l - p_i P p_i P p_l) & \text{if } i \text{ is odd} \\ (-1)^{k-1}(-p_i P p_i P p_l) & \text{if } i \text{ is even} \end{cases} \\ &= \begin{cases} (-1)^{k-1}(I - X)p_i P p_l & \text{if } i \text{ is odd} \\ (-1)^{k-1}(-1)(I - X)p_i P p_i P p_l & \text{if } i \text{ is even} \end{cases} \end{aligned}$$

(again take into account (10)). Replacing k by $j - i$ yields assertion (a). The proof for (b) and (c) is analogous. □

4.3 *Localization, and Identification of the Local Algebras*

The element X belongs to the center of the algebra $\mathcal{B}^0$ (Proposition 1) and thus to the center of $\mathcal{B}$ itself. Hence, the smallest closed subalgebra C of $\mathcal{B}$ which contains the identity element I and the element X is in the center of $\mathcal{B}$, and this offers the possibility of localizing $\mathcal{B}$ over C by the local principle of Allan and Douglas (Theorem 2). It is well known that the maximal ideal space of the singly (by X) generated Banach algebra C is homeomorphic to the spectrum $\sigma_C(X)$ of its generator (see [12], 15.3.6) and that under this homeomorphism the point

$x \in \sigma_C(X)$ corresponds to the smallest closed ideal of C which contains $X - xI$. In accordance with Theorem 2, we introduce ideals I_x in $\mathcal{B}$ for all $x \in \sigma_C(X)$.

Proposition 4 *(a) If $x \in \sigma_{\mathcal{B}}(X)(\subseteq \sigma_C(X))$ then $I_x \neq \mathcal{B}$.*
(b) If $x \in \sigma_C(X) \setminus \sigma_{\mathcal{B}}(X)$ then $I_x = \mathcal{B}$.

For a proof see [18]. Thus, by Theorem 2, an element $b \in \mathcal{B}$ is invertible if and only if the cosets $b + I_x$ are invertible for all $x \in \sigma_{\mathcal{B}}(X)$.

For $x \in \sigma_{\mathcal{B}}(X)$, let

$$\mathcal{B}_x := \mathcal{B}/I_x$$

denote the local algebra associated with x and let

$$\Phi_x : \mathcal{B} \to \mathcal{B}_x$$

be the canonical homomorphism. Let us remark once more that, by Proposition 4, each algebra $\mathcal{B}_x$ contains at least two different elements (the zero and the identity). Our next goal is the explicit description of the local algebras $\mathcal{B}_x$.

Proposition 5 *If $x \in \sigma_{\mathcal{B}}(X) \setminus \{0, 1\}$, then $\mathcal{B}_x$ is isomorphic to $\mathbb{C}^{2N \times 2N}$.*

Proof Consider the image $\Phi_x(\mathcal{B}^0)$ of the algebra $\mathcal{B}^0$ in $\mathcal{B}_x$. Since each element of $\mathcal{B}^0$ can be written in the form (9) and since

$$\Phi_x(X) = x\Phi_x(I)$$

by definition, it follows that

$$\Phi_x(R(X)) = R(x)\Phi_x(I)$$

for each polynomial R. Consequently, we conclude that each element of $\Phi_x(\mathcal{B}^0)$ is a complex linear combination of the elements

$$\Phi_x(p_i) \quad (i = 1, \ldots, 2N) \quad \text{and} \quad \Phi_x(p_i P p_j) \quad (i, j = 1, \ldots, 2N, i \neq j). \tag{15}$$

Conversely, each linear combination of the elements (15) is in $\Phi_x(\mathcal{B}^0)$. Thus, $\Phi_x(\mathcal{B}^0)$ is a finite dimensional linear space (of dimension $\leq (2N)^2$). In particular, $\Phi_x(\mathcal{B}^0)$ is closed in $\mathcal{B}_x$. On the other hand, $\mathcal{B}^0$ is dense in $\mathcal{B}$ and, hence, $\Phi_x(\mathcal{B}^0)$ is dense in $\Phi_x(\mathcal{B}) = \mathcal{B}_x$. Thus, $\mathcal{B}_x = \Phi_x(\mathcal{B}^0)$, and $\mathcal{B}_x$ is a linear space of dimension $\leq (2N)^2$.

We claim that the dimension of $\mathcal{B}_x$ is exactly $(2N)^2$ and that the elements (15) form a basis of this space. Given $i, j = 1, \ldots, 2N$, define $a_{ij} \in \mathcal{B}_x$ by

$$a_{ij} = \begin{cases} (-1)^{i-1}(x-1)^{-1}\Phi_x(p_i P p_j) \text{ if } i < j \\ (-1)^{i-1}x^{-1}\Phi_x(p_i P p_j) \text{ if } i > j \\ \Phi_x(p_i) \text{ if } i = j. \end{cases}$$

(This definition is correct since $x \neq 0$ and $x \neq 1$.) Proposition 3 implies that

$$a_{ij}a_{kl} = \delta_{jk} \cdot a_{il} \quad \text{for all} \quad 1 \leq i, j, k, l \leq 2N. \tag{16}$$

We check (16), for example, in case $j = k$ and $j > i > l$:

$$\begin{aligned} a_{ij}a_{jl} &= (-1)^{i-1}(x-1)^{-1}\Phi_x(p_i P p_j) \cdot (-1)^{j-1}x^{-1}\Phi_x(p_j P p_l) \\ &= (-1)^{i-1}(-1)^{j-1}x^{-1}(x-1)^{-1}\Phi_x(p_i P p_j P p_l) \\ &= (-1)^{i-1}(-1)^{j-1}x^{-1}(x-1)^{-1}\Phi_x((-1)^{j-1}(X-I)p_i P p_l) \\ &= (-1)^{i-1}x^{-1}\Phi_x(p_i P p_l) = a_{il}. \end{aligned}$$

The other cases can be disposed of analogously.

Now suppose the elements a_{ij} are linearly dependent. Then there are numbers c_{ij} with

$$\sum_{i,j=1}^{2N} c_{ij}a_{ij} = 0 \tag{17}$$

but $c_{i_0 j_0} \neq 0$ for certain i_0, j_0. Multiplying (17) from the left by a_{ki_0} and from the right by $a_{j_0 k}$ yields that

$$c_{i_0 j_0} a_{k i_0} a_{i_0 j_0} a_{j_0 k} = c_{i_0 j_0} a_{kk} = 0$$

and hence, $a_{kk} = 0$ for all $k = 1, \ldots, 2N$. Consequently,

$$\Phi_x(I) = \Phi_x\left(\sum_{k=1}^{2N} p_k\right) = \sum_{k=1}^{2N} a_{kk} = \Phi_x(0)$$

which contradicts Proposition 4(a) (see also the remark following this proposition).

Thus, the elements $(a_{ij})_{i,j=1}^{2N}$ are linearly independent. It follows that so are also the elements (15), and therefore both sets of elements form a basis of $\mathcal{B}_x$. Finally, it is immediate from (16) that the mapping

$$\Psi_x : (a_{ij})_{i,j=1}^{2N} \to \mathbb{C}^{2N \times 2N}, \quad a_{ij} \mapsto E_{ij},$$

where E_{ij} refers to the $2N \times 2N$ matrix where i, j entry is 1 and all other entries of which are zero, extends to an algebra isomorphism from $\mathcal{B}_x$ onto $\mathbb{C}^{2N \times 2N}$. □

Here are the images of the generating elements of the algebra $\mathcal{B}$ under the homomorphism $F_x := \Psi_x \circ \Phi_x : \mathcal{B} \to \mathbb{C}^{2N \times 2N}$.

Corollary 1 *Let $x \in \sigma_{\mathcal{B}}(X) \setminus \{0, 1\}$. Then*

$$F_x(p_i) = \text{diag}(0, \dots, 0, 1, 0, \dots, 0), \tag{18}$$

the 1 *standing at the ith place, and*

$$\begin{aligned} F_x(P) =& \text{diag}(1, -1, 1, -1, \dots, 1, -1)\times \\ &\times \begin{pmatrix} x & x-1 & x-1 & x-1 & \dots & x-1 & x-1 \\ x & x-1 & x-1 & x-1 & \dots & x-1 & x-1 \\ x & x & x & x-1 & \dots & x-1 & x-1 \\ x & x & x & x-1 & \dots & x-1 & x-1 \\ \vdots & \vdots & \vdots & \vdots & \ddots & \vdots & \vdots \\ x & x & x & x & \dots & x & x-1 \\ x & x & x & x & \dots & x & x-1 \end{pmatrix}. \end{aligned} \tag{19}$$

Proof To verify (18) recall that $\Phi_x(p_i) = a_{ii}$, and to get (19) observe that

$$\begin{aligned} F_x(P) =& F_x\left(\sum_{i,j=1}^{2N} p_i P p_j\right) \\ =& (\Psi_x \circ \Phi_x)\left(\sum_{\substack{j,\,j=1\\ i<j}}^{2N} p_i P p_j\right) + (\Psi_x \circ \Phi_x)\left(\sum_{\substack{j,\,j=1\\ i>j}}^{2N} p_i P p_j\right) \\ &+ (\Psi_x \circ \Phi_x)\left(\sum_{j=1}^{2N} p_i P p_i\right) \\ =& \Psi_x\left(\sum_{\substack{i,\,j=1\\ i<j}}^{2N} (-1)^{i-1}(x-1)a_{ij}\right) + \Psi_x\left(\sum_{\substack{i,\,j=1\\ i>j}}^{2N} (-1)^{i-1} x a_{ij}\right) \\ &+ (\Psi_x \circ \Phi_x)\left(\sum_{j=1}^{2N} p_i P p_i\right) \end{aligned}$$

and take into account (10). □

Our next subject is the local algebras $\mathcal{B}_x$ associated with the points in $\sigma_{\mathcal{B}}(X) \cap \{0, 1\}$. These algebras will not be identified completely; we will only show that all irreducible representations are one-dimensional and will compute them.

Proposition 6 *If $x \in \sigma_{\mathcal{B}}(X) \cap \{0, 1\}$ then $\mathcal{B}_x$ is an F_2^{N+1}-algebra.*

Proof Instead of working with the polynomial

$$F_2^{N+1}(a, b) = (ab - ba)^{N+1}$$

in two variables, which is non-linear, let us consider the polynomial

$$F_2^{(N+1)}(a_1, b_1, \dots, a_{N+1}, b_{N+1}) := \prod_{k=1}^{N+1} (a_k b_k - b_k a_k)$$

in $2(N+1)$ variables, which is linear in each variable. Notice that if $\mathcal{B}_x$ is an $F_2^{(N+1)}$-algebra, then it is also and F_2^{N+1}-algebra.

Since $\mathcal{B}_X$ is a linear space (recall the proof of Proposition 5) and since $F_2^{(N+1)}$ is multilinear, it remains to prove that

$$\prod_{k=1}^{N+1} (a_k b_k - b_k a_k) = 0$$

for all choices of cosets a_k, b_k $(k = 1, \dots, N+1)$ among the (possible) basis elements of the algebra $\mathcal{B}_x$:

$$\Phi_x(p_i) \quad (i = 1, \dots, 2N) \quad \text{and} \quad \Phi_x(p_i P p_j) \quad (i, j = 1, \dots, 2N, \ i \neq j).$$

Proposition 3 entails that each commutant $a_k b_k - b_k a_k$ can be written as

$$c_k \Phi_x(p_{i_k} P p_{j_k})$$

where $i_k, j_k \in \{1, \dots, N+1\}$ and $c_k \in \mathbb{C}$ can be zero. Hence,

$$\prod_{k=1}^{N+1} (a_k b_k - b_k a_k) - c\Phi_x\left(\prod_{k=1}^{N+1} p_{i_k} P p_{j_k}\right).$$

Since the partition of unity into projections (p_i) consists of $2N$ elements, there are two of the elements p_{i_k} and p_{j_k} with $k = 1, \dots, N+1$ which coincide. Thus, $\prod_{k=1}^{N+1} p_{i_k} P p_{j_k}$ contains at least one subproduct of the form

$$p_i P p_{l_1} P p_{l_2} \dots P p_{l_r} P p_i$$

with $r \geq 1$, and invoking Proposition 3 once more, one easily gets

$$\Phi_x(p_i P p_{l_1} P p_{l_2} \dots P p_{l_r} P p_i) = 0.$$

Thus,

$$\Phi_x \left(\prod_{k=1}^{N+1} p_{i_k} P p_{j_k} \right) = 0 \quad \text{for} \quad x \in \sigma_{\mathcal{B}}(X) \cap \{0, 1\},$$

which concludes the proof. □

By Theorem 3 (extended version), the algebras $\mathcal{B}_x$ possess matrix symbols of order 1, i.e. scalar-valued symbols. Since each algebra homomorphism

$$\Psi : \mathcal{B}_x \to \mathbb{C}$$

gives rise to an algebra homomorphism $\Psi \circ \Phi_x : \mathcal{B} \to \mathbb{C}$, we proceed with determining the one-dimensional representations of the algebra $\mathcal{B}$.

Clearly, each homomorphism $G : \mathcal{B} \to \mathbb{C}$ maps idempotents to idempotents. Thus, if $p \in \mathcal{B}$ is idempotent, then $G(p)$ is either 0 or 1. Moreover, since $G(I) = 1$ for each non-zero homomorphism G, we conclude that, given a partition of unity into projections $(p_i)_{i=1}^{2N}$, there is an i_0 such that $G(p_{i_0}) = 1$ and $G(p_i) = 0$ for all $i \neq i_0$. Hence, the restriction of a non-zero homomorphism $G : \mathcal{B} \to \mathbb{C}$ to the set $\{P, p_1, p_2, \ldots, p_{2N}\}$ coincides with one of the following mappings G_n with $n \in \{1, 2, \ldots, 4N\}$:

$$\begin{aligned}
G_{4m}(p_i) &= \begin{cases} 1 & \text{if } i = 2m \\ 0 & \text{if } i \neq 2m, \end{cases} & G_{4m}(P) &= 0, \\
G_{4m-1}(p_i) &= \begin{cases} 1 & \text{if } i = 2m \\ 0 & \text{if } i \neq 2m, \end{cases} & G_{4m-1}(P) &= 1, \\
G_{4m-2}(p_i) &= \begin{cases} 1 & \text{if } i = 2m-1 \\ 0 & \text{if } i \neq 2m-1, \end{cases} & G_{4m-2}(P) &= 1, \\
G_{4m-3}(p_i) &= \begin{cases} 1 & \text{if } i = 2m-1 \\ 0 & \text{if } i \neq 2m-1, \end{cases} & G_{4m-3}(P) &= 0,
\end{aligned} \tag{20}$$

where $m = 1, \ldots, N$. Set

$$Y := \sum_{i=1}^{N} (p_{2i-1} P + p_{2i} Q) + \sum_{i=1}^{2N} (2i - 1) p_i .$$

Proposition 7 *If $m \in \sigma_{\mathcal{B}}(Y) \cap \{1, 2, \ldots, 4N\}$ then the mapping*

$$G_m : \{P, p_1, p_2, \ldots, p_{2N}\} \to \mathbb{C},$$

given by (20)*, extends to an algebra homomorphism from $\mathcal{B}$ onto $\mathbb{C}$.*

Proof First of all notice that if G_m extends to an algebra homomorphism, then

$$G_m(Y) = m. \tag{21}$$

We claim that, for $m \in \sigma_{\mathcal{B}}(Y) \cap \{1, 2, \ldots, 4N\}$ and $x \in \sigma_{\mathcal{B}}(X) \setminus \{0, 1\}$,

$$m \notin \sigma_{\mathcal{B}_x}(\Phi_x(Y)). \tag{22}$$

What we have to prove is, by Corollary 1, that the $2N \times 2N$ matrices

$$(\Psi_x \circ \Phi_x)(Y) - \operatorname{diag}(m, m, \ldots, m) =$$

$$= \begin{pmatrix} x & x-1 & x-1 & x-1 & \ldots & x-1 & x-1 \\ x & x & x-1 & x-1 & \ldots & x-1 & x-1 \\ x & x & x & x-1 & \ldots & x-1 & x-1 \\ x & x & x & x & \ldots & x-1 & x-1 \\ \vdots & \vdots & \vdots & \vdots & \ddots & \vdots & \vdots \\ x & x & x & x & \ldots & x & x-1 \\ x & x & x & x & \ldots & x & x \end{pmatrix}$$

$$+ \operatorname{diag}(1-m, 3-m, \ldots, 4N-1-m)$$

are invertible. For this goal we compute the determinant of the slightly more general $M \times M$ matrix

$$\begin{pmatrix} x+\lambda_1 & x-1 & x-1 & x-1 & \ldots & x-1 & x-1 \\ x & x+\lambda_2 & x-1 & x-1 & \ldots & x-1 & x-1 \\ x & x & x+\lambda_3 & x-1 & \ldots & x-1 & x-1 \\ x & x & x & x+\lambda_4 & \ldots & x-1 & x-1 \\ \vdots & \vdots & \vdots & \vdots & \ddots & \vdots & \vdots \\ x & x & x & x & \ldots & x+\lambda_{M-1} & x-1 \\ x & x & x & x & \ldots & x & x+\lambda_M \end{pmatrix} \tag{23}$$

where $\lambda_1, \ldots, \lambda_M, x \in \mathbb{C}$. Consider x as being variable and denote the determinant of the matrix (23) by $D(x)$. Subtracting in (23) the first row from all other rows, and then the last column from all other columns, one gets a matrix the $1, N$ entry of which is $x - 1$ while all other entries are independent of x. Thus, $D(x)$ is a polynomial of first degree in x and, since $D(0) = \prod_{i=1}^{M} \lambda_i$ and $D(1) = \prod_{i=1}^{M}(1 + \lambda_i)$, one has

$$D(x) = x \prod_{i=1}^{M} (1 + \lambda_i) + (1 - x) \prod_{i=1}^{M} \lambda_i. \tag{24}$$

Now let $m \in \{1, \ldots, 4N\}$, $M = 2N$, and $\lambda_i = 2i - 1 - m$ for $i = 1, \ldots, 2N$. If m is odd, then one of the numbers λ_i is equal to zero, but

$$\prod_{i=1}^{M}(1 + \lambda_i) \neq 0.$$

If m is even, then one of the numbers $1 + \lambda_i$ is zero, but

$$\prod_{i=1}^{M}\lambda_i \neq 0.$$

Hence, in any case,

$$x\prod_{i=1}^{M}(1 + \lambda_i) + (1 - x)\prod_{i=1}^{M}\lambda_i \neq 0$$

whenever $x \notin \{0, 1\}$. This proves our claim (22).

Now the assertion can be obtained as follows. Let $m \in \sigma_{\mathcal{B}}(Y) \cap \{1, 2, \ldots, 4N\}$. Then, by the local principle,

$$m \in \bigcup_{x \in \sigma_{\mathcal{B}}(X)} \sigma_{\mathcal{B}_x}(\Phi_x(Y))$$

whereas, by (22),

$$m \notin \bigcup_{x \in \sigma_{\mathcal{B}}(X) \setminus \{0,1\}} \sigma_{\mathcal{B}_x}(\Phi_x(Y)).$$

Hence,

$$m \in \bigcup_{x \in \sigma_{\mathcal{B}}(X) \cap \{0,1\}} \sigma_{\mathcal{B}_x}(\Phi_x(Y)).$$

But the algebras $\mathcal{B}_x$ with $x \in \sigma_{\mathcal{B}}(X) \cap \{0, 1\}$ possess a scalar-valued symbol (Proposition 6 and Theorem 3). Thus, if $m \in \sigma_{\mathcal{B}_{x_0}}(\Phi_{x_0}(Y))$ with a certain $x_0 \in \sigma_{\mathcal{B}}(X) \cap \{0, 1\}$ then there is an algebra homomorphism G' from $\mathcal{B}_{x_0}$ onto $\mathbb{C}$ with $G'(\Phi_{x_0}(Y)) = m$. Then $G := G' \circ \Phi_{x_0}$ is an algebra homomorphism from $\mathcal{B}$ onto $\mathbb{C}$ with $G(Y) = m$. The restriction of G to the set $\{P, p_1, \ldots, p_{2N}\}$ coincides with one of the mappings G_n introduced in (20) and, by (21), this restriction is just G_m. In other words, G_m extends to an (evidently continuous) algebra homomorphism from $\mathcal{B}$ onto $\mathbb{C}$. □

For $m \in \sigma_{\mathcal{B}}(Y) \cap \{1, 2, \ldots, 4N\}$, let us denote the extension of G_m by G_m again. One easily checks that

$$G_m(X) = \begin{cases} 0 \text{ if } m \text{ is odd} \\ 1 \text{ if } m \text{ is even.} \end{cases}$$

Thus, if $0 \in \sigma_{\mathcal{B}}(X)$ and m is odd, then the local ideal I_0 lies in the kernel of G_m and consequently, for each $A \in \mathcal{B}$ the number $G_m(A)$ depends on the coset $\Phi_0(A)$ only. So the quotient mapping

$$G'_m : \mathcal{B}_0 \to \mathbb{C}, \quad \Phi_0(A) \mapsto G_m(A)$$

is correctly defined, and it is an algebra homomorphism from $\mathcal{B}_0$ onto $\mathbb{C}$. Analogously, if $0 \in \sigma_{\mathcal{B}}(X)$ and m is even, then

$$G'_m : \mathcal{B}_1 \to \mathbb{C}, \quad \Phi_1(A) \mapsto G_m(A)$$

is a correctly defined and non-trivial algebra homomorphism.

Proposition 8 *(a) If $0 \in \sigma_{\mathcal{B}}(X)$ then the set $\{G_m\}$, consisting of all mappings G_m with $m \in \sigma_{\mathcal{B}}(Y) \cap \{1, 2, \ldots, 4N\}$ and m odd, forms a scalar-valued symbol for $\mathcal{B}_0$. (b) If $1 \in \sigma_{\mathcal{B}}(X)$ then the set $\{G_m\}$, consisting of all mappings G_m with $m \in \sigma_{\mathcal{B}}(Y) \cap \{1, 2, \ldots, 4N\}$ and m even, forms a scalar-valued symbol for $\mathcal{B}_1$.*

Proof The mappings G'_m with m odd (even) are the only non-trivial algebra homomorphisms from $\mathcal{B}_0$ (resp. $\mathcal{B}_1$) into $\mathbb{C}$. But since the algebras $\mathcal{B}_0$ (resp. $\mathcal{B}_1$) possess a scalar-valued symbol by Theorem 3 and Proposition 6, we conclude that for all $A \in \mathcal{B}$ the coset $\Phi_0(A)$ (resp. $\Phi_1(A)$) is invertible whenever all

$$G'_m(\Phi_0(A)) = G_m(A)$$

with m odd (resp. even) are invertible. □

4.4 The N Projections Theorem

Now we are in a position to state our main result.

Theorem 9 *Let $\mathcal{A}$ be a Banach algebra with identity I. Let $p_1, p_2, \ldots, p_{2N}$ and P be nonzero elements of $\mathcal{A}$ satisfying*

$$p_i \cdot p_j = \delta_{ij} p_i \quad \text{for all} \quad i, j \quad \text{and} \quad p_1 + p_2 + \cdots + p_{2N} = I,$$

where δ_{ij} is the Kronecker delta, and

$$P(p_{2i-1} + p_{2i})P = (p_{2i-1} + p_{2i})P \quad \text{and} \quad Q(p_{2i} + p_{2i+1})Q = p_{2i} + p_{2i+1})Q$$

for all $i = 1, \ldots, N$, where

$$Q := I - P \text{ and } p_{2N+1} := p_1.$$

Let further $\mathcal{B}$ stand for the smallest closed subalgebra of $\mathcal{A}$ containing the elements P and $p_1, \ldots, p_{2N}$. Then the following assertions hold.

(a) If $x \in \sigma_{\mathcal{B}}(X) \setminus \{0, 1\}$ where

$$X = \sum_{i=1}^{N} (p_{2i-1} P p_{2i-1} + p_{2i} Q p_{2i}),$$

then the mapping $F_x : \{P, p_1, \ldots, p_{2N}\} \to \mathbb{C}^{2N\times 2N}$ given by

$$F_x(p_i) = \operatorname{diag}(0, \ldots, 0, 1, 0, \ldots, 0),$$

with the 1 standing at the ith place, and

$$F_x(P) = \operatorname{diag}(1, -1, 1, -1, \ldots, 1, -1)\times$$
$$\times \begin{pmatrix} x & x-1 & x-1 & x-1 & \ldots & x-1 & x-1 \\ x & x-1 & x-1 & x-1 & \ldots & x-1 & x-1 \\ x & x & x & x-1 & \ldots & x-1 & x-1 \\ x & x & x & x-1 & \ldots & x-1 & x-1 \\ \vdots & \vdots & \vdots & \vdots & \ddots & \vdots & \vdots \\ x & x & x & x & \ldots & x & x-1 \\ x & x & x & x & \ldots & x & x-1 \end{pmatrix}$$

extends to a continuous algebra homomorphism from $\mathcal{B}$ onto $\mathbb{C}^{2N\times 2N}$.

(b) If $m \in \sigma_{\mathcal{B}}(Y) \cap \{1, \ldots, 4N\}$ where

$$Y := \sum_{i=1}^{N} (p_{2i-1} P + p_{2i} Q) + \sum_{i=1}^{2N} (2i-1) p_i,$$

then the mapping $G_m : \{P, p_1, \ldots, p_{2N}\} \to \mathbb{C}$ defined by

$$G_{4m}(p_i) = \begin{cases} 1 & \text{if } i = 2m \\ 0 & \text{if } i \neq 2m, \end{cases} \qquad G_{4m}(P) = 0,$$
$$G_{4m-1}(p_i) = \begin{cases} 1 & \text{if } i = 2m \\ 0 & \text{if } i \neq 2m, \end{cases} \qquad G_{4m-1}(P) = 1,$$
$$G_{4m-2}(p_i) = \begin{cases} 1 & \text{if } i = 2m-1 \\ 0 & \text{if } i \neq 2m-1, \end{cases} \qquad G_{4m-2}(P) = 1,$$
$$G_{4m-3}(p_i) = \begin{cases} 1 & \text{if } i = 2m-1 \\ 0 & \text{if } i \neq 2m-1, \end{cases} \qquad G_{4m-3}(P) = 0$$

where $m = 1, \ldots, N$, extends to a continuous algebra homomorphism from $\mathcal{B}$ onto $\mathbb{C}$.

(c) An element $B \in \mathcal{B}$ is invertible in $\mathcal{B}$ if and only if the matrices $F_x(B)$ are invertible for all $x \in \sigma_{\mathcal{B}}(X) \setminus \{0, 1\}$ and the numbers $G_m(B)$ are non-zero for all $m \in \sigma_{\mathcal{B}}(Y) \cap \{1, \ldots, 4N\}$.

(d) An element $B \in \mathcal{B}$ is invertible in $\mathcal{A}$ if and only if the matrices $F_x(B)$ are invertible for all $x \in \sigma_{\mathcal{A}}(X) \setminus \{0, 1\}$ and the numbers $G_m(B)$ are non-zero for all $m \in \sigma_{\mathcal{A}}(Y) \cap \{1, \ldots, 4N\}$.

Proof The proof of assertions (a), (b) and (c) is immediate from the local principle in combination with the description of the local algebras given in the preceding subsection. Concerning the continuity of the mappings F_x and G_m we refer to a general result by Johnson (see, e.g., [28], Chapter 6, Theorem 2.65) stating that an algebra homomorphism from a Banach algebra onto a semi-simple Banach algebra is always continuous.

For a proof of assertion (d) recall that the algebra $\mathcal{B}^0$ is a $(2N)^2$ dimensional module over its center. Thus, Corollary 1.2 in [22] tells us that there is a set $\{\nu_t\}$, $t \in T$, of representations of $\mathcal{B}$ such that $\operatorname{Im} \nu_t = \mathbb{C}^{l \times l}$ with $l = l(t) \leq 2N$ and such that an element B of $\mathcal{B}$ is invertible in $\mathcal{A}$ if and only if $\det \nu_t(B) \neq 0$ for all $t \in T$. The very same arguments as in the proof of assertion (c) entail that each of these representations is of the form F_x (with an $x \in \mathbb{C} \setminus \{0, 1\}$) as defined in Corollary 1 or G_m (with an $m \in \{1, 2, \ldots, 4N\}$) as defined after Proposition 6. Hence, there exist two sets $\xi = \xi(\mathcal{A}, \mathcal{B}) \subset \mathbb{C} \setminus \{0, 1\}$ and $\mu = \mu(\mathcal{A}, \mathcal{B}) \subseteq \{1, 2, \ldots, 4N\}$ such that

$$\sigma_{\mathcal{A}}(B) = \bigcup_{x \in \xi} \sigma(F_x(B)) \cup \{G_m(B) : m \in \mu\} \tag{25}$$

for all $B \in \mathcal{B}$. We claim that $\xi = \sigma_{\mathcal{A}}(X) \setminus \{0, 1\}$ and $\mu = \sigma_{\mathcal{A}}(Y) \cap \{1, \ldots, 4N\}$.

Since $G_m(X) \in \{0, 1\}$ and $\xi \cap \{0, 1\} = \emptyset$, one has

$$\sigma_{\mathcal{A}}(X) \setminus \{0, 1\} = \bigcup_{x \in \xi} \sigma(F_x(X)) \cup \{G_m(X) : m \in \mu\} \setminus \{0, 1\} = \bigcup_{x \in \xi} \{x\} = \xi. \tag{26}$$

For the second claim note that, for any $\lambda \in \mathbb{C}$, the matrix $F_x(Y - \lambda I)$ coincides with the matrix (21) with the λ_i in (21) replaced by $2i - 1 - \lambda$. It follows from the explicit form (24) of the determinant of this matrix that every eigenvalue λ of $F_x(Y)$ solves the equation

$$x \prod_{i=1}^{2N} (2i - \lambda) + (1 - x) \prod_{i=1}^{2N} (2i - 1 - \lambda) = 0.$$

But, if $x \notin \{0, 1\}$, then $\sigma(F_x(Y)) \cap \{1, 2, \ldots, 2N\} = \emptyset$. Thus,

$$\sigma_{\mathcal{A}}(Y) \cap \{1, \ldots, 4N\} = \{G_m(Y) : m \in \mu\} \cap \{1, \ldots, 4N\} = \{m\}_{m \in \mu} = \mu. \tag{27}$$

Now assertion (d) follows immediately from (25), (26) and (27). □

Observe that assertion (d) is evident in case the algebra $\mathcal{B}$ is inverse closed in $\mathcal{A}$. However, this is not always satisfied as the following example indicates.

Example Let $\mathbb{T}$ be the unit circle $\{t \in \mathbb{C} : |t| = 1\}$ and consider the algebra $\mathcal{A}$ of all continuous 2×2 matrix functions on $\mathbb{T}$. If t denotes the identical mapping of $\mathbb{T}$ then

$$P = \begin{pmatrix} t & 1-t \\ t & 1-t \end{pmatrix}, \quad p_1 = \begin{pmatrix} 1 & 0 \\ 0 & 0 \end{pmatrix}, \quad \text{and} \quad p_2 = \begin{pmatrix} 0 & 0 \\ 0 & 1 \end{pmatrix}$$

are elements of $\mathcal{A}$ which satisfy the assumptions of Theorem 9 (with $N = 1$). The element

$$X = p_1 P p_1 + p_2(I - P)p_2 = \begin{pmatrix} t & 0 \\ 0 & t \end{pmatrix}$$

is invertible in $\mathcal{A}$ but not invertible in $\mathcal{B}$ since the latter algebra consists of matrix functions holomorphic in the unit disk only. □

In this connection, let us emphasize an evident consequence of assertions (c) and (d) of the previous theorem.

Corollary 2 *If $\sigma_{\mathcal{B}}(X) = \sigma_{\mathcal{A}}(X)$ and $\sigma_{\mathcal{B}}(Y) = \sigma_{\mathcal{A}}(Y)$, then the algebra $\mathcal{B}$ is inverse closed in $\mathcal{A}$.*

The following additional assertions are often useful.

Proposition 9 *(a) If $0 \notin \sigma_{\mathcal{B}}(X)$ and $1 \notin \sigma_{\mathcal{B}}(X)$ then $\sigma_{\mathcal{B}}(Y) \cap \{1, \ldots, 4N\} = \emptyset$.*
(b) If $0 \in \sigma_{\mathcal{B}}(X)$ and $1 \in \sigma_{\mathcal{B}}(X)$, and if both points are not isolated in $\sigma_{\mathcal{B}}(X)$, then the family (F_x) with $x \in \sigma_{\mathcal{B}}(X)$ is a matrix symbol for $\mathcal{B}$.
(c) If $0 \notin \sigma_{\mathcal{A}}(X)$ and $1 \notin \sigma_{\mathcal{A}}(X)$ then $\sigma_{\mathcal{A}}(Y) \cap \{1, \ldots, 4N\} = \emptyset$.
(d) If $0 \in \sigma_{\mathcal{A}}(X)$ and $1 \in \sigma_{\mathcal{A}}(X)$, and if both points are not isolated in $\sigma_{\mathcal{A}}(X)$, then the family (F_x) with $x \in \sigma_{\mathcal{A}}(X)$ is a matrix symbol for the invertibility of the elements of $\mathcal{B}$ in the algebra $\mathcal{A}$.

Proof (a) Observe that $G_m(X) \in \{0, 1\}$ in any case. Thus, if

$$\sigma_{\mathcal{B}}(X) \cap \{0, 1\} = \emptyset,$$

one-dimensional representations cannot exist.

(b) Let M be a mapping from $\mathbb{C}$ into the set of all subsets of $\mathbb{C}$. Given a sequence $(x_n) \subseteq \mathbb{C}$ with $x_n \to 0$ as $n \to \infty$, we consider the set $L(x_n)$ of all limiting points of all sequences (λ_{x_n}) with $\lambda_{x_n} \in M(x_n)$, and we define the *limiting set* $\lim_{x \to 0} M(x)$ as $\cup L(x_n)$ where the union is taken over all sequences (x_n) with $x_n \to 0$ but $x_n \neq 0$ for all n. Analogously, we define $\lim_{x \to 1} M(x)$.

The function $x \mapsto F_x(Y)$ is continuous on $\sigma_{\mathcal{B}}(X)$. Due to the continuous dependence of the eigenvalues of a matrix on the matrix itself (see [27], Appendix D), one has

$$\sigma(F_0(Y)) = \lim_{x \to 0} \sigma(F_x(Y))$$

and consequently,

$$\sigma(F_0(Y)) = \lim_{x \to 0} \sigma(\Phi_x(Y)). \tag{28}$$

We claim that

$$\lim_{x \to 0} \sigma(\Phi_x(Y)) \subseteq \sigma(\Phi_0(Y)). \tag{29}$$

To prove (29), we need the following supplement to the local principle. Let the notation be as in Theorem 2. □

Proposition 10 *Let $a \in \mathcal{A}$ and suppose $a + I_x$ to be invertible for some x. Then there is an open neighborhood U of x such that the cosets $a + I_y$ are invertible and*

$$\|(a + I_y)^{-1}\| \le 4\|(a + I_x)^{-1}\| \quad \textit{for all} \quad y \in U.$$

Proof Set $\phi_x(a) := a + I_x$ and let $\phi_x(a)$ be invertible. Then there is a $b \in \mathcal{A}$ such that

$$\phi_x(ab - e) = \phi_x(ba - e) = 0.$$

As shown in [1], or [6], Theorem 1.34, or [25], Theorem 1.5, the mappings

$$y \mapsto \|\phi_y(ab - e)\| \quad \text{and} \quad y \mapsto \|\phi_y(ba - e)\|,$$

defined on the maximal ideal space of C, are upper semi-continuous. Hence

$$\|\phi_y(ab - e)\| < 1/2 \quad \text{and} \quad \|\phi_y(ba - e)\| < 1/2$$

for all maximal ideals y in a certain neighborhood U' of x. Since

$$\phi_y(a)\phi_y(b) = \phi_y(e) + \phi_y(ab - e) \quad \text{and} \quad \phi_y(b)\phi_y(a) = \phi_y(e) + \phi_y(ba - e),$$

and since $\phi_y(e)$ is the identity element in $\mathcal{A}/I_y$, this implies (Neumann's series) that $\phi_y(a)$ is invertible in $\mathcal{A}/I_y$ and that

$$\|\phi_y(a)^{-1}\| \le 2\|\phi_y(b)\| \quad \text{for all} \quad y \in U'.$$

Invoking upper semi-continuity once more we get

$$\|\phi_y(b)\| \leq 2\|\phi_x(b)\| = 2\|\phi_x(a)^{-1}\|$$

for all y in a neighborhood U'' of x, which proves Proposition 10. □

Continuation of the Proof of Proposition 9 Now, in order to prove our claim (29), assume there are sequences $(x_n) \subseteq \sigma_{\mathcal{B}}(X)$ and (λ_n) with $\lambda_n \in \sigma(\Phi_{x_n}(Y))$ such that $x_n \to 0$, $x_n \neq 0$, $\lambda_n \to \lambda$, but $\lambda \notin \sigma(\Phi_0(Y))$. Then $\Phi_0(Y - \lambda I)$ is invertible and, by Proposition 10, $\Phi_{x_n}(Y - \lambda I)$ is invertible and

$$\|\Phi_{x_n}(Y - \lambda I)^{-1}\| \leq 4\|\Phi_0(Y - \lambda I)^{-1}\|$$

for all n large enough. Thus,

$$\operatorname{dist}(\lambda, \sigma(\Phi_{x_n}(Y))) \geq \frac{1}{4\|\Phi_0(Y - \lambda I)^{-1}\|},$$

which contradicts our assumption since

$$|\lambda - \lambda_{x_n}| \geq \operatorname{dist}(\lambda, \sigma(\Phi_{x_n}(Y))).$$

This proves our claim (29).

From (28) and (29) we see that $\sigma(F_0(Y)) \subseteq \sigma(\Phi_0(Y))$ and, analogously, $\sigma(F_1(Y)) \subseteq \sigma(\Phi_1(Y))$. Hence,

$$\sigma(F_0(Y)) \cup \sigma(F_1(Y)) \subseteq \sigma(\Phi_0(Y)) \cup \sigma(\Phi_1(Y)) \subseteq \sigma_{\mathcal{B}}(Y).$$

Evidently,

$$\sigma(F_0(Y)) \cup \sigma(F_1(Y)) = \{1, 2, \ldots, 4N\},$$

and consequently,

$$\sigma_{\mathcal{B}}(Y) \cap \{1, 2, \ldots, 4N\} = \{1, 2, \ldots, 4N\}.$$

In other words, all possible one-dimensional representations occur.

It remains to observe that, for each $A \in \mathcal{B}$, the matrices $F_0(A)$ and $F_1(A)$ are triangular and that the diagonal of $F_0(A)$ equals $(G_1(A), G_3(A), \ldots, G_{2N-1}(A))$, while the diagonal of $F_1(A)$ is $(G_2(A), G_4(A), \ldots, G_{2N}(A))$.

The proof of assertions (c) and (d) can be given in a completely analogous manner. □

5 Examples

5.1 Abstract Analogues of Singular Integral Operators

Let T be a non-empty proper subset of $\{1, 2, \ldots, 2N\}$, set $P := \sum_{i \in T} p_i$ and $q := I - p$. Elements of the form $A := pPp + q (\in \mathcal{B})$ are called *abstract analogues of singular integrals*. Our first concern is to demonstrate how Theorem 9 can be used to compute the spectrum of abstract singular integrals in case the spectrum of X is known. From Theorem 9 we conclude that this spectrum equals

$$\bigcup_{x \in \sigma_{\mathcal{F}}(X) \setminus \{0,1\}} \sigma(F_x(A)) \cup \bigcup_{m \in \sigma_{\mathcal{F}}(Y) \cap \{1,\ldots,2N\}} \sigma(G_m(A)),$$

where $\mathcal{F} \in \{\mathcal{A}, \mathcal{B}\}$ depends on whether we want to know the spectrum of A in $\mathcal{F} = \mathcal{A}$ or in $\mathcal{F} = \mathcal{B}$. Let us first determine the spectrum of $F_x(A)$ for $x \in \sigma_{\mathcal{F}}(X) \setminus \{0, 1\}$. Let $\lambda \in \mathbb{C}$ and set $D(x) := \det(F_x(A - \lambda I))$. Further, let t, t_o, and t_e refer to the number of the elements of the sets T, $T \cap \{1, 3, \ldots, 2N-1\}$, and $T \cap \{2, 4, \ldots, 2N\}$, respectively. Also put $\upsilon := t_o - t_e$. Changing the rows and columns of $F_x(A)$ in an appropriate way produces a matrix of the form

$$\begin{pmatrix} F_{11} & 0 \\ 0 & I \end{pmatrix} \tag{30}$$

where F_{11} is a $t \times t$ matrix and I is the $(2N - t) \times (2N - t)$ identity matrix. The determinant $D(x)$ of (30) is a polynomial of first degree in x (see the proof of Proposition 7), and

$$D(0) = (-\lambda)^{t_o}(1-\lambda)^{t_e}(1-\lambda)^{2N-t}, \quad D(1) = (1-\lambda)^{t_o}(-\lambda)^{t_e}(1-\lambda)^{2N-t},$$

the factors $(1 - \lambda)^{2N-t}$ coming from the lower right corner in (30) and the other factors resulting from the upper left one. Thus,

$$D(x) = (1-\lambda)^{2N-t}[x(1-\lambda)^{t_o}(-\lambda)^{t_e} + (1-x)(-\lambda)^{t_o}(1-\lambda)^{t_e}].$$

Depending on whether $\upsilon > 0$, $\upsilon = 0$, or $\upsilon < 0$, this equals

$$D(x) = (1-\lambda)^{2N-t}(1-\lambda)^{t_e}(-\lambda)^{t_e}[x(1-\lambda)^{\upsilon} + (1-x)(-\lambda)^{\upsilon}],$$

$$D(x) = (1-\lambda)^{2N-t}(1-\lambda)^{t_o}(-\lambda)^{t_o},$$

$$D(x) = (1-\lambda)^{2N-t}(1-\lambda)^{t_o}(-\lambda)^{t_o}[x(-\lambda)^{|\upsilon|} + (1-x)(1-\lambda)^{|\upsilon|}],$$

respectively. Thus, if $\upsilon = 0$, $\sigma(F_x(A)) = \{0, 1\}$. In case $\upsilon > 0$, we have

$$x(1-\lambda)^{\upsilon} + (1-x)(-\lambda)^{\upsilon} = 0 \tag{31}$$

if and only if

$$\left(\frac{\lambda}{\lambda-1}\right)^{\upsilon} = \frac{x}{x-1} \tag{32}$$

(observe that $x \neq 1$ by assumption and that (31) cannot vanish if $\lambda = 1$). Hence, on denoting by

$$\zeta_0(x), \ldots, \zeta_{\upsilon-1}(x)$$

the υ roots of $x/(x-1)$, we infer form (32) that the spectrum of $F_x(A)$ equals

$$\{0, 1\} \cup \left\{\frac{\zeta_0(x)}{\zeta_0(x)-1}, \ldots, \frac{\zeta_{\upsilon-1}(x)}{\zeta_{\upsilon-1}(x)-1}\right\} \quad \text{for} \quad t_e > 0, \tag{33}$$

$$\{1\} \cup \left\{\frac{\zeta_0(x)}{\zeta_0(x)-1}, \ldots, \frac{\zeta_{\upsilon-1}(x)}{\zeta_{\upsilon-1}(x)-1}\right\} \quad \text{for} \quad t_e = 0. \tag{34}$$

In the case $\upsilon < 0$ we obtain analogously that $\sigma(F_x(A))$ is

$$\{0, 1\} \cup \left\{\frac{-1}{\zeta_0(x)-1}, \ldots, \frac{-1}{\zeta_{|\upsilon|-1}(x)-1}\right\} \quad \text{for} \quad t_o > 0, \tag{35}$$

$$\{1\} \cup \left\{\frac{-1}{\zeta_0(x)-1}, \ldots, \frac{-1}{\zeta_{|\upsilon|-1}(x)-1}\right\} \quad \text{for} \quad t_o = 0. \tag{36}$$

Finally, it is evident that $G_m(A) \in \{0, 1\}$ for all m, and it is clear which value is actually assumed.

The case where $\sigma_{\mathcal{F}}(X)$ is a logarithmic double spiral is of particular interest for applications. For $\delta \in \mathbb{R}$ and $\nu \in (0, 1)$, put

$$S_{\delta,\nu} := \{re^{i\delta \log r} e^{2\pi i \nu} : r \in (0, \infty)\},$$

and given two distinct numbers $z, w \in \mathbb{C}$, let

$$\mathcal{S}(z, w; \delta; \nu) := \{(w\zeta - z)/(\zeta - 1) : \zeta \in S_{\delta,\nu}\} \cup \{z, w\}.$$

If $\delta = 0$, then $S_{\delta,\nu}$ is a ray and hence, $\mathcal{S}(z, w; \delta, \nu)$ is a circular arc between z and w, which degenerates to the line segment $[z, w]$ in case $\nu = 1/2$. If $\delta \neq 0$, then $S_{\delta,\nu}$ is a logarithmic spiral and therefore $\mathcal{S}(z, w; \delta; \nu)$ is a double spiral wriggling out of z and scrolling up at w. We call a set a *logarithmic double spiral* (between z and w) if it is of the form $\mathcal{S}(z, w; \delta; \nu)$ with some $\delta \in \mathbb{R}$ and $\nu \in (0, 1)$. Notice that segments and circular arcs are logarithmic double spirals in this sense.

Now suppose $\sigma_{\mathcal{F}}(X) = \mathcal{S}(0, 1; \delta; \nu)$ and let $x \in \mathcal{S}(0, 1; \delta; \nu) \setminus \{0, 1\}$. Assume first that $\upsilon := t_o - t_e > 0$ and $t_e > 0$. Then $\sigma(F_x(A))$ is given by (33). If

$$x = re^{i\delta \log r} e^{2\pi i \nu} / (re^{i\delta \log r} e^{2\pi i \nu} - 1),$$

then

$$x/(x-1) = re^{i\delta \log r}e^{2\pi i \nu} = re^{i\delta \log r}e^{2\pi i(\nu+k)}$$

and consequently, the υ roots of $\zeta^{\upsilon} = x/(x-1)$ are

$$\zeta_k(x) = r^{1/\upsilon}e^{i\delta(\log r)/\upsilon}e^{2\pi i(\nu+k)/\upsilon} = se^{i\delta \log s}e^{2\pi i(\nu+k)/\upsilon}$$

where $s := r^{1/\upsilon}$ and $k = 0, \ldots, \upsilon - 1$. Thus, if x traces out $\mathcal{S}(0, 1; \delta; \nu) \setminus \{0, 1\}$ then $\zeta_k(x)/(\zeta_k(x)-1)$ describes the logarithmic double spiral $\mathcal{S}(0, 1; \delta; (\nu+k)/\upsilon) \setminus \{0, 1\}$. In the case where $\upsilon < 0$ we similarly see that if x ranges over $\mathcal{S}(0, 1; \delta; \nu) \setminus \{0, 1\}$ then $-1/(\zeta_k(x) - 1)$ moves along the logarithmic double spiral

$$\mathcal{S}(1, 0; \delta; (\nu + k)/|\upsilon|) \setminus \{0, 1\}.$$

Taking into account that spectra are closed we so obtain from (33)–(36) the following result, a concrete version of which was by means of slightly different methods already proved and explicitly stated in [9] (Theorem 2.2.2).

Theorem 10 *Let $\mathcal{F}$ be $\mathcal{A}$ or $\mathcal{B}$. If $\sigma_{\mathcal{F}}(X)$ is the logarithmic double spiral $\mathcal{S}(0, 1; \delta; \nu)$, then the spectrum of $A := pPp + q$ in $\mathcal{F}$ equals*

$$\{0, 1\} \quad \textit{for} \quad \upsilon = 0,$$

$$\bigcup_{k=0}^{\upsilon-1} \mathcal{S}(0, 1; \delta; (\nu + k)/\upsilon) \quad \textit{for} \quad \upsilon > 0, \tag{37}$$

$$\bigcup_{k=0}^{|\upsilon|-1} \mathcal{S}(0, 1; \delta; (\nu + k)/|\upsilon|) \quad \textit{for} \quad \upsilon < 0. \tag{38}$$

Clearly, if $\sigma_{\mathcal{F}}(X)$ is a union of logarithmic spirals,

$$\sigma_{\mathcal{F}}(X) = \bigcup_{\delta \in [\delta_1, \delta_2]} \mathcal{S}(0, 1; \delta; \nu),$$

then the conclusion of Theorem 10 remains true with (37) and (38) replaced by

$$\bigcup_{k=0}^{\upsilon-1} \bigcup_{\delta \in [\delta_1, \delta_2]} \mathcal{S}(0, 1; \delta; (\nu + k)/\upsilon) \quad \text{for} \quad \upsilon > 0,$$

$$\bigcup_{k=0}^{|\upsilon|-1} \bigcup_{\delta \in [\delta_1, \delta_2]} \mathcal{S}(0, 1; \delta; (\nu + k)/|\upsilon|) \quad \text{for} \quad \upsilon < 0.$$

5.2 Applications to Singular Integral Operators

The results of this paper yield a symbol calculus for the closed algebra generated by singular integral operators with piecewise continuous coefficients. This symbol calculus reduces the question of deciding whether an operator is Fredholm to the problem of finding out whether a family of matrix functions consists of invertible matrices only. The simplest nontrivial operator in the algebra mentioned is the Cauchy singular integral operator S_Γ, and we now apply the results of Sect. 5.1 to the operator S_Γ. To avoid complications that go beyond the scope of this paper, we will not study the problem in full generality. We abstain in particular from considering the operators on spaces with general (Muckenhoupt) weights [5, 45].

A *simple arc* is an oriented rectifiable curve in the plane which is homeomorphic to a line segment. The union of finitely many simple arcs each pair of which have at most endpoints in common is called a *composed curve*. If Γ is a composed curve and $z \in \Gamma$, then in a small neighborhood of z the curve is locally comprised by a finite number of simple arcs. This number is referred to as the *multiplicity* of z and is denoted by $t := t(z)$. At a point z of multiplicity t, the curve has $t_o := t_o(z)$ outgoing and $t_e := t_e(z)$ incoming simple arcs, where $t_o \geq 0$, $t_e \geq 0$, and $t_o + t_e = t$. We call $\upsilon := \upsilon(z) := t_o - t_e$ the *valency* of the point z.

Let Γ be a composed curve. The curve Γ is said to be a *Carleson curve* (or to be *Ahlfors-David regular*) if

$$\sup_{z \in \Gamma} \sup_{\varepsilon > 0} |\Gamma(z, \varepsilon)|/\varepsilon < \infty$$

where $|\Gamma(z, \varepsilon)|$ denotes the (length) measure of the portion $\Gamma(z, \varepsilon) := \{\zeta \in \Gamma : |\zeta - z| < \varepsilon\}$. David [10, 11] proved that the Cauchy singular integral operator S_Γ,

$$(S_\Gamma f)(z) := \lim_{\varepsilon \to 0} \frac{1}{\pi i} \int_{\Gamma \setminus \Gamma(z,\varepsilon)} \frac{f(\zeta)}{\zeta - z} d\zeta \quad (z \in \Gamma),$$

is a well-defined and bounded operator on $L^p(\Gamma)$ $(1 < p < \infty)$ if and only if Γ is a Carleson curve (see also [33] for the "only if" portion). So let us henceforth suppose that Γ is Carleson. Our aim is to determine the essential spectrum of S_Γ on $L^p(\Gamma)$, i.e. to determine the set

$$\sigma_{\text{ess}}(S_\Gamma) := \{\lambda \in \mathbb{C} : S_\Gamma - \lambda I \text{ is not Fredholm on } L^p(\Gamma)\}.$$

Recall that an operator $A \in L(L^p(\Gamma))$ is Fredholm if and only if it is invertible modulo the ideal $K(L^p(\Gamma))$ of the compact operators, that is, if and only if the coset $\pi(A) := A + K(L^p(\Gamma))$ is invertible in the Calkin algebra $L(L^p(\Gamma))/K(L^p(\Gamma))$.

For $a \in L^\infty(\Gamma)$, let $aI : L^p(\Gamma) \to L^p(\Gamma)$ be the multiplication operator $f \mapsto af$. We denote by $C(\Gamma)$ the continuous functions on Γ and by $PC(\Gamma)$ the closure in $L^\infty(\Gamma)$ of all piecewise constant functions on Γ.

An operator $A \in L(L^p(\Gamma))$ is said to be of *local type* if $AcI - cA$ is a compact operator for every $c \in C(\Gamma)$. Clearly, compact operators as well as multiplication operators are of local type. It is well known that S_Γ is also of local type (see [20], Vol. I, Chap. 1, Theorem 4.3 and [3], Lemma 5.1). One can easily see that the set OLT of all operators of local type is a closed subalgebra of $L(L^p(\Gamma))$ and that an operator $A \in \text{OLT}$ is Fredholm if and only if the coset $\pi(A)$ is invertible in $\pi(\text{OLT}) := \text{OLT}/K(L^p(\Gamma))$. For $z \in \Gamma$, let J_z be the smallest closed two-sided ideal of $\pi(\text{OLT})$ containing $\{\pi(cI) : c \in C(\Gamma), c(z) = 0\}$. Put $\mathcal{A}_z := \pi(\text{OLT})/J_z$ and denote the coset $\pi(A) + J_z$ by $\pi_z(A)$. Allan's local principle (Theorem 2) with $\mathcal{A} := \text{OLT}$ and $C := \{\pi(cI) : c \in C(\Gamma)\}$ so implies that an operator $A \in \text{OLT}$ is Fredholm on $L^p(\Gamma)$ if and only if $\pi_z(A)$ is invertible in $\mathcal{A}_z$ for every $z \in \Gamma$.

The algebra $\mathcal{A}_z$ contains $P := \pi_z(P_\Gamma)$ and $p_j := \pi_z(\chi_j I)$ $(j = 1, \ldots, t)$ where $P_\Gamma = (I + S_\Gamma)/2$ and $\chi_1, \ldots, \chi_t$ are the characteristic functions of the t connected components of $(\Gamma \cap U) \setminus \{z\}$ (U sufficiently small). Let $\mathcal{B}_z$ stand for the closed subalgebra of $\mathcal{A}_z$ which is generated by $P, p_1, p_2, \ldots, p_t$.

Clearly, $\mathcal{B}_z$ is of much better structure than $\mathcal{A}_z$. It is obvious that $p_1, p_2, \ldots, p_t$ are idempotents whose sum is the identity and which satisfy $p_i p_j = \delta_{ij} p_i$. Unfortunately, in general P is not an idempotent, by virtue of which Theorem 9 is not immediately applicable. We therefore construct two other "local algebras" $\mathcal{A}_z^* \supset \mathcal{B}_z^*$ and identify $\pi_z(P_\Gamma)$ as an abstract singular integral (in the sense of Sect. 5.1) in these algebras.

A counter-clockwise oriented curve homeomorphic to a circle is called a Jordan curve. A composed curve consisting of a finite number $N \geq 2$ of Jordan curves which have exactly one point in common is referred to as a *flower*. All points of a flower have valency zero, exactly one point, the center of the flower, has multiplicity $2N$, while the remaining points have multiplicity 2.

Suppose Γ^* is both a flower and a Carleson curve. Denote the center of Γ^* by z, and let $\mathcal{A}_z^*$ and $\mathcal{B}_z^*$ be the algebras that arise from the above construction with Γ^* in place of Γ. If $\varepsilon > 0$ is sufficiently small, then the connected component of the portion $\Gamma^*(z, \varepsilon)$ containing z may be written in the form

$$\bigcup_{i=1}^{N} (\Gamma_{2i}^* \cup \Gamma_{2i-1}^*) \tag{39}$$

where Γ_{2i}^* and Γ_{2i-1}^* $(i = 1, \ldots, N)$ are outgoing and incoming simple arcs, respectively. The algebra $\mathcal{B}_z^*$ is generated by $P := \pi_z(P_{\Gamma^*})$ and $p_j := \pi_z(\chi_j I)$ $(j = 1, \ldots, 2N)$ where χ_j is the characteristic function of Γ_j^*. One can show that now P is idempotent and that (3), (4) hold. Thus, Theorem 9 is applicable to the pair of algebras $\mathcal{A} := \mathcal{A}_z^*$, $\mathcal{B} := \mathcal{B}_z^*$, and we may use the results of Sect. 5.1 to compute the local spectrum of the singular integral operator

$$A := \left(\sum_{j \in T} \chi_j I \right) P_{\Gamma^*} \left(\sum_{j \in T} \chi_j I \right) + \left(\sum_{j \notin T} \chi_j I \right), \tag{40}$$

i.e. the spectrum of the abstract singular integral $\pi_z(A) = pPp + q$, where T is a nonempty proper subset of $\{1, 2, \ldots, 2N\}$ (note that $\sigma_{\mathcal{A}_z^*}\pi_z(A) = \sigma_{\mathcal{B}_z^*}\pi_z(A) = \{0, 1\}$ for $T = \{1, 2, \ldots, 2N\}$).

What we need is the spectrum $\sigma_{\mathcal{A}_z^*}(X)$ of

$$X := \sum_{i=1}^{N}(p_{2i-1}Pp_{2i-1} + p_{2i}Qp_{2i}).$$

It is easily seen that

$$\sigma_{\mathcal{A}_z^*}(X) = \bigcup_{i=1}^{N}\sigma_{\mathcal{A}_z^*}(p_{2i-1}Pp_{2i-1} + p_{2i}Qp_{2i}),$$

which reduces the problem to finding the local spectrum of singular integral operators with piecewise continuous coefficients on Carleson Jordan curves. These spectra were completely determined in [4]. In order to illustrate the basic phenomena, let us for the sake of simplicity assume that the arcs Γ_j^* of the flower may be parametrized as

$$\Gamma_j^* = \{\zeta = z + re^{i\phi(r)+b_j(r))} : 0 \le r < \varepsilon\} \quad (j = 1, \ldots, 2N) \tag{41}$$

where $\varepsilon \in (0, 1)$, ϕ is a real-valued function of the form

$$\phi(r) = h(\log(-\log r))(-\log r)$$

with a function $h \in C^2(\mathbb{R})$ for which h, h', h'' are bounded on $\mathbb{R}$, and b_j are real-valued functions in $C^1[0, \varepsilon]$ such that

$$0 \le b_1(r) < b_2(r) < \cdots < b_{2N}(r) < 2\pi \quad \text{for} \quad r \in (0, \varepsilon).$$

We remark that the ansatz $h(\log(-\log r))$ guarantees that $r\dot{\phi}(r)$ is bounded for $r \in (0, 1)$, which in turn implies that Γ^* is a Carleson curve (see e.g. [4]). Clearly, every piecewise C^1 flower can be parametrized in this way with $h = 0$. If h and $b_l, \ldots, b_{2N}$ are constant functions, then Γ locally consists of $2N$ logarithmic spirals scrolling up at z. The choice

$$h(x) = \delta + \mu \sin x, \quad b_j(r) = b_j = \text{constant} \tag{42}$$

gives $2N$ "oscillating spirals" terminating at z. In accordance with [4], the spirality indices δ_z^- and δ_z^+ of Γ^* are defined by

$$\delta_z^- = \liminf_{x\to\infty}(h(x) + h'(x)) \, (= \liminf_{r\to 0}(-r\dot{\phi}(r))),$$

$$\delta_z^+ = \limsup_{x\to\infty}(h(x) + h'(x)) \;(= \limsup_{r\to 0}(-r\dot{\phi}(r))).$$

In case $h = 0$, i.e. for piecewise C^1 flowers, we have $\delta_z^- = \delta_z^+ = 0$. If h is as in (42), then

$$\delta_z^- = \delta - |\mu|\sqrt{2}, \quad \delta_z^+ = \delta + |\mu|\sqrt{2}.$$

The symbol calculus of [4] implies that

$$\sigma_{\mathcal{A}_z^*}(p_{2i-1}Pp_{2i-1} + p_{2i}Qp_{2i}) = \bigcup_{\delta\in[\delta_z^-,\delta_z^+]} \mathcal{S}(0, 1; \delta; 1/p)$$

for every $i = 1, \dots, N$, whence

$$\sigma_{\mathcal{A}_z^*}(X) = \bigcup_{\delta\in[\delta_z^-,\delta_z^+]} \mathcal{S}(0, 1; \delta; 1/p). \tag{43}$$

The set on the right of (43) is a union of logarithmic double spirals; such sets were called skew spiralic horns in [4] and are logarithmic leaves with a separating point in the terminology of [5]. Clearly, for piecewise C^1 flowers or, more generally, for flowers whose spirality indices are both zero, the set (43) is a circular arc.

Since the set (43) does not separate the complex plane (i.e., does not contain "holes"), a standard result from the theory of Banach algebras implies that

$$\sigma_{\mathcal{A}_z^*}(X) = \sigma_{\mathcal{B}_z^*}(X).$$

By a *substar* of the flower Γ^* we understand a set Γ of the form $\Gamma = \cup_{j\in T}\Gamma_j^*$ where the simple arcs Γ_j^* are given by (39) and T is a non-empty subset of $\{1, 2, \dots, 2N\}$. Obviously, the operator $P_\Gamma = (I + S_\Gamma)/2$ may be identified with the singular integral operator (40). Thus, combining Theorem 10 (and the remark after it) with (43) we arrive at the following result for $S_\Gamma = 2P_\Gamma - I$.

Theorem 11 *Let Γ^* be a Carleson flower with the center z and let Γ be a substar of Γ^*. Denote the valency of $z \in \Gamma$ by $\upsilon(z)$ and let δ_z^-, δ_z^+ be the spirality indices of z. Then the local spectra $\sigma_{\mathcal{B}_z}(S_\Gamma)$ and $\sigma_{\mathcal{A}_z}(S_\Gamma)$ of S_Γ at z coincide and are equal to*

$$\{-1, 1\} \quad \textit{if} \quad \upsilon(z) = 0,$$

$$\bigcup_{k=0}^{\upsilon(z)-1} \bigcup_{\delta\in[\delta_z^-,\delta_z^+]} \mathcal{S}(-1, 1; \delta; (1/p + k)/\upsilon(z)) \quad \textit{if} \quad \upsilon(z) > 0,$$

$$\bigcup_{k=0}^{|\upsilon(z)|-1} \bigcup_{\delta\in[\delta_z^-,\delta_z^+]} \mathcal{S}(-1, 1; \delta; (1/p + k)/|\upsilon(z)|) \quad \textit{if} \quad \upsilon(z) < 0.$$

We finally return to the case of an arbitrary composed Carleson curve Γ. A point $z \in \Gamma$ is called a *bud* if there exists an $\varepsilon > 0$ such that the connected component of

$$\Gamma \cap \{\zeta \in \mathbb{C} : |\zeta - z| < \varepsilon\}$$

containing z is a substar of some Carleson flower. It is easily seen that composed Carleson curves which may locally be "parametrized by the radius" as in (41) consist entirely of buds. We conjecture that every point of a composed Carleson curve is a bud; three of the authors are planning to devote a forthcoming paper to this problem. The following theorem is immediate from the preceding discussion.

Theorem 12 *Let Γ be a composed Carleson curve each point of which is a bud. Then the essential spectrum of S_Γ on $L^p(\Gamma)$ is*

$$\sigma_{\text{ess}}(S_\Gamma) = \bigcup_{z\in\Gamma} \sigma(S_\Gamma)$$

where $\sigma(S_\Gamma) := \sigma_{\mathcal{A}_z}(S_\Gamma) = \sigma_{\mathcal{B}_z}(S_\Gamma)$ is as in Theorem 11.

6 Miscellanea

6.1 Other Partitions of Unity into Projections

Besides the (obvious) partition of unity into projections (p_i), there are other partitions in $\mathcal{B}$. Set, for example,

$$w_{2i} = (p_{2i} + p_{2i+1})Q \quad \text{and} \quad w_{2i-1} = (p_{2i-1} + p_{2i})P.$$

Proposition 11 *The set $(w_i)_{i=1}^{2n}$ is a partition of unity into projections in $\mathcal{B}$.*

The consideration of this partition is motivated by [22]. To get another one, set

$$a_i = p_{2i-1} + p_{2i}$$

and

$$q_i = \begin{cases} a_i P a_i & \text{if } i = 1, \ldots, N \\ a_{2N+1-i} Q a_{2N+1-i} & \text{if } i = N+1, \ldots, 2N. \end{cases} \quad v$$

Proposition 12 *The set $(q_i)_{i=1}^{2n}$ is a partition of unity into projections in $\mathcal{B}$.*

The proofs of the preceding propositions are straightforward.

Clearly, the use of other partitions of unity into projections than (p_i) yields other descriptions of the local algebras at points

$$x \in \sigma_{\mathcal{B}}(X) \setminus \{0, 1\}.$$

We shall illustrate this for the partition (w_i). Here are the analogues of Propositions 2 and 3.

Proposition 13 *Considered as module over its center, the algebra $\mathcal{B}^0$ is generated by the $(2N)^2$ elements $(w_i)_{i=1}^{2N}$ and $(w_i Y w_j)_{i,j=1}^{2N}$ with $i \neq j$. To be more precise, given $A \in \mathcal{B}^0$, there are polynomials R_{ij} in X such that*

$$A = \sum_{i=1}^{2N} R_{ii}(X) w_i + \sum_{\substack{i,j=1 \\ i \neq j}}^{2N} R_{ij}(X) w_i Y w_j. \tag{44}$$

Proposition 14 *(a) If $l > j > i$ or $j > i > l$ or $i > l > j$ then*

$$w_i Y w_j Y w_l = (X - I) w_i Y w_l.$$

(b) If $l > i > j$ or $j > l > i$ or $i > j > l$ then

$$w_i Y w_j Y w_l = X w_i Y w_l.$$

(c) If $i = l$ and $i \neq j$ then

$$w_i Y w_j Y w_i = X(X - I) w_i.$$

The proofs are omitted.

As in the proof of Proposition 5 one can show that, for $x \in \sigma_{\mathcal{B}}(X) \setminus \{0, 1\}$, the elements

$$b_{ij} = \begin{cases} (x-1)^{-1} \Phi_x(w_i Y w_j) & \text{if } i < j \\ x^{-1} \Phi_x(w_i Y w_j) & \text{if } i > j \\ \Phi_x(w_i) & \text{if } i = j \end{cases}$$

form a basis of the linear space $\mathcal{B}_x$ which, moreover, satisfies $b_{ij} b_{kl} = \delta_{jk} b_{il}$. Thus, there is an algebra homomorphism

$$\Psi'_x : \mathcal{B}_x \to \mathbb{C}^{2N \times 2N}$$

with $\Psi'_x(b_{ij}) = E_{ij}$. Set $H_x = \Psi'_x \circ \Phi_x$. Then

$$H_x(w_i) = \operatorname{diag}(0, \ldots, 0, 1, 0, \ldots, 0) \tag{45}$$

the 1 standing at the ith place, and

$$H_x(Y) = \begin{pmatrix} x & x-1 & x-1 & x-1 & \dots & x-1 & x-1 \\ x & x & x-1 & x-1 & \dots & x-1 & x-1 \\ x & x & x & x-1 & \dots & x-1 & x-1 \\ x & x & x & x & \dots & x-1 & x-1 \\ \vdots & \vdots & \vdots & \vdots & \ddots & \vdots & \vdots \\ x & x & x & x & \dots & x & x-1 \\ x & x & x & x & \dots & x & x \end{pmatrix}. \tag{46}$$

Here is the analogue of Corollary 1.

Corollary 3 *Let $x \in \sigma_{\mathcal{B}}(X) \setminus \{0, 1\}$. Then*

$$H_x(P) = \mathrm{diag}(1, 0, 1, 0, \dots, 1, 0),$$

and $H_x(p_i)$ is the matrix with $(i-1)$st column

$$(1-x, 1-x, \dots, 1-x, -x, -x, \dots, -x),$$

ith column

$$(x-1, x-1, \dots, x-1, x, x, \dots, x)$$

(the entries $1-x$ and $x-1$ both appear $i-1$ times), and all other columns are zero.

Proof The proof is based on checking that $P = \sum_{i=1}^{N} w_{2i-1}$, that

$$p_i = (I - X)w_{i-1} + Xw_i - \sum_{\substack{k=1 \\ k \neq i-1}}^{2N} w_k Y w_{i-1} + \sum_{\substack{k=1 \\ k \neq i}}^{2N} w_k Y w_i,$$

and on employing (45) and (46). □

We renounce to give an explicit formulation of Theorem 9 based on the partition (w_i).

6.2 Other Indicator Elements

The elements X and Y indicate which matrix representations of the algebra $\mathcal{B}$ actually appear. While X is distinguished by the fact that it belongs to the center of

$\mathcal{B}$, there is some latitude to choose Y. For example, one can show that the element

$$Z := P + \sum_{i=1}^{2N} 2i p_i$$

can play the role of Y in the determination of all one-dimensional representations. Indeed, consider the mappings K_m given by $K_{2i}(P) = 0$, $K_{2i+1}(P) = 1$,

$$K_{2i}(p_i) = \begin{cases} 1 & \text{if } i = j \\ 0 & \text{if } i \neq j, \end{cases} \qquad K_{2i+1}(p_i) = \begin{cases} 1 & \text{if } i = j \\ 0 & \text{if } i \neq j, \end{cases}$$

where $i = 1, \ldots, 2N$. The analogue of Proposition 7 reads as follows.

Proposition 15 *If $m \in \sigma_{\mathcal{B}}(Z) \cap \{2, 3, 4, \ldots, 4N + 1\}$ then the complex-valued mapping K_m defined on $\{P, p_1, \ldots, p_{2N}\}$ extends to an algebra homomorphism from $\mathcal{B}$ onto $\mathbb{C}$.*

The proof runs as that of Proposition 7.

The following observation is often useful in order to determine the spectrum of X. For $i = 1, 2, \ldots, 2N$ let $\mathcal{B}_i$ denote the algebra

$$p_i \mathcal{B} p_i = \{p_i b p_i, b \in \mathcal{B}\}.$$

Proposition 16 *If $\{0, 1\} \subseteq \sigma_{\mathcal{B}_i}(p_i X p_i)$ for some i then $\sigma_{\mathcal{B}}(X) = \sigma_{\mathcal{B}_i}(p_i X p_i)$.*

Proof Since (p_i) is a partition of unity into projections and X is in the center of $\mathcal{B}$, we have

$$\sigma_{\mathcal{B}}(X) = \bigcup_{j=1}^{2N} \sigma_{\mathcal{B}_j}(p_j X p_j). \tag{47}$$

We claim that

$$\sigma_{\mathcal{B}_j}(p_j X p_j) \setminus \{0, 1\} = \sigma_{\mathcal{B}_k}(p_k X p_k) \setminus \{0, 1\} \tag{48}$$

for all $j, k = 1, \ldots, 2N$. Indeed, let $\lambda \notin \sigma_{\mathcal{B}_j}(p_j X p_j)$. Then there is an a in $\mathcal{B}$ such that

$$p_j a p_j (p_j X p_j - \lambda p_j) = p_j.$$

Multiplying this identity from the left hand side by $p_k P p_j$ and from the right hand side by $p_j P p_k$ with some $k \neq j$ yields

$$p_k P p_j a p_j (p_j X p_j - \lambda p_j) p_j P p_k = p_k P p_j P p_k$$

and, by Proposition 3,

$$p_k P p_j a p_j P p_k (p_k X p_k - \lambda p_k) = (-1)^{j-k} X(X - I) p_k.$$

The element $p_k X p_k$ lies in the center of the algebra $\mathcal{B}_k$. Thus, localizing $\mathcal{B}_k$ over its smallest closed subalgebra which contains p_k and $p_k X p_k$ via Theorem 2 yields that at the point $\mu \in \sigma_{\mathcal{B}_k}(p_k X p_k)$ (where Ω_μ refers to the canonical homomorphism from $\mathcal{B}_k$ onto its local algebra at μ) the following equality holds:

$$(\mu - \lambda)\Omega_\mu(p_k P p_j a p_j P p_k) = (-1)^{j-k} \mu(\mu - 1)\Omega_\mu(I).$$

Thus, if $\mu \notin \{0, 1\}$ then $\mu - \lambda \neq 0$ and, hence,

$$\lambda \notin \sigma_{\mathcal{B}_k}(p_k X p_k) \setminus \{0, 1\}.$$

This gives our claim (48). Clearly, (48) in combination with (47) proves the assertion. □

6.3 The Two Projections Theorem

If $N = 1$ in Theorem 9, then the partition (p_i) consists of two elements p_1 and p_2 with $p_2 = I - p_1$. Moreover, the axioms (3) and (4) reduce to $P^2 = P$ and $Q^2 = Q$, respectively. Thus, $\mathcal{B}$ is nothing but the (general) algebra generated by two idempotents (P and p_1) and the identity.

Obviously, there are some differences between the specification of Theorem 9 to the case $N = 1$ and Theorem 1. In case $N = 1$, set $p := p_1$ and $q := P$ in Theorem 1.

The first difference concerns the indicator element for the one-dimensional representations. In Theorem 1, it is the element $p + 2q$, whereas it is

$$Y = pq + (I - p)(I - q) + P + 3(I - p) = 2pq + 4I - 3p - q$$

in Theorem 9, which seems to be much more complicated. But if Y is replaced by the element Z from preceding remark, then

$$Z = q + 2p + 4(I - p) = q - 2p + 4I$$

which is as simple as $P + 2q$.

The second difference concerns the explicit form of the 2×2 matrices. In Theorem 1, the matrix associated with q at the point $x \in \sigma_{\mathcal{B}}(X) \setminus \{0, 1\}$ is

$$\begin{pmatrix} x & \sqrt{x(1-x)} \\ \sqrt{x(1-x)} & 1-x \end{pmatrix}, \tag{49}$$

whereas the corresponding matrix from Theorem 9 is

$$\begin{pmatrix} x & 1-x \\ x & 1-x \end{pmatrix}.$$

But, for $x \in \mathbb{C} \setminus \{0, 1\}$,

$$\begin{pmatrix} x & \sqrt{x(1-x)} \\ \sqrt{x(1-x)} & 1-x \end{pmatrix} = \begin{pmatrix} \sqrt[4]{\frac{x}{1-x}} & 0 \\ 0 & \sqrt[4]{\frac{1-x}{x}} \end{pmatrix} \begin{pmatrix} x & 1-x \\ x & 1-x \end{pmatrix} \begin{pmatrix} \sqrt[4]{\frac{1-x}{x}} & 0 \\ 0 & \sqrt[4]{\frac{x}{1-x}} \end{pmatrix}$$

and, moreover,

$$\begin{pmatrix} 1 & 0 \\ 0 & 0 \end{pmatrix} = \begin{pmatrix} \sqrt[4]{\frac{x}{1-x}} & 0 \\ 0 & \sqrt[4]{\frac{1-x}{x}} \end{pmatrix} \begin{pmatrix} 1 & 0 \\ 0 & 0 \end{pmatrix} \begin{pmatrix} \sqrt[4]{\frac{1-x}{x}} & 0 \\ 0 & \sqrt[4]{\frac{x}{1-x}} \end{pmatrix}$$

where $\sqrt[4]{\frac{x}{1-x}}$ is any number with

$$\left(\sqrt[4]{\frac{x}{1-x}}\right)^4 = \frac{x}{1-x} \quad \text{and} \quad \sqrt[4]{\frac{1-x}{x}} \quad \text{is} \quad \left(\sqrt[4]{\frac{x}{1-x}}\right)^{-1}.$$

Thus, both representations are equivalent.

6.4 Symmetric Representations in Case $N > 1$

In case $N = 1$, the $2N \times 2N$ dimensional representations of p_i and P can be chosen to be symmetric (and even self-adjoint in case $\sigma_{\mathcal{B}}(X) \subseteq \mathbb{R}$, which is of particular interest in many applications) (compare the matrix (49)). This observation suggests the following question: Is there a symmetric representation in case $N > 1$, too? To be more precise, is there an invertible matrix D such that again

$$D^{-1} F_x(p_i) D = \operatorname{diag}(0, \dots, 0, 1, 0, \dots, 0) = F_x(p_i) \tag{50}$$

(which is desirable for symmetry) but, moreover,

$$D^{-1} F_x(P) D = (D^{-1} F_x(P) D)^T \tag{51}$$

where ^{T} marks the transposed matrix?

In general, a matrix D with these properties does not exist in case $N > 1$! Indeed, (50) involves that D itself is a diagonal matrix, say

$$D = \mathrm{diag}(d_1, d_2, \ldots, d_{2N}).$$

Then identity (51) yields for the 2, 1, the 4, 1, and the 4, 2 entry

$$d_2 d_1^{-1} x = d_1 d_2^{-1}(1 - x), \tag{52}$$

$$d_4 d_1^{-1} x = d_1 d_4^{-1}(1 - x), \tag{53}$$

$$d_4 d_2^{-1}(-x) = d_2 d_4^{-1}(1 - x), \tag{54}$$

respectively. Identities (52) and (53) imply that

$$d_2^2 = d_4^2,$$

which contradicts (54).

6.5 *Coefficient Algebras*

Again let $\mathcal{A}$ be a Banach algebra with identity I, let $(p_i)_{i=1}^{2N}$ be a partition of unity into projections and P be an idempotent in $\mathcal{A}$ such that the axioms (3) and (4) hold. The smallest closed subalgebra of $\mathcal{A}$ containing the partition (p_i) as well as the element P will be denoted by $\mathcal{B}$ again. Suppose $\mathcal{G}$ is a closed subalgebra of $\mathcal{A}$ containing I and having the property that

$$p_i g = g p_i \quad \text{and} \quad gP = Pg \quad \text{for all} \quad i = 1, \ldots, 2N \quad \text{and} \quad g \in \mathcal{G}.$$

The algebra $\mathcal{G}$ is referred to as a coefficient algebra. As in [18], one can derive a version of Theorem 9 which provides us with an invertibility symbol for the smallest closed subalgebra C of $\mathcal{A}$ which contains the partition (p_i), the idempotent P, and the algebra $\mathcal{G}$. Here is the formulation of this version under the stronger condition that $\mathcal{G}$ be a simple algebra.

Theorem 13 *Let C be as above and let $\mathcal{G}$ be simple.*

(a) If $x \in \sigma_{\mathcal{B}}(X) \setminus \{0, 1\}$, then the mapping

$$F_x : \{P, p_1, \ldots, p_{2N}\} \cup \mathcal{G} \to \mathcal{G}^{2N \times 2N}$$

given by

$$F_x(p_i) = \mathrm{diag}(0, \ldots, 0, I, 0, \ldots, 0),$$

the I standing at the ith place,

$$F_x(P) = \mathrm{diag}(I, -I, I, -I, \ldots, I, -I) \times$$
$$\times \begin{pmatrix} x & x-1 & x-1 & x-1 & \ldots & x-1 & x-1 \\ x & x-1 & x-1 & x-1 & \ldots & x-1 & x-1 \\ x & x & x & x-1 & \ldots & x-1 & x-1 \\ x & x & x & x-1 & \ldots & x-1 & x-1 \\ \vdots & \vdots & \vdots & \vdots & \ddots & \vdots & \vdots \\ x & x & x & x & \ldots & x & x-1 \\ x & x & x & x & \ldots & x & x-1 \end{pmatrix},$$
$$F_x(g) = \mathrm{diag}(g, g, \ldots, g),$$

extends to a continuous algebra homomorphism from $\mathcal{C}$ onto $\mathcal{G}^{2N\times 2N}$.

(b) If $m \in \sigma_{\mathcal{B}}(Y) \cap \{1, \ldots, 4N\}$, then the mapping

$$G_m : \{P, p_1, \ldots, p_{2N}\} \cup \mathcal{G} \to \mathcal{G}$$

given by

$$G_{4m}(p_i) = \begin{cases} I & \text{if } i = 2m \\ 0 & \text{if } i \neq 2m, \end{cases} \qquad G_{4m}(P) = 0,$$
$$G_{4m-1}(p_i) = \begin{cases} I & \text{if } i = 2m \\ 0 & \text{if } i \neq 2m, \end{cases} \qquad G_{4m-1}(P) = I,$$
$$G_{4m-2}(p_i) = \begin{cases} I & \text{if } i = 2m-1 \\ 0 & \text{if } i \neq 2m-1, \end{cases} \qquad G_{4m-2}(P) = I,$$
$$G_{4m-3}(p_i) = \begin{cases} I & \text{if } i = 2m-1 \\ 0 & \text{if } i \neq 2m-1, \end{cases} \qquad G_{4m-3}(P) = 0,$$

where $m = 1, \ldots, N$, and by $G_m(g) = g$ extends to a continuous algebra homomorphism from $\mathcal{C}$ onto $\mathcal{G}$.

(c) An element $C \in \mathcal{C}$ is invertible in $\mathcal{C}$ if and only if the matrices $F_x(C)$ are invertible for all $x \in \sigma_{\mathcal{B}}(X) \setminus \{0, 1\}$ and the elements $G_m(C)$ are invertible for all $m \in \sigma_{\mathcal{B}}(Y) \cap \{1, \ldots, 4N\}$.

(d) An element $C \in \mathcal{C}$ is invertible in $\mathcal{A}$ if and only if the matrices $F_x(C)$ are invertible for all $x \in \sigma_{\mathcal{A}}(X) \setminus \{0, 1\}$ and the elements $G_m(C)$ are invertible for all $m \in \sigma_{\mathcal{A}}(Y) \cap \{1, \ldots, 4N\}$.

Observe that the conditions of the theorem are satisfied if, for example, $\mathcal{G}$ is the algebra $\mathbb{C}^{n\times n}$ which yields just the matrix version of Theorem 9.

References

1. Allan, G.R.: Ideals of vector-valued functions. Proc. Lond. Math. Soc. (3) **18**, 193–216 (1968)
2. Böttcher, A.: Toeplitz operators with piecewise continuous symbols—a neverending story? Jahresber. Deutsch. Math.-Verein. **97**(4), 115–129 (1995)
3. Böttcher, A., Karlovich, Yu.I.: Toeplitz and singular integral operators on Carleson curves with logarithmic whirl points. Integral Equ. Oper. Theory **22**(2), 127–161 (1995)
4. Böttcher, A., Karlovich, Yu.I.: Toeplitz and singular integral operators on general Carleson Jordan curves. In: Singular Integral Operators and Related Topics (Tel Aviv, 1995). Operator Theory: Advances and Applications, vol. 90, pp. 119–152. Birkhäuser, Basel (1996)
5. Böttcher, A., Karlovich, Yu.I.: Toeplitz operators with PC symbols on general Carleson Jordan curves with arbitrary Muckenhoupt weights. Trans. Amer. Math. Soc. **351**(8), 3143–3196 (1999)
6. Böttcher, A., Silbermann, B.: Analysis of Toeplitz Operators. Springer, Berlin (1990)
7. Böttcher, A., Spitkovsky, I.M.: Pseudodifferential operators with heavy spectrum. Integral Equ. Oper. Theory **19**(3), 251–269 (1994)
8. Böttcher, A., Roch, S., Silbermann, B., Spitkovskiĭ, I.M.: A Gohberg-Krupnik-Sarason symbol calculus for algebras of Toeplitz, Hankel, Cauchy, and Carleman operators. In: Topics in Operator Theory: Ernst D. Hellinger Memorial Volume. Operator Theory: Advances and Applications, vol. 48, pp. 189–234. Birkhäuser, Basel (1990)
9. Böttcher, A., Karlovich, Yu.I., Rabinovich, V.S.: Emergence, persistence, and disappearance of logarithmic spirals in the spectra of singular integral operators. Integral Equ. Oper. Theory **25**(4), 406–444 (1996)
10. David, G.: L'integrale de Cauchy sur les courbes rectifiables. Prepublication Univ. Paris-Sud, Dept. Math. 82T05 (1982)
11. David, G.: Opérateurs intégraux singuliers sur certaines courbes du plan complexe. Ann. Sci. École Norm. Sup. (4) **17**(1), 157–189 (1984)
12. Dieudonné, J.: Grundzüge der modernen Analysis. Band 2. VEB Deutscher Verlag der Wissenschaften, Berlin (1975)
13. Douglas, R.G.: Banach Algebra Techniques in Operator Theory. Pure and Applied Mathematics, vol. 49. Academic Press, New York (1972)
14. Douglas, R.G.: Local Toeplitz operators. Proc. Lond. Math. Soc. (3) **36**(2), 243–272 (1978)
15. Ehrhardt, T., Roch, S., Silbermann, B.: Symbol calculus for singular integrals with operator-valued PQC-coefficients. In: Singular Integral Operators and Related Topics (Tel Aviv, 1995). Operator Theory: Advances and Applications, vol. 90, pp. 182–203. Birkhäuser, Basel (1996)
16. Ehrhardt, T., Roch, S., Silbermann, B.: Finite section method for singular integrals with operator-valued PQC-coefficients. In: Singular Integral Operators and Related Topics (Tel Aviv, 1995). Operator Theory: Advances and Applications, vol. 90, pp. 204–243. Birkhäuser, Basel (1996)
17. Finck, T., Roch, S.: Banach algebras with matrix symbol of bounded order. Integral Equ. Oper. Theory **18**(4), 427–434 (1994)
18. Finck, T., Roch, S., Silbermann, B.: Two projection theorems and symbol calculus for operators with massive local spectra. Math. Nachr. **162**, 167–185 (1993)
19. Gohberg, I., Krupnik, N.: On singular integral operators on a composed curve. Soobshch. Akad. Nauk Gruz. SSR **64**, 21–24 (1971), in Russian
20. Gohberg, I., Krupnik, N.: One-dimensional Linear Singular Integral Equations. Vols. I, II. Operator Theory: Advances and Applications, vols. 53, 54. Birkhäuser, Basel (1992)
21. Gohberg, I., Krupnik, N.: Extension theorems for invertibility symbols in Banach algebras. Integral Equ. Oper. Theory **15**(6), 991–1010 (1992)
22. Gohberg, I., Krupnik, N.: Extension theorems for Fredholm and invertibility symbols. Integral Equ. Oper. Theory **16**(4), 514–529 (1993)
23. Gohberg, I., Krupnik, N., Spitkovsky, I.: Banach algebras of singular integral operators with piecewise continuous coefficients. General contour and weight. Integral Equ. Oper. Theory **17**(3), 322–337 (1993)

24. Hagen, R., Silbermann, B.: A Banach algebra approach to the stability of projection methods for singular integral equations. Math. Nachr. **140**, 285–297 (1989)
25. Hagen, R., Roch, S., Silbermann, B.: Spectral Theory of Approximation Methods for Convolution Equations. Operator Theory: Advances and Applications, vol. 74. Birkhäuser, Basel (1995)
26. Halmos, P.R.: Two subspaces. Trans. Amer. Math. Soc. **144**, 381–389 (1969)
27. Horn, R.A., Johnson, C.R.: Matrix Analysis. Cambridge University Press, Cambridge (1985)
28. Khelemskiĭ, A.Y.: Banach and Polynormed Algebras: General Theory, Representations, Homology. Nauka, Moscow (1989), in Russian
29. Krupnik, N.: Banach Algebras with Symbol and Singular Integral Operators. Operator Theory: Advances and Applications, vol. 26. Birkhäuser, Basel (1987)
30. Krupnik, N.: Minimal number of idempotent generators of matrix algebras over arbitrary field. Comm. Algebra **20**(11), 3251–3257 (1992)
31. Krupnik, N., Spigel, E.: Invertibility symbol for a Banach algebra generated by two idempotents and a shift. Integral Equ. Oper. Theory **17**(4), 567–578 (1993)
32. Krupnik, N., Roch, S., Silbermann, B.: On C^*-algebras generated by idempotents. J. Funct. Anal. **137**(2), 303–319 (1996)
33. Paatashvili, V.A., Khuskivadze, G.A.: Boundedness of a singular Cauchy operator in Lebesgue spaces in the case of nonsmooth contours. Trudy Tbiliss. Mat. Inst. Razmadze Akad. Nauk Gruzin. SSR **69**, 93–107 (1982), in Russian
34. Pedersen, G.K.: Measure theory for C^* algebras. II. Math. Scand. **22**, 63–74 (1968)
35. Power, S.C.: C^*-algebras generated by Hankel operators and Toeplitz operators. J. Funct. Anal. **31**, 52–68 (1979)
36. Power, S.C.: Essential spectra of piecewise continuous Fourier integral operators. Proc. Roy. Irish Acad. Sect. A **81**(1), 1–7 (1981)
37. Power, S.C.: Hankel Operators on Hilbert Space. Research Notes in Mathematics, vol. 64. Pitman, Boston (1982)
38. Raeburn, I., Sinclair, A.M.: The C^*-algebra generated by two projections. Math. Scand. **65**(2), 278–290 (1989)
39. Roch, S.: Spline approximation methods for Wiener-Hopf operators. In: Recent Developments in Operator Theory and Its Applications (Winnipeg, MB, 1994). Operator Theory: Advances and Applications, vol. 87, pp. 282–308. Birkhäuser, Basel (1996)
40. Roch, S., Silbermann, B.: Algebras generated by idempotents and the symbol calculus for singular integral operators. Integral Equ. Oper. Theory **11**(3), 385–419 (1988)
41. Roch, S., Silbermann, B.: A symbol calculus for finite sections of singular integral operators with shift and piecewise continuous coefficients. J. Funct. Anal. **78**(2), 365–389 (1988)
42. Roch, S., Silbermann, B.: Algebras of Convolution Operators and their Image in the Calkin Algebra. Report MATH, vol. 90. Akademie der Wissenschaften der DDR, Karl-Weierstrass-Institut für Mathematik, Berlin (1990)
43. Roch, S., Silbermann, B.: Asymptotic Moore-Penrose invertibility of singular integral operators. Integral Equ. Oper. Theory **26**(1), 81–101 (1996)
44. Silbermann, B.: The C^*-algebra generated by Toeplitz and Hankel operators with piecewise quasicontinuous symbols. Integral Equ. Oper. Theory **10**(5), 730–738 (1987)
45. Spitkovsky, I.: Singular integral operators on the spaces with general weights. J. Funct. Anal. **105**, 129–143 (1992)
46. Spitkovsky, I.: Once more on algebras generated by two projections. Linear Algebra Appl. **208/209**, 377–395 (1994)
47. Sunder, V.S.: N subspaces. Canad. J. Math. **40**(1), 38–54 (1988)
48. Vasilevskiĭ, N.L., Spitkovskiĭ, I.M. On the algebra generated by two projections. Dokl. Akad. Nauk Ukrain. SSR Ser. A **8**, 10–13 (1981), in Russian
49. Weiss, Y.: On algebras generated by two idempotents. In: Seminar Analysis: Operator Eq. and Numer. Anal., pp. 139–145. Karl-Weierstrass-Institut für Mathematik, Berlin (1987/88)

Part II
Invited Contributions

M-Local Type Conditions for the C^*-Crossed Product and Local Trajectories

M. Amélia Bastos, Catarina C. Carvalho, and Manuel G. Dias

Dedicated to Professor Yuri Karlovich on the occasion of his 75th birthday.

Abstract The local trajectories method establishes invertibility in algebras $\mathcal{B} = \text{alg}(\mathcal{A}, U_G)$ for a unital C^*-algebra $\mathcal{A}$ with a non-trivial center and a unitary group U_g, $g \in G$, with G a discrete group, assuming that G is amenable and the action $a \mapsto U_g a U_g^*$ is topologically free. It is applicable in particular to C^*-algebras associated with convolution type operators with amenable groups of shifts. We introduce here an M-local type condition that allows to establish an isomorphism between $\mathcal{B}$ and a C^*-crossed product, which is fundamental for the local trajectories method to work. We replace amenability of G by the more general condition that the action is amenable. The influence of the structure of the fixed points of the group action is analysed and a condition is introduced that applies when the action is not topologically free. If $\mathcal{A}$ is commutative, the referred conditions are related to the subalgebra $\text{alg}(U_G)$ yielding, in particular, a sufficient condition that depends essentially on U_G. It is shown that in $\pi(\mathcal{B}) = \text{alg}(\pi(\mathcal{A}), \pi(U_G))$, with π the local trajectories representation, the M-local type condition is satisfied, which allows establishing the isomorphism essential for the local trajectories method.

1 Introduction

The study of invertibility criteria in algebras of operators plays an important role in operator theory, with wide applications in many areas. One approach that has been fruitful is to use suitable families of representations, so that to reduce to invertibility

M. A. Bastos (✉) · C. C. Carvalho · M. G. Dias
Instituto Superior Técnico, University of Lisbon, Lisbon, Portugal
e-mail: amelia.bastos@tecnico.ulisboa.pt; catarinaccarvalho@tecnico.ulisboa.pt; manuel.g.dias@tecnico.ulisboa.pt

A. Böttcher et al. (eds.), *Achievements and Challenges in the Field of Convolution Operators*, Operator Theory: Advances and Applications 306,
https://doi.org/10.1007/978-3-031-80486-1_5

of the representatives in 'nicer', so-called local, algebras. The Allan-Douglas local principle applies to C^*-algebras $\mathcal{A}$ with a non-trivial center, in that if $\mathcal{Z}$ is a central subalgebra then using the isomorphism $\mathcal{Z} \cong C(M)$ given by Gelfand theory, with M the maximal ideal space of $\mathcal{Z}$, we have local algebras

$$\mathcal{A}_m = \mathcal{A}/J_m, \quad \pi_m : \mathcal{A} \to \mathcal{A}_m, \quad m \in M,$$

where J_m is the closed *-ideal of $\mathcal{A}$ generated by m, π_m is the quotient map, and in this case, $a \in \mathcal{A}$ is invertible in $\mathcal{A}$ if, and only if, $\pi_m(a)$ is invertible in $\mathcal{A}_m$ for all $m \in M$.

In this paper, we are interested in the case when we have a unital C^*-algebra of operators $\mathcal{A}$, a discrete group G and a group of unitary operators U_G defining an action of G on $\mathcal{A}$, that is, if we have a C^*-dynamical system. One is then bound to study invertibility in the algebra of operators generated by $\mathcal{A}$ and U_g, $g \in G$,

$$\mathcal{B} := \text{alg}(\mathcal{A}, U_G), \quad \text{with} \quad \mathcal{A} \subset B(H), \quad U : G \to B(H) \text{ unitary},$$

with $B(H)$ being the C^*-algebra of all bounded linear operators acting on some Hilbert space H. The issue is that even when $\mathcal{A}$ is commutative, $\mathcal{B}$ typically has a trivial center, so local principles do not apply directly.

The study of C^*-algebras of operators associated with C^*-dynamical systems was developed by Antonevich when the initial C^*-algebra is commutative and the group is subexponential or admissible (see [3, 4]). The extension of the Allan-Douglas local principle to C^*-algebras induced by an action of a discrete amenable group was developed by Karlovich [16, 18], relying on a given arbitrary central subalgebra of $\mathcal{A}$ and representations of C^*-crossed product algebras, originating what we call the *local trajectories method* (see below for details). An alternative approach was also developed by Antonevich, Lebedev, Brenner (see [5] and references therein).

The local trajectories method gives a powerful machinery for studying invertibility and Fredholmness in C^*-algebras of nonlocal type operators with discontinuous data. A first example of this application can be found in the paper by Karlovich [17] in which he analysed the Fredholm theory of convolution type operators with discrete groups of displacements and coefficients admitting discontinuities of semi almost periodic type. It has also been applied in the Fredholm analysis of C^*-algebras with amenable groups of shifts, algebras of convolution type operators with oscillating coefficients, algebras of singular integral operators with piecewise quasi-continuous and semi almost periodic coefficients. Examples of these applications can be found for instance in [8–11, 17, 21].

More precisely, the local trajectories method is applicable to study invertibility in C^*-algebras of the form $\mathcal{B} = \text{alg}(\mathcal{A}, U_G)$, where $\mathcal{A} \subset B(H)$ is a C^*-algebra with non-trivial center and $U_G := U(G)$, where $U : G \to B(H)$ is a unitary representation of an amenable discrete group G. We assume that we have an action $\alpha : G \to \text{Aut}(\mathcal{A})$ given by $\alpha_g(a) = U_g a U_g^*$, so we have a dynamical system $(\mathcal{A}, G, \alpha)$, with (id, U) a covariant representation on H, and we can consider the crossed product algebra $\mathcal{A} \rtimes_\alpha G$. We assume that G also acts on a given central subalgebra $\mathcal{Z} \cong C(M) \subset \mathcal{A}$, so that we have an induced action of G on the maximal ideal space M.

The idea behind the local trajectories method is to consider first localization in $\mathcal{A}$ as in Allan-Douglas and then to consider the associated regular representations of $\mathcal{A} \rtimes_\alpha G$ on ℓ^2-spaces. If we let Ω be the set of orbits of the action on M, then for points in the same orbit the respective regular representations are unitarily equivalent, so the local trajectories representations reduce to a family of regular representations $\{\pi_\omega\}_{\omega\in\Omega}$ of the crossed product algebra $\mathcal{A} \rtimes_\alpha G$. (We review these definitions in Sect. 3.)

In order to study invertibility in $\mathcal{B}$ through this family of representations, it is therefore a fundamental step to establish an isomorphism between the algebra $\mathcal{B}$ and the crossed product $\mathcal{A} \rtimes_\alpha G$. Conditions for this isomorphism, as well as for the faithfulness of the local trajectories family, typically assume that the group G is amenable. The approaches in [4, 5, 16, 18] rely on proving suitable isomorphism theorems giving sufficient conditions namely through the set of fixed points and the crucial notion of a *topologically free* action (see also [7]).

In the setting of local trajectories, following [18], the notion of being topologically free relies on the topology of the pure state space. If we let $\mathcal{P}_\mathcal{A}$ and $\mathcal{P}_\mathcal{Z}$ be the classes of pure states of $\mathcal{A}$ and $\mathcal{Z}$, then since $\mathcal{Z} \cong C(M)$, we have $\mathcal{P}_\mathcal{Z} \cong \hat{\mathcal{Z}} \cong M$, and since $\mathcal{Z}$ is central, there is a well-defined, surjective, restriction map

$$\psi : \mathcal{P}_\mathcal{A} \to \mathcal{P}_\mathcal{Z} \cong \hat{\mathcal{Z}} \cong M.$$

Then the action is said to be *topologically free* if for any finite set $G_0 \subset G$, and for any non-empty open set $W \subset \mathcal{P}_\mathcal{A}$, there exists $m \in \psi(W)$ with $\beta_g(m) \neq m$ for all $g \neq e \in G_0$, with $\beta : G \times M \to M$ the induced action on M; this is referred to as condition (A3) (see Sect. 3).

The main result in [18] (see also Theorem 1) can then be written as:

Local Trajectories Method: If G is amenable and the action is topologically free, then $\mathcal{B} \cong \mathcal{A} \rtimes_\alpha G$ and the family $\{\pi_\omega\}_{\omega\in\Omega}$ is faithful, so that $b \in \mathcal{B}$ is invertible if and only if $\oplus_\omega \pi_\omega(b)$ is invertible if and only if $\pi_\omega(b)$ is invertible for all orbits ω and $\sup_{\omega\in\Omega} \|\pi_\omega^{-1}(b)\| < \infty$.

If the action is not topologically free, the situation is typically much harder to analyse, but in some situations there are still methods to reduce things to this case, see [8, 18] (and references therein), and also [4, 5].

The purpose of the present article is to explore the relation between the notion of topologically freeness and the isomorphism with the crossed product algebra, as well as alternative conditions for the local trajectories method to hold in order to better understand its domain of applicability.

We do this by going back to a basic global condition, known to be equivalent to the existence of an isomorphism $\mathcal{B} \cong \mathcal{A} \rtimes_\alpha G$—referred to as *condition (B0)*—and establishing similar conditions but of *M-local type*, that is, on open sets of the space M of maximal ideals in $\mathcal{Z}$—referred to as *condition (B1)*. This basically amounts to considering 'local' elements of the form zb, where $b \in \mathcal{B}$ and $z \in \mathcal{Z} = C(M)$ is a bump function with support in V (the precise definitions are in Sect. 4.1).

We note that a necessary condition for condition (B0) to yield the isomorphism of $\mathcal{B}$ with the crossed product is that the full and reduced crossed product algebras coincide. In previous works, this identification typically comes from assuming that the group G is amenable. Here this condition is replaced by the more general notion of *amenability of the action* $\alpha : G \to \mathrm{Aut}(\mathcal{A})$ (see for instance [1, 12]), which we assume in Sect. 4. That is, instead of assuming that the group is 'nice', we assume that it acts 'nicely', and we still have $\mathcal{A} \rtimes_\alpha G \cong \mathcal{A} \rtimes_\alpha^r G$.

We show that M-localization applies to a class of algebras $\mathcal{A}$ where open sets of $\mathcal{P}_\mathcal{A}$ always determine an open set in M, in that the restriction map ψ defined above is open. We say that such an algebra $\mathcal{A}$ is *M-localizable*, or satisfies condition (C). This class includes all commutative algebras and algebras of matrices of continuous functions, as well as the class of algebras considered in [4].

The point in using M-local type conditions is that we can split between open sets $V \subset M$ where the action is 'well-behaved', in that there are points in V that are not fixed by a finite $G_0 \subset G$, and the open sets V that are fixed by some non-trivial finite $G_0 \subset G$ such that $(\beta_g)_{|V} = id_{|V}$, $g \in G_0$ (see Sect. 4.2). This is made possible by Lemma 2 and Proposition 4, which imply on one hand that if the action is topologically free then the M-local condition (B1) is automatically satisfied, so that we have an isomorphism with the crossed product, and on the other hand, that in the general case, to check (B1) it suffices to consider the class of open sets where the action of some finite subset is trivial - referred to as *condition (B2)*. As a result, in Corollary 4, we obtain an isomorphism between $\mathcal{B}$ and $\mathcal{A} \rtimes_\alpha G$ as long as we replace the assumption of the action being topologically free by our M-local condition being satisfied on such sets, should they exist, condition (B2).

We use condition (B2) to show, in Examples 1 and 2, that in case the action is not topologically free, then the fact that $\mathcal{B}$ is isomorphic to the crossed product may depend on the way $\mathcal{A}$ and $\mathrm{alg}(U_G)$ 'sit' inside $B(H)$ and how they interact, which cannot happen in the presence of topological freeness, due to the isomorphim theorems in [5, 7, 18].

This work then explores the local type conditions in two directions. The first one, developed in Sect. 4.3, applies directly to commutative algebras, and has to do with a C^*-subalgebra $\mathcal{A}'$ that is M-locally 'arbitrarily close' to $\mathcal{A}$ and where G also acts; in this case, assuming the generated algebras are also M-locally 'arbitrarily close', then the local condition (B2) needs only be verified on $\mathcal{A}'$ and $\mathcal{B}'$. We say that $\mathcal{A}'$ and $\mathcal{B}'$ are *M-locally dense in $\mathcal{A}$ and $\mathcal{B}$*.

It turns out that when the algebra $\mathcal{A}$ is commutative, and we see that this is the case with $\mathcal{A}' = \mathbb{C}$ and $\mathcal{B}' = \mathrm{alg}(U_G)$, so we obtain an M-local condition in $\mathrm{alg}(U_G)$ sufficient for our local condition (B2) to be satisfied in $\mathcal{B}$, and for the isomorphism with the crossed product algebra; this is Theorem 2.

Moreover, under an additional non-degeneracy condition, we can use (B2) to get rid of the M-local element, and we find a global condition on $\mathrm{alg}(U_G)$, namely that $\mathrm{alg}(U_G) \cong C^*(G)$, the group algebra, which guarantees (B2) and the isomorphism with the crossed product algebra. We point out that this condition depends only on

U_G, not on localization, and effectively substitutes the requirement of topological freeness in this situation, see Theorems 3 and 8.

Another application of the M-local conditions (B1) and (B2) is to guarantee conditions for the applicability of the method of local trajectories to M-localizable algebras, that is, those satisfying (C). As it is pointed out in Sect. 3, in general, the local trajectories method is applicable to $\mathcal{B}$, and yields a faithful family if there are isomorphisms

$$\mathcal{B} \cong \mathcal{A} \rtimes_{\alpha} G, \quad \pi(\mathcal{B}) \cong \pi(\mathcal{A}) \rtimes_{\alpha'} G$$

where $\pi = \oplus \pi_{\omega}$ and $\{\pi_{\omega}\}_{\omega \in \Omega}$ is the local trajectories family, π_{ω} regular representations. For the second isomorphism, we have representations on C^*-algebras of the form $B(\ell^2(G, H_{\omega}))$ and we can explore their norm properties to show that the M-local type condition (B1) is in fact always satisfied in $\pi(\mathcal{B}) = \text{alg}(\pi(\mathcal{A}), U'_G)$. We conclude that if $\mathcal{A}$ is M-localizable, then the conditions obtained before to ensure the isomorphism $\mathcal{B} \cong \mathcal{A} \rtimes_{\alpha} G$ are also sufficient for the method of local trajectories to work on $\mathcal{B}$; this leads to Theorems 6, 7 and 8.

In this article, we chose to formulate our results for a 'concrete' algebra of operators $\mathcal{A} \subset B(H)$, for some fixed Hilbert space H, together with an action of a discrete group on $\mathcal{A}$, and a group of unitary operators $U_G \subset B(H)$. The condition that the action satisfies $\alpha_g(a) = U_g a U_g^*$ is nothing more than the pair (id, U) being a covariant representation for $(\mathcal{A}, G, \alpha)$. We point out that in fact our results also apply to the setting when we start with an 'abstract' unital C^*-algebra $\mathcal{A}$ and consider an arbitrary faithful representation $\phi : \mathcal{A} \to B(H_{\phi})$ on some Hilbert space H_{ϕ}, and a covariant representation (ϕ, U), with $U_G \subset B(H_{\phi})$ unitary operators. Defining $\mathcal{B}_{\phi,U} = \text{alg}(\mathcal{A}, U_G)$, conditions (B0), (B1) and (B2) can be easily written (see Remark 3) in a way such that they yield an isomorphism $\mathcal{B}_{\phi,U} \cong \mathcal{A} \rtimes_{\alpha} G$, and similar conditions can be given such that the locally trajectories method works on $\mathcal{B}_{\phi,U}$.

We now give an outline of the paper. In Sect. 2, we review some objects and concepts needed throughout the paper. We also establish the setting we will work on and define the relevant C^*-algebras associated with our structures.

In Sect. 3, we present the local trajectories method. We follow the lines of [18], while focusing on pinpointing the main steps and in particular, on the role of the isomorphisms with crossed product algebras.

In Sect. 4, we explore conditions to guarantee the isomorphism with the crossed product, using what we call M-localization, that is, reduction to open subsets of M and assuming that the action is amenable.

We start, in Sect. 4.1, by introducing the classical condition to guarantee the isomorphism of $\mathcal{B}$ with the respective crossed product algebra, called here condition (B0). We then give a similar condition of M-local type, on open sets of the space of the maximal ideals of $\mathcal{Z}$, called condition (B1). Under an assumption on $\mathcal{A}$, condition (C) that $\mathcal{A}$ is M-localizable, we prove the equivalence between these two notions in Proposition 3.

In Sect. 4.2, we see how this localized condition can be used together with the structure of the fixed point set and the notion of topologically free action to guarantee the isomorphism with the crossed product algebra, and we will arrive at condition (B2) that only involves open sets of M fixed by a finite subset of G (Proposition 4).

In Sect. 4.3, we show that in the commutative case we need check conditions for isomorphism on $\mathrm{alg}(U_G)$ and the group algebra $C^*(G)$ depending only on G. We prove first that this algebra is what we call M-locally dense in $\mathcal{B}$ and then abstract these results to the general case.

Then in Sect. 5, we tackle the conditions for applicability of the local trajectories method, applying our results to the local trajectories representation $\pi = \oplus\pi_\omega$. We show that the image $\pi(\mathcal{B})$ of $\mathcal{B}$ is always isomorphic to a crossed product, proving that if $\mathcal{A}$ is M-localizable, the local trajectories method works on $\mathcal{B}$ as long as there is an isomorphism of $\mathcal{B}$ with the crossed product. In the commutative case, we give a sufficient condition depending only on U_G.

2 Preliminaries

Throughout the paper, $\mathcal{A}$ will always denote a unital C^*-algebra. We review in this section some concepts and results on C^*-algebras and representation theory that will be needed in the paper. For general references, see for instance [14, 15, 20].

2.1 Representations and States

By a *representation* of a C^*-algebra $\mathcal{A}$ on a Hilbert space H_π we always mean a non-degenerate $*$-homomorphism

$$\pi : \mathcal{A} \to B(H_\pi),$$

with $B(H_\pi)$ the C^*-algebra of bounded linear operators on $\mathcal{H}_\pi$. Non-degenerate in the unital case means that $\pi(1_\mathcal{A}) = I$. We say that π is *irreducible* if its only invariant subspaces are trivial. An injective representation is called *faithful*.

A *state* in $\mathcal{A}$ is a positive linear functional μ on $\mathcal{A}$ with $\|\mu\| = 1$, or, equivalently, $\mu(1_\mathcal{A}) = 1$. The state space $\mathcal{S}_\mathcal{A} \subset \mathcal{A}^*$ is convex and compact in the weak-$*$ topology, and as such, has extreme points that are called *pure states*. We denote the pure state space of $\mathcal{A}$ by $\mathcal{P}_\mathcal{A}$, and always endow $\mathcal{S}_\mathcal{A}$ and $\mathcal{P}_\mathcal{A}$ with the weak-$*$ topology, that is the topology of pointwise convergence.

States are related to representations in a fundamental way: $\pi : \mathcal{A} \to B(H)$ is an irreducible representation on a Hilbert space H, and x is a unit vector in H, then

$$\mu(a) := \langle \pi(a)x, x \rangle$$

is a pure state, and, by the GNS construction, any pure state defines an irreducible representation π_μ satisfying the above.

By the Gelfand-Naimark theorem, for every C^*-algebra $\mathcal{A}$ there exists a faithful representation $\pi : \mathcal{A} \to B(H)$ given by the direct sum

$$\pi = \bigoplus \pi_\nu, \qquad \pi_\nu : \mathcal{A} \to B(H_\nu)$$

where π_ν are irreducible representations associated to pure states.

We will make extensive use of the following properties (see [15], Proposition 4.3.1 and Theorems 4.3.8 and 4.3.14):

(i) Pure states separate points in $\mathcal{A}$: if $\mu(a) = 0$ for all $\mu \in \mathcal{P}_\mathcal{A}$ then $a = 0$.
(ii) For any state μ on $\mathcal{A}$, and $a, b \in \mathcal{A}$, we have the Cauchy-Schwarz inequality for states:

$$|\mu(a^*b)|^2 \le \mu(a^*a)\mu(b^*b).$$

In particular, since $\mu(1_\mathcal{A}) = 1$, we get $|\mu(a)| \le \sqrt{\mu(a^*a)} \le \|a\|$.
(iii) If $a \in \mathcal{A}$ is normal, that is, if $a^*a = aa^*$, then there exists a pure state $\mu \in \mathcal{P}_\mathcal{A}$ such that

$$\|a\| = |\mu(a)|.$$

In general, for $a \in \mathcal{A}$,

$$\|a\| = \max_{\mu \in \mathcal{P}_\mathcal{A}} \sqrt{\mu(a^*a)}.$$

In particular, if a is a positive element, then $\|a\| = \max_{\mu \in \mathcal{P}_\mathcal{A}} \mu(a)$.
(iv) If $\mathcal{Z}$ is a central subalgebra of $\mathcal{A}$, then for any $\mu \in \mathcal{P}_\mathcal{A}$, we have

$$\mu(za) = \mu(z)\mu(a), \qquad \text{for } a \in \mathcal{A},\ z \in \mathcal{Z}.$$

It follows that $\mu_{|\mathcal{Z}}$ is a multiplicative linear functional.

It follows from (iii) and the GNS construction that for any $a \in \mathcal{A}$ there exists an irreducible representation ϕ of $\mathcal{A}$ such that

$$\|a\| = \|\phi(a)\|.$$

We shall also need results on extension of states. Let $\mathcal{Z}$ be a closed C^*-subalgebra of $\mathcal{A}$, containing the identity. Then any state on $\mathcal{Z}$ can be extended to a state on $\mathcal{A}$ (see [15], Theorem 4.3.13); for each state in $S_\mathcal{Z}$, the set of its extensions to $S_\mathcal{A}$ is weak-$*$ compact and convex. Moreover, pure states can be extended to pure states. On the other hand, if $\mathcal{Z}$ is a *central* subalgebra of $\mathcal{A}$, pure states restrict to pure states (by (iv) and Proposition 4.4.1 in [15]), so restriction yields then a surjective map

$$\psi : \mathcal{P}_{\mathcal{A}} \to \mathcal{P}_{\mathcal{Z}}, \quad \mu \mapsto \mu_{|\mathcal{Z}}, \tag{1}$$

which is moreover continuous, since $\mathcal{P}_{\mathcal{A}}, \mathcal{P}_{\mathcal{Z}}$ have the weak-$*$ topology. Since $\mathcal{Z}$ is a central subalgebra, then from (iv) above, $\mu_{|\mathcal{Z}}$ is a multiplicative linear functional and $\mathcal{P}_{\mathcal{Z}} = \hat{\mathcal{Z}}$, the character space of $\mathcal{Z}$. Moreover, by Gelfand's theorem, since $\mathcal{Z}$ is commutative, $\hat{\mathcal{Z}} \cong M$, the space of maximal ideals of $\mathcal{Z}$, and the Gelfand transform yields an isomorphism $\mathcal{Z} \cong C(M)$.

2.2 Algebras Associated to an Unitary Action

In what follows, we let $\mathcal{A}$ be a unital C^*-algebra and G be a discrete group. For details on the constructions below, see for instance [12, 20, 22] and references therein.

An *action* of G on $\mathcal{A}$ is a homomorphism $\alpha : G \to \mathrm{Aut}(\mathcal{A})$, where $\mathrm{Aut}(\mathcal{A})$ is the group of $*$-automorphisms of $\mathcal{A}$. Given such an action we call $(\mathcal{A}, G, \alpha)$ a *C^*-dynamical system.*

Given a C^*-dynamical system $(\mathcal{A}, G, \alpha)$, we denote by $C_c(G, \mathcal{A})$ the linear space of finitely supported functions in G,

$$C_c(G, \mathcal{A}) = \{f : G \to \mathcal{A} \mid f(s) = 0, s \notin G_0 \text{ finite }\}.$$

We use the action α to define a α-twisted convolution product on $C_c(G, \mathcal{A})$, as well as an involution:

$$(f * g)(s) := \sum_{t \in G} f(t)\alpha_t(g(t^{-1}s)), \quad f^*(s) := \alpha_s(f^*(s^{-1})).$$

We call the $*$-algebra $C_c(G, \mathcal{A})$ the *convolution algebra* of $(\mathcal{A}, G, \alpha)$.

If $\mathcal{A}$ is commutative, $\mathcal{A} \cong C(M)$, with M some compact Hausdorff space, then C^*- dynamical systems are in one-to-one correspondence with group actions on M: if G acts on compact space M, $G \times M \to M, (g, m) \to g \cdot m$, then$(C(M), G, \alpha)$ is a C^*-dynamical system with

$$\alpha_s(f)(m) = f(s^{-1} \cdot m), \quad m \in M, \quad s \in G.$$

Conversely, if $\mathcal{A} = C(M)$ is a commutative algebra, then G acts on M and (G, M) is a transformation group.

We are interested in studying invertibility in C^*-algebras associated to dynamical systems.

Definition 1 Assume that $\mathcal{A} \subset B(H)$, for some Hilbert space H, and let $U : G \to B(H)$, $g \mapsto U_g$ be a unitary representation. We denote by

$$\mathcal{B} := \mathrm{alg}(\mathcal{A}, U_G)$$

the C^*-subalgebra of $B(H)$ generated by $\mathcal{A}$ and $U_G = \{U_g, g \in G\}$. Assume also that $U_g a U_g^*$ is a $*$-automorphism of $\mathcal{A}$ for all $g \in G$, that is, that we have an action

$$\alpha : G \to \mathrm{Aut}(\mathcal{A}), \quad \alpha_g(a) := U_g a U_g^*,$$

then $\mathcal{B}$ is the closure in $B(H)$ of the $*$-subalgebra

$$\mathcal{B}_0 := \left\{ \sum_{g \in G_0} a_g U_g : a_g \in \mathcal{A}, G_0 \subset G \text{ finite} \right\} = \left\{ \sum_{g \in G} a(g) U_g : a \in C_c(G, \mathcal{A}) \right\}.$$

Given an arbitrary dynamical system $(\mathcal{A}, G, \alpha)$, there is always a universal object that encodes both the original C^*-algebra $\mathcal{A}$ and the group action.

Definition 2 The *crossed product algebra* $\mathcal{A} \rtimes_\alpha G$ is the completion of $C_c(G, \mathcal{A})$ with respect to the universal norm

$$\|f\|_u = \sup_\pi \|\pi(f)\|,$$

where π ranges over all $*$-homomorphisms $\pi : C_c(G, \mathcal{A}) \to B(H)$, with H a Hilbert space.

When $\mathcal{A} = \mathbb{C}$, the crossed product algebra yields the group algebra $C^*(G)$.

If $\mathcal{B}$ is as in Definition 1, one of our goals is to discuss conditions under which $\mathcal{B} \cong \mathcal{A} \rtimes_\alpha G$ (see Sect. 4). One can check that there is always a $*$-homomorphism $\Phi : C_c(G, \mathcal{A}) \to \mathcal{B}_0$, surjective, given by

$$\Phi(f) = \sum_{g \in G} f(g) U_g, \tag{2}$$

that extends to a surjection $\Phi : \mathcal{A} \rtimes_\alpha G \to \mathcal{B}$, so the algebra $\mathcal{B}$ is always a quotient of the crossed product algebra. To study representations of such algebras $\mathcal{B}$, we consider first representations of the crossed product.

Let $(\mathcal{A}, G, \alpha)$ be a dynamical system. Given a Hilbert space H, consider a representation $\pi : \mathcal{A} \to B(H)$ and a unitary representation $U : G \to B(H)$, then the pair (π, U) is said to be a *covariant representation* of $(\mathcal{A}, G, \alpha)$ if

$$\pi(\alpha_s(a)) = U_s \pi(a) U_s^*.$$

If (π, U) is a covariant representation on some Hilbert space H, we can define the so-called *integrated representation* of $\mathcal{A} \rtimes_\alpha G$ by

$$\pi \rtimes U : \mathcal{A} \rtimes_\alpha G \to B(H), \quad (\pi \rtimes U)(f) = \sum_{t \in G} \pi(f(g))U_g, \quad f \in C_c(G, \mathcal{A}).$$

We have that any covariant representation defines a $*$-representation of $C_c(G, \mathcal{A})$ and, conversely, every non-degenerate $*$-representation of $C_c(G, \mathcal{A})$ is induced by some covariant representation of $(\mathcal{A}, G, \alpha)$, so that

$$\|f\|_u = \sup_{(\pi, U)} \|\pi \rtimes U(f)\|,$$

where (π, U) ranges over all covariant representations of $(\mathcal{A}, G, \alpha)$. Note that in Definition 1, we are simply assuming that (id, U) is a covariant representation, and the map Φ in (2) is given by $\Phi = id \rtimes U$.

In fact, the definition of the crossed product algebra yields the following *universal property*: for every covariant representation (π, U), there is a $*$-homomorphism $\sigma : \mathcal{A} \rtimes_\alpha G \to \operatorname{alg}(\pi(\mathcal{A}), U_G) \subset B(H)$ such that

$$\sigma(f) = \sum_{g \in G} \pi(f(g))\, U_g \quad \text{for all} \quad f \in C_c(G, \mathcal{A}).$$

We are interested in a particular class of integrated representations that correspond to taking the *left regular representations of* G: given a Hilbert space H, one defines the representation

$$\lambda : G \to B(\ell^2(G, H)), \quad \lambda_g \xi(s) = \xi(g^{-1}s).$$

Given a representation $\pi : \mathcal{A} \to B(H)$ of $\mathcal{A}$, define also

$$\tilde{\pi} : \mathcal{A} \to B(\ell^2(G, H)), \quad \tilde{\pi}(a)\xi(s) = \pi(\alpha_s^{-1}(a))\xi(s).$$

Then one can see that $(\tilde{\pi}, \lambda)$ is covariant representation.

Definition 3 The *regular representation of* $\mathcal{A} \rtimes_\alpha G$ *induced by the representation* π is given by the integrated representation induced by $(\tilde{\pi}, \lambda)$, that is,

$$\tilde{\pi} \rtimes \lambda : \mathcal{A} \rtimes_\alpha G \to B(\ell^2(G, H))$$

such that for $f \in C_c(G, \mathcal{A})$,

$$[(\tilde{\pi} \rtimes \lambda)(f)\xi](g) = \sum_{s \in G} \tilde{\pi}\left(\alpha_{g^{-1}}[f(s)]\right) \xi\left(s^{-1}g\right), \quad \xi \in \ell^2(G, H). \tag{3}$$

The *reduced crossed product algebra* $\mathcal{A} \rtimes_\alpha^r G$ is the completion of $C_c(G, \mathcal{A})$ in the norm

$$\|f\|_r = \sup_{(\tilde{\pi},\lambda)} \|(\tilde{\pi} \rtimes \lambda)(f)\|,$$

where $(\tilde{\pi} \rtimes \lambda)$ ranges over all regular representations of $\mathcal{A} \rtimes_\alpha G$. If $\mathcal{A} = \mathbb{C}$ then $\mathcal{A} \rtimes_\alpha^r G$ is the reduced group algebra $C_r^*(G)$.

In this paper, we shall work in the setting where the full and reduced crossed product algebras coincide, so that

$$\|f\|_u = \|f\|_r = \sup_{(\tilde{\pi},\lambda)} \|(\tilde{\pi} \rtimes \lambda)(f)\|, \quad \text{with } (\tilde{\pi} \rtimes \lambda) \text{ a regular representation.}$$

A discrete group G is said to be *amenable* if there exists a state μ on $\ell^\infty(G)$ which is invariant under the left translation action: for all $s \in G$ and $f \in \ell^\infty(G)$,

$$\mu(sf) = \mu(f).$$

The state μ is called an invariant mean. In this case, we have $C_r^*(G) \cong C^*(G)$ and it follows that $\mathcal{A} \rtimes_\alpha G \cong \mathcal{A} \rtimes_\alpha^r G$, for any dynamical system (A, G, α). The class of amenable groups includes all compact groups, abelian groups, solvable groups and finitely generated groups of subexponential growth. On the other hand, if G contains a copy of the free group in two generators, then G is not amenable.

We will consider here a more general notion of amenability that suffices for our purposes, that of an *amenable action* (see for instance [12], Section 4.3). Define a norm in $C_c(G, \mathcal{A})$ by

$$\|f\|_2 := \left\| \sum_{g \in G} f(g) f(g)^* \right\|^{1/2}, \quad f \in C_c(G, \mathcal{A}).$$

Let $\mathcal{Z}(\mathcal{A})$ denote the center of $\mathcal{A}$. For $s \in G$, let $\delta_s \in C_c(G, \mathcal{A})$ be such that $\delta_s(s) = 1_{\mathcal{A}}$ and $\delta_s(g) = 0$, $g \neq s$. Then $(\delta_s * f)(g) = \alpha_s(f(s^{-1}g))$, $g \in G$.

Definition 4 An action $\alpha : G \to \mathrm{Aut}(\mathcal{A})$ is *amenable* if there exist finitely supported functions $x_i : G \to \mathcal{Z}(\mathcal{A}) \subset \mathcal{A}$, $i \in \mathbb{N}$, with the following properties:

1. $x_i(g) \geq 0$ for all $i \in \mathbb{N}$ and $g \in G$;
2. $\sum_{g \in G} x_i(g)^2 = 1_{\mathcal{A}}$ for all $i \in \mathbb{N}$;
3. $\|\delta_s * x_i - x_i\|_2 \to 0$ for all $s \in G$.

Given a dynamical system $(\mathcal{A}, G, \alpha)$ where the action is amenable, we always have $\mathcal{A} \rtimes_\alpha G \cong \mathcal{A} \rtimes_\alpha^r G$ (Theorem 4.3.4 in [12]).

Moreover, every action by an amenable discrete group is amenable. This can be seen using the following equivalent definition of amenability: a discrete group G

is amenable if, and only if, for any finite set $G_0 \subset G$ there exist nonnegative unit vectors $x_i \in \ell^2(G)$, $i \in \mathbb{N}$, such that $\|\lambda_s x_i - x_i\|_2 \to 0$ for all $s \in G_0$ (see for instance [12], Theorem 2.6.8). In this case, for any action α of an amenable discrete group G on a C^*-algebra $\mathcal{A}$, identifying $\mathbb{C}$ with $\mathbb{C}1_{\mathcal{A}} \subset \mathcal{Z}(\mathcal{A})$, we have

$$(\delta_s * x_i)(g) = \alpha_s(x_i(s^{-1}g)) = x_i(s^{-1}g)\alpha_s(1_{\mathcal{A}}) = \lambda_s x_i(g)1_{\mathcal{A}}, \quad g \in G.$$

On the other hand, there are many relevant amenable actions of non-amenable groups, such as the action of a free group on its Gromov boundary, see [12].

3 Local Trajectories Method

We let $\mathcal{A} \subset B(H)$ be a unital C^*-algebra and $U : G \to B(H)$ be a unitary representation. We review in this section the method of local trajectories, whose goal is to establish an invertibility criterion for operators in $\mathcal{B} = \mathrm{alg}(\mathcal{A}, U_G)$ in terms of the invertibility of local representatives associated to the orbits of an action of G on some compact space. We follow here the approach and notation in [18] (see also [4]).

From now on, we will let $\mathcal{Z}$ be a central subalgebra of $\mathcal{A}$, with $1_{\mathcal{A}} \in \mathcal{Z}$. In this case, the Gelfand transform yields an isomorphism $\mathcal{Z} \cong C(M)$, where M is the compact Hausdorff space of maximal ideals of $\mathcal{Z}$ or equivalently, the class of non-zero multiplicative linear functionals with the weak-$*$ topology. We typically identify $z \in \mathcal{Z}$ with its Gelfand transform in $C(M)$, that is, we regard z as a continuous function.

Following the terminology in [18], we consider conditions (A1) and (A2):

Condition (A1): For every $g \in G$, the mapping $\alpha_g : a \mapsto U_g a U_g^*$ is a $*$-automorphism of the C^*-algebras $\mathcal{A}$ and $\mathcal{Z}$.

Condition (A2): G is an amenable discrete group.

Assume condition (A1), that is, we assume that (id, U) is a covariant representation of $(\mathcal{A}, G, \alpha)$ where α is an action on $\mathcal{A}$ that maps $\mathcal{Z} = C(M)$ to $\mathcal{Z} = C(M)$. We have then an action β of G on M given by $\beta_g : M \to M$ such that for $z \in \mathcal{Z}$,

$$z\left(\beta_g(m)\right) = \left(\alpha_g(z)\right)(m), \quad m \in M, \quad g \in G.$$

For each $m \in M$, let $G(m) = \{\beta_g(m) : g \in G\}$ be the G-orbit of m and Ω be the orbit space, that is, $\Omega = M/\sim$ with $m \sim n \Leftrightarrow G(m) = G(n)$. Let also:

- J_m be the closed two-sided ideal of $\mathcal{A}$ generated by maximal ideal $m \in M$ of $\mathcal{Z} \subset \mathcal{A}$;
- $\mathcal{A}_m := \mathcal{A}/J_m$, and $\rho_m : \mathcal{A} \to \mathcal{A}/J_m$ be the quotient map. Then by (A1), $J_{\beta_g(m)} = \alpha_g^{-1} J_m$, so it follows that

$$\mathcal{A}_{m'} \cong \mathcal{A}_m, \quad m' \in G(m).$$

We define now a family of representations of $\mathcal{A}$. For each orbit $\omega \in \Omega$, choose a representative $m_\omega \in \omega$ and a faithful representation $\phi_\omega : \mathcal{A}_{m_\omega} \to B(H_\omega)$ on some Hilbert space H_ω, and define

$$\pi'_\omega : \mathcal{A} \to B(H_\omega), \quad \pi'_\omega = \phi_\omega \circ \rho_m$$

Recall that to π'_ω we associate $\tilde{\pi}'_\omega : \mathcal{A} \to B(\ell^2(G, H_\omega))$ such that

$$(\tilde{\pi}'_\omega(a)\xi)(g) := \pi'_\omega(\alpha_g^{-1}(a))(\xi(g)).$$

Definition 5 The *local trajectories family on* $\mathcal{A} \rtimes_\alpha G$ is the family of regular representations $\{\pi_\omega\}_{\omega\in\Omega}$ induced by π'_ω, $\omega \in \Omega$, that is, induced by the covariant representation $(\pi'_\omega, \lambda_\omega)$, such that

$$\pi_\omega : \mathcal{A} \rtimes_\alpha G \to B(\ell^2(G, H_\omega)), \quad \pi_\omega := \tilde{\pi}'_\omega \rtimes \lambda_\omega,$$

with

$$[\pi_\omega(a)\xi](t) = \pi'_{m_\omega}\left(\alpha_t^{-1}(a)\right)\xi(t), \quad \left[\pi_\omega\left(U_g\right)\xi\right](t) = \xi\left(g^{-1}t\right),$$

for $\xi \in \ell^2(G, H_\omega)$, $t \in G$. Let $\pi = \oplus_{\omega\in\Omega}\pi_\omega$ be the direct sum representation of $\mathcal{A} \rtimes_\alpha G$ on $B(H_\Omega)$, with $H_\Omega = \sum_{\omega\in\Omega} \ell^2(G, H_\omega)$.

In order for the local trajectories maps to be well-defined on the algebra $\mathcal{B} = \mathrm{alg}(\mathcal{A}, U_G)$, we need extra conditions, namely to guarantee that there is uniqueness of representation of elements in $\mathcal{B}_0$. One such condition has to do with the structure of fixed points of the action. Recall that we say that G acts freely on M if the group $\left\{\beta_g : g \in G\right\}$ of homeomorphisms of M onto itself acts freely on M, that is, if $\beta_g(m) \neq m$ for all $g \in G\backslash\{e\}$ and all $m \in M$. One considers here a more general notion of freeness that relies on the topology of the state space.

Let $\mathcal{P}_\mathcal{A}$ be the set of all pure states on $\mathcal{A}$, equipped with the weak-* topology. As we have noted in (1), we have a map

$$\psi : \mathcal{P}_\mathcal{A} \to \hat{\mathcal{Z}} \cong M, \quad \mu \mapsto \mu_{|\mathcal{Z}}.$$

We often write $m = m_\mu$ if $m = \mu_{|\mathcal{Z}} = \mathcal{Z} \cap \ker\mu$.

If $J_m \subset \mathcal{A}$ is the ideal generated by the maximal ideal $m \in M$ of $\mathcal{Z}$ and $m = \mu_{|\mathcal{Z}}$, for some $\mu \in \mathcal{P}_\mathcal{A}$, then $\ker\mu \supset J_m$ and therefore (see [11], Lemma 4.1)

$$\mathcal{P}_\mathcal{A} = \bigcup_{m\in M} \mathcal{P}_m, \quad \mathcal{P}_m := \{\mu \in \mathcal{P}_\mathcal{A} : \ker\mu \supset J_m\}. \tag{4}$$

Writing $m = \mu_{|\mathcal{Z}}$, then $\mathcal{P}_m \subset \mathcal{P}_\mathcal{A}$ is the class of extensions of the pure state $\mu_{|\mathcal{Z}}$ to a pure state on $\mathcal{A}$.

As in [18], we adopt the following notion of *topologically free action*:

Condition (A3): For every finite set $G_0 \subset G$ and every open set $W \subset \mathcal{P}_{\mathcal{A}}$ there exists $\nu \in W$ such that $\beta_g(m_\nu) \neq m_\nu$ for all $g \in G_0 \backslash \{e\}$, where $m_\nu := \nu_{|\mathcal{Z}} = \mathcal{Z} \cap \ker \nu \in M$.

Note that (A3) guarantees that the set of fixed points has empty interior. If the C^*-algebra $\mathcal{A}$ is commutative, then taking $\mathcal{Z} = \mathcal{A} \cong C(M)$, we have $\mathcal{P}_{\mathcal{A}} \cong M$, so we can rewrite (A3):

(Commutative (A3)) For every finite set $G_0 \subset G$ and every open set $V \subset M$ there exists $m_0 \in V$ such that $\beta_g(m_0) \neq m_0$ for all $g \in G_0 \backslash \{e\}$.

Under assumptions (A1)-(A3), we have an isomorphism $\mathcal{B} \cong \mathcal{A} \rtimes_\alpha G$ (see [18, Theorem 3.2]). In particular, the local trajectories family is well-defined in $\mathcal{B}$. Moreover, we have:

Theorem 1 (Local Trajectories Method: [18, Theorem 4.1]) *If (A1)-(A3) hold, then the local trajectories representation $\pi = \oplus_{\omega\in\Omega}\pi_\omega$ is faithful in $\mathcal{B}$. Hence, $b \in \mathcal{B}$ is invertible if and only if $\pi_\omega(b)$ is invertible in $B(\ell^2(G, H_\omega))$ for all $\omega \in \Omega$ and*

$$\sup\left\{\left\|(\pi_\omega(b))^{-1}\right\| : \omega \in \Omega\right\} < \infty.$$

If the number of orbits is finite, then the bound on the norms of the inverse elements always holds, so that $b \in \mathcal{B}$ is invertible if, and only if, $\pi_\omega(b)$ is invertible in $B(\ell^2(G, H_\omega))$ for all $\omega \in \Omega$. In this case, local trajectories family is said to be *sufficient*.

A crucial step in the proof of the above criterion is that the local trajectory family is always injective over $\mathcal{A}$, that is, $\pi(\mathcal{A}) \cong \mathcal{A}$. This result relies on the structure of pure states of $\mathcal{A}$ as in (4), we give here a short proof for completeness (see also [18] for a direct proof of the equality of norms).

Proposition 1 *Assume (A1) is satisfied. Then $\{(\pi_\omega)_{|A}\}_{\omega\in\Omega}$ is a faithful family in $\mathcal{A}$, i.e., $\pi_{|A} := \oplus(\pi_\omega)_{|A}$ is injective and*

$$\|\pi(a)\| = \sup_{\omega\in\Omega} \|\pi_\omega(a)\| = \|a\|_{\mathcal{A}} \quad \textit{for all} \quad a \in \mathcal{A}.$$

Proof Let $\pi(a) = 0$, $a \in \mathcal{A}$, so that $\pi_\omega(a) = 0$ for all $\omega \in \Omega$. We show that

$$\ker \pi_\omega \cap \mathcal{A} = \bigcap_{g\in G} J_{\beta_g(m)}, \qquad m \in \omega.$$

We have $\pi_\omega(a) = 0 \Leftrightarrow \pi'_\omega(\alpha_g(a)) = 0$, for any $g \in G \Leftrightarrow \rho_m(\alpha_g(a))) = 0$, for $g \in G \Leftrightarrow \alpha_g(a) \in J_m$, for $g \in G$. Since, by (A1), $J_m = \alpha_g\left(J_{\beta_g(m)}\right)$, the equality above holds.

Hence, $\pi(a) = 0$ yields that for any $m \in M$ and $\mu \in \mathcal{P}_m$, we have $\mu(a) = 0$, since $J_m \subset \ker \mu$. By (4), we have $\mu(a) = 0$, for any $\mu \in \mathcal{P}_{\mathcal{A}}$, so $a = 0$. □

Assume (A1) and suppose that $\Phi := id \rtimes U : \mathcal{A} \rtimes_\alpha G \to \mathcal{B}$ is an isomorphism, so the representations $\pi_w : \mathcal{B} \to B\left(\ell^2(G, H_w)\right)$ defined by:

$$[\pi_\omega(a)\xi](t) = \pi'_{m_\omega}\left(\alpha_t^{-1}(a)\right)\xi(t), \quad \left[\pi_\omega\left(U_g\right)\xi\right](t) = \xi\left(g^{-1}t\right)$$

for $\xi \in \ell^2(G, H_\omega)$ are well defined. As before, let $\pi = \sum_{\omega\in\Omega} \pi_\omega$.

Now since, by Proposition 1, $\pi : \mathcal{A} \to \pi(\mathcal{A})$ is an isomorphism, we can define $\Psi : C_c(G, \mathcal{A}) \to C_c(G, \pi(\mathcal{A}))$ given by

$$\Psi(f)(s) = \pi(f(s)), \quad f \in C_c(G, \mathcal{A}),$$

which extends to an isomorphism $\Psi : \mathcal{A} \rtimes_\alpha G \to \pi(\mathcal{A}) \rtimes_{\alpha'} G$, with

$$\alpha'_g(\pi(a)) := \pi(\alpha_g(a)) = \pi(U_g a U_g^*) = \pi(U_g)\pi(a)\pi(U_g^*).$$

Thus, by definition, $(id, \pi(U))$ is a covariant representation of $(\pi(\mathcal{A}), G, \alpha')$. We have also that $\pi(\mathcal{B}) = \text{alg}\,(\pi(\mathcal{A}), \pi\,(U_G))$.

We then obtain the following commutative diagram:

$$\begin{array}{ccc} \mathcal{A} \rtimes_\alpha G & \xrightarrow{id \rtimes_\alpha U} & \mathcal{B} \\ \downarrow{\cong} & & \downarrow{\pi} \\ \pi(\mathcal{A}) \rtimes_{\alpha'} G & \xrightarrow{id \rtimes_{\alpha'} \pi(U)} & \pi(\mathcal{B}). \end{array} \tag{5}$$

The idea of the proof of Theorem 1 in [18] can be summarized roughly as follows: assuming (A1)–(A3) then $\mathcal{A} \rtimes_\alpha G \cong \mathcal{B}$ and moreover $(\pi(\mathcal{A}), G, \alpha')$ also satisfies (A1)-(A3), since $(id, \pi(U))$ is a covariant representation of $(\pi(\mathcal{A}), G, \alpha')$, α' maps $\pi(\mathcal{Z})$ to $\pi(\mathcal{Z})$, and the condition (A3) of topological freeness relies on the sets of pure states of $\mathcal{A}$ and $\pi(\mathcal{A})$, which are isomorphic. This then implies that π is an isomorphism due to the commutative diagram above, since all the other arrows are isomorphisms.

Remark 1 It follows from the discussion above that, assuming (A1), and therefore knowing that $\pi(\mathcal{A}) \cong \mathcal{A}$, with $\pi = \sum_{\omega\in\Omega} \pi_\omega$, then the conclusion of the local trajectories method, Theorem 1, holds in $\mathcal{B}$ as long as one can show that

$$id \rtimes_\alpha U : \mathcal{A} \rtimes_\alpha G \to \mathcal{B}, \quad id \rtimes_{\alpha'} \pi(U) : \pi(\mathcal{A}) \rtimes_{\alpha'} G \to \pi(\mathcal{B})$$

are isomorphisms, with $\alpha' = \pi \circ \alpha$. The proof of injectivity of π on $\mathcal{B}$ will then follow by the commutativity of the diagram (5). Recall that $\mathcal{B} = \text{alg}(\mathcal{A}, U_G)$ and $\pi(\mathcal{B}) = \text{alg}(\pi(\mathcal{A}), \pi(U_G))$.

4 Isomorphism with the Crossed Product

As we have noted, one important step to establish the local trajectory method, Theorem 1, in the non-local algebra $\mathcal{B} = \text{alg}(\mathcal{A}, U_G)$ is to give conditions such that $\mathcal{B}$ is isomorphic to a crossed product algebra. Our goal in this section is to give conditions that guarantee that such an isomorphism exists, without assuming topological freeness of the action.

We assume throughout this section that $\mathcal{A} \subset B(H)$, for some Hilbert space H, α is an action by automorphisms of $\mathcal{A}$ and that $U : G \to B(H)$, $g \mapsto U_g$ is a unitary representation satisfying

$$\alpha_g(a) = U_g a U_g^*,$$

for $g \in G, a \in \mathcal{A}$, that is, that (id, U) is a covariant representation of the dynamical system $(\mathcal{A}, G, \alpha)$ on $B(H)$. As in Definition 1, we let $\mathcal{B} := \text{alg}(\mathcal{A}, U_G)$ be the C^*-algebra generated by $\mathcal{A}$ and U, which coincides with the closure in $B(H)$ of the $*$-subalgebra

$$\mathcal{B}_0 = \left\{ \sum_{g \in G_0} a_g U_g : a_g \in \mathcal{A}, G_0 \subset G \text{ finite} \right\}.$$

The condition that the group G is amenable will be replaced by the weaker condition that the action α is amenable. Note that in the case we have $\mathcal{A} \rtimes_\alpha G \cong \mathcal{A} \rtimes_\alpha^r G$.

4.1 General Conditions and M-Localization

We start with noting that by the universal property of the crossed product, one has a surjection $\Phi : C_c(G, \mathcal{A}) \to \mathcal{B}_0$ given by $\Phi(f) = \sum_{s \in G} f(s) U_s$ that is bounded in the universal norm and therefore extends to $\mathcal{A} \rtimes_\alpha G$, so we obtain a surjective map

$$\Phi := id \rtimes U : \mathcal{A} \rtimes_\alpha G \to \mathcal{B}.$$

Hence we have $\mathcal{B} \cong \mathcal{A} \rtimes_\alpha G / \ker \Phi$, so in order to establish that $\mathcal{B} \cong \mathcal{A} \rtimes_\alpha G$, it suffices to give conditions such that $\ker \Phi = 0$. As it is known, this condition can be written as the boundedness of a family of 'evaluation maps' [4, 5, 18, 19].

For $s \in G$, let

$$E_s(f) := f(s), \quad f \in C_c(G, \mathcal{A}). \tag{6}$$

Then E_s is bounded in the universal norm so that it extends to $E_s : \mathcal{A} \rtimes_\alpha G \to \mathcal{A}$. We can write, by continuity,

$$\Phi(f) = \sum_{s \in G} E_s(f) U_s, \quad f \in \mathcal{A} \rtimes_\alpha G.$$

We will need the following simple lemma.

Lemma 1 *If $\mathcal{A} \rtimes_\alpha G \cong \mathcal{A} \rtimes_\alpha^r G$ then $\bigcap_{s \in G} \ker E_s = \{0\}$.*

Proof Let $\pi : \mathcal{A} \to B(H_\mathcal{A})$ be an arbitrary $*$-representation of $\mathcal{A}$ and $\tilde{\pi} \rtimes \lambda : \mathcal{A} \rtimes_\alpha G \to B(\ell^2(G, H_\mathcal{A}))$ be the induced regular representation. Extending (3) by continuity, we can write, for $b \in \mathcal{A} \rtimes_\alpha G$,

$$[(\tilde{\pi} \rtimes \lambda)(b)\xi](g) = \sum_{s \in G} \pi \left(\alpha_{g^{-1}}[E_s(b)]\right) \xi \left(s^{-1} g\right)$$

for $\xi \in \ell^2(G, H_\mathcal{A})$, $g \in G$. If $b \in \bigcap_{s \in G} \ker E_s$, then $\pi \left(\alpha_{g^{-1}}[E_s(b)]\right) = 0$ for all $s, g \in G$, hence $(\tilde{\pi} \rtimes \lambda)(b) = 0$. Since π is arbitrary, it follows that for the reduced norm we have $\|b\|_r = 0$, and by the isomorphism $\mathcal{A} \rtimes_\alpha G \cong \mathcal{A} \rtimes_\alpha^r G$, also $b = 0$. □

Assume that also $E_s' : \mathcal{B}_0 \to \mathcal{A}$ such that

$$E_s' \left(\sum_{g \in G_0} a_g U_g \right) = a_s$$

is well defined, that is, we have uniqueness of representation of elements in $\mathcal{B}_0$. Then we have a bijection between $\mathcal{B}_0$ and $C_c(G, \mathcal{A})$ and $E_s(f) = E_s' \circ \Phi(f)$, for $f \in C_c(G, \mathcal{A})$.

Proposition 2 *Assume that the action $\alpha : G \to \mathrm{Aut}(\mathcal{A})$ is amenable, in particular, $\mathcal{A} \rtimes_\alpha G \cong \mathcal{A} \rtimes_\alpha^r G$. Then the following statements are equivalent:*

(i) $\Phi = id \rtimes U : \mathcal{A} \rtimes_\alpha G \to \mathcal{B}$ *is an isomorphism;*
(ii) $E_s' : \mathcal{B}_0 \to \mathcal{A}$ *is well defined and bounded for all* $s \in G$*;*
(iii) For any finite set $G_0 \subset G$ *and* $a_g \in \mathcal{A}$, $g \in G_0$, *we have*

$$\|a_e\| \leq \left\| \sum_{g \in G_0} a_g U_g \right\|.$$

Proof If $\Phi = id \rtimes U$ is an isomorphism on $\mathcal{A} \rtimes_\alpha G$, then $E_s' = E_s \circ \Phi^{-1}$ is well defined and bounded. Conversely, if E_s' bounded then $E_s = E_s' \circ \Phi$ extends to $\mathcal{A} \rtimes_\alpha G$. If $b \in \ker \Phi$ then also $E_s(b) = 0$ for all s, and since the action of G

is amenable, it follows from Lemma 1 that $b = 0$. So (i) $\Leftrightarrow$ (ii), and (ii) clearly gives (iii). Assuming (iii), we have $\sum_{g \in G_0} a_g U_g = 0 \Rightarrow a_e = 0$ and also $a_s = 0$, for $s \in G_0$, so we have uniqueness of representation and E'_s is well defined. The boundedess of the maps E'_s for any $s \in G$ is equivalent to boundedness for $s = e$, since $E'_s \left(\sum_{g \in G_0} a_g U_g \right) = E'_e \left(\sum_{g \in G_0} a_g U_{gs^{-1}} \right)$. □

This result can be found in [4, 5, 18], with the assumption that G is amenable.

Remark 2 The reduced C^*-algebra $\mathcal{A} \rtimes^r_\alpha G$ always has a canonical faithful conditional expectation to $\mathcal{A}$, that is, a contractive projection onto $\mathcal{A}$, given by the extension of E_e as in (6) (see for instance [12], Proposition 4.1.9). Faithful here means that $E_e(b^*b) = 0$ if, and only if, $b = 0$ (or equivalently that the kernel of E_e does not contain any non-zero ideal). Assuming $\mathcal{A} \rtimes_\alpha G \cong \mathcal{A} \rtimes^r_\alpha G$, we see that the equivalent conditions (ii) and (iii) just mean that there exists a canonical faithful conditional expectation from $\mathcal{B} = \text{alg}(\mathcal{A}, U_G)$ to $\mathcal{A}$ given by E'_e.

According to the previous proposition, our aim is to give conditions so that we have:

Condition (B0): For any finite set $G_0 \subset G$ and $a_g \in \mathcal{A}$, $g \in G_0$, we have

$$\|a_e\| \leq \left\| \sum_{g \in G_0} a_g U_g \right\|. \tag{7}$$

When (7) holds, we say that $(B0)$ holds for $b = \sum_{g \in G_0} a_g U_g$.

When $\mathcal{A} = \mathbb{C}$, condition $(B0)$ gives conditions such that the group C^*-algebra $C^*(G)$ and $\text{alg}(U_G)$, the C^*-algebra generated by the unitary representations U_g, $g \in G$, are isomorphic.

Remark 3 We are working in a concrete setting where $A \subset B(H)$ for some fixed Hilbert space H and (id, U) is a covariant representation. However, condition (B0) and the result above can be used more generally. For an 'abstract' C^*-algebra $\mathcal{A}$ and action $\alpha : G \to \text{Aut}(\mathcal{A})$, suppose that we are given any faithful representation $\pi : \mathcal{A} \to B(H_\pi)$ on a Hilbert space H_π and $U : G \to B(H_\pi)$ is a unitary representation. Let $\mathcal{B}_{\pi,U}$ be the C^*-algebra generated by $\pi(\mathcal{A})$ and U_g, $g \in G$.

Then (π, U) is a covariant representation of $(\mathcal{A}, G, \alpha)$ on $B(H_\pi)$ if, and only if, (id, U) is a covariant representation of $(\pi(\mathcal{A}), G, \alpha')$, with $\alpha' := \pi \circ \alpha \circ \pi^{-1}$. Moreover α is amenable if, and only if, α' is amenable. There is an isomorphism $\mathcal{A} \rtimes_\alpha G \cong \pi(\mathcal{A}) \rtimes_{\alpha'} G$ that intertwines $\pi \rtimes U$ and $id \rtimes U$.

We conclude from Proposition 2 that if (π, U) is a faithful covariant representation of $(\mathcal{A}, G, \alpha)$ and α is amenable, then $\pi \rtimes U : \mathcal{A} \rtimes_\alpha G \to \mathcal{B}_{\pi,U}$ is an isomorphism if, and only if, for any finite set $G_0 \subset G$ and $a_g \in \mathcal{A}$, $g \in G_0$, we have

$$\left\| \sum_{g \in G_0} \pi(a_g) U_g \right\| \geq \|\pi(a_e)\| = \|a_e\|.$$

In order to estimate norms, it is often useful to consider positive elements, by property (iii) in Sect. 2.1. We let $b = \sum_{g \in G_0} a_g U_g$, where $G_0 \subset G$ is a finite set. In terms of pure states, (B0) becomes:

$$\|a_e\|^2 = \|a_e^* a_e\| = \max_{\mu \in \mathcal{P}_{\mathcal{A}}} \mu(a_e^* a_e) \leq \max_{\nu_0 \in \mathcal{P}_{B(H)}} \nu_0(b^* b) = \|b^* b\| = \|b\|^2.$$

We will write often

$$b^* b = \tilde{a} + \sum_{s \neq t \in G_0} \alpha_s^{-1}\left(a_s^* a_t\right) U_{s^{-1}t}, \quad \text{with} \quad \tilde{a} := \sum_{s \in G_0} \alpha_s^{-1}\left(a_s^* a_s\right) \in \mathcal{A}. \tag{8}$$

Then $\tilde{a}$ is positive and we can write $\tilde{a} = a_e^* a_e + a'$, with $a_e^* a_e$ and a' positive elements. Hence, $\mu(a_e^* a_e) \leq \mu(\tilde{a})$, for $\mu \in \mathcal{P}_{\mathcal{A}}$, so that also $\|a_e^* a_e\| \leq \|\tilde{a}\|$. It follows that if (B0) holds for all positive elements, that is, if

$$\|\tilde{a}\| \leq \|b^* b\| \quad \text{for all} \quad b \in \mathcal{B}_0,$$

then $\|a_e\|^2 = \|a_e^* a_e\| \leq \|b^* b\| = \|b\|^2$ and (B0) holds in general. We conclude that to guarantee that Φ is an isomorphism, one needs only check (B0) for positive elements b and a_e.

We now give a localized version of (B0). We let $\mathcal{Z}$ be a central subalgebra of $\mathcal{A}$, as in Sect. 3. From now on, we assume condition (A1) of the previous section, that is, that α acts by automorphisms both of $\mathcal{A}$ and of $\mathcal{Z}$ and (id, U) is a covariant representation of $(\mathcal{A}, G, \alpha)$. We will replace condition (A2) of amenability of the group G by the weaker assumption that the action α is amenable.

As before, we let M be the maximal ideal space of the commutative C^*-algebra $\mathcal{Z}$, and identify $\mathcal{Z}$ with $C(M)$ through the Gelfand transform. Recall from (1) that we have a surjective, continuous map

$$\psi : \mathcal{P}_{\mathcal{A}} \to \mathcal{P}_{\mathcal{Z}} = \hat{\mathcal{Z}} \cong M, \quad \mu \mapsto \mu_{|\mathcal{Z}},$$

where $\mathcal{P}_{\mathcal{A}}$, $\hat{\mathcal{Z}} \cong M$ have the weak-$*$ topology. We have $\mathcal{P}_{\mathcal{A}} = \bigcup_{m \in M} \mathcal{P}_m$, where $\mathcal{P}_m$ are the pure states that restrict to m and for $V \subset M$, we write $\mathcal{P}_V := \psi^{-1}(V)$.

We make the following assumption on the algebra $\mathcal{A}$ of an M-localization property:

Condition (C): For any open set $W \subset \mathcal{P}_A$, the set

$$V = \psi(W) = \{m \in M : m = \mu_{|\mathcal{Z}}, \mu \in W\} \tag{9}$$

is open in $M \cong \hat{\mathcal{Z}}$.

We say that such an algebra $\mathcal{A}$ is *M-localizable*. Clearly any commutative algebra satisfies (C), since in this case $\mathcal{P}_{\mathcal{A}}$ is homeomorphic to M. Also matrix algebras $[C(M)]_{N \times N}$, $N \in \mathbb{N}$, satisfy (C), and more generally also algebras of the form $HOM(E, F)$, where E, F are vector bundles, as the ones considered in [4].

Moreover, all C^*-algebras that have the uniqueness of extension property [2, 6, 13]), that is, if ψ is injective in that any pure state in $\mathcal{Z}$ has a unique extension to a pure state in $\mathcal{A}$, also satisfy (C), since in this case the extension in $\mathcal{S}_{\mathcal{A}}$ is also unique and the extension map $\mathcal{P}_{\mathcal{Z}} \to \mathcal{P}_{\mathcal{A}} \subset \mathcal{S}_{\mathcal{A}}$ is continuous (see [13], Lemma 1, or directly).

Note that if $\mathcal{A}$ satisfies (C), then topological freeness, condition (A3), becomes equivalent to requiring that no non-empty open set of M is fixed by a finite, non-trivial, subset $G_0 \subset G$, similarly to the commutative case and to the definition adopted in [4].

According to (8), (B0) for the positive element b^*b comes down to $\|b^*b\| \geq \|\tilde{a}\|$. The next result shows that, assuming our algebra satisfies (C), if this inequality holds locally, for any positive element, then also (B0) holds in $\mathcal{B}$.

We first introduce the following notation that we shall use throughout the paper: for $\emptyset \neq V \subset M$ open, let $\mathcal{Z}(V) \subset \mathcal{Z}$ be those $z \in C(M)$ with $0 \leq z \leq 1$, $\|z\| = 1$, $\operatorname{supp} z \subset V$. We think of elements zb, $z \in \mathcal{Z}(V)$, $b \in \mathcal{B}$, as being a localization of b to V.

If we let $\mu \in \mathcal{P}_{\mathcal{A}}$, with $m_\mu = \mu_{|\mathcal{Z}} \in M$ and $z \in \mathcal{Z}$, then if $z(m_\mu) = 0$ then $\mu_0(zb) = 0$ for all $b \in \mathcal{B}$, μ_0 any extension of μ to a state in $\mathcal{B}$, since

$$|\mu_0(zb)|^2 \leq \mu(zz^*)\mu_0(b^*b) = z^2(m_\mu)\mu_0(b^*b) = 0.$$

In particular, if $V \subset M$ is open and $z_V \in \mathcal{Z}(V)$, then for μ_0 an extension of μ with $\mu \notin \mathcal{P}_V$, that is, $m_\mu \notin V$, we have

$$\mu_0(z_V b) = 0, \qquad \text{for } b \in \mathcal{B}. \tag{10}$$

Moreover, if $a \in \mathcal{A}$ with $a \geq 0$ then, since a and $z_V \in \mathcal{Z}$ commute and $z_V \geq 0$, we have $z_V a \geq 0$ and

$$\|z_V a\| = \max_{\mu \in \mathcal{P}_V} \mu(z_V a) = \max_{\mu \in \mathcal{P}_V} z_V(\mu_{|\mathcal{Z}})\mu(a).$$

We now prove:

Proposition 3 *Assume $\mathcal{A}$ satisfies (C). Let $G_0 \subset G$ be finite and $b = \sum_{g \in G_0} a_g U_g$, $a_g \in \mathcal{A}$, and $\tilde{a} := \sum_{g \in G_0} \alpha_g^{-1}(a_g^* a_g)$. If for any non-empty open set $V \subset M$, there is a central element $z_V \in \mathcal{Z}(V)$ satisfying*

$$\|z_V b^* b\| \geq \|z_V \tilde{a}\| \tag{11}$$

then (B0) holds for b. In fact, it suffices that $\|z_V b^* b\| \geq \nu(z_V \tilde{a})$, for $\nu \in \mathcal{P}_{\mathcal{A}}$ such that $z_V(m_\nu) = 1$, with $m_\nu = \nu_{|\mathcal{Z}}$.

Proof Let $\phi \in \mathcal{P}_{\mathcal{A}}$ be such that $\|a_e^* a_e\| = \phi(a_e^* a_e) \leq \phi(\tilde{a})$ and consider the open sets

$$W = \{\mu \in \mathcal{P}_{\mathcal{A}} : |\mu(\tilde{a}) - \phi(\tilde{a})| < \varepsilon\} \quad \text{and} \quad V = \{\mu_{|\mathcal{Z}} : \mu \in W\} \subset M.$$

Let $z_V \in \mathcal{Z}(V)$. Then since $z_V\tilde{a}$ is positive, writing $m_\mu = \mu_{|\mathcal{Z}}$, $\mu \in \mathcal{P}_\mathcal{A}$,

$$\begin{aligned}\|z_V\tilde{a}\| &= \max_{\mu\in\mathcal{P}_\mathcal{A}} \mu(z_V)\mu(\tilde{a}) \geq \max_{\mu\in W} \mu(z_V)\mu(\tilde{a})\\ &> (\phi(\tilde{a}) - \varepsilon) \max_{m_\mu\in V} z_V(m_\mu) = \phi(\tilde{a}) - \varepsilon \geq \|a_e\|^2 - \varepsilon.\end{aligned}$$

Taking z_V satisfying the assumption, and noting that $\|z_V b^* b\| \leq \|b\|^2$, we get $\|b\|^2 \geq \|a_e\|^2 - \varepsilon$ for all $\varepsilon > 0$, so (B0) follows.

For the last assertion, in this case, taking $m \in V$ with $z(m) = 1$ and $\nu \in W$ with $\nu_{|\mathcal{Z}} = m$, we get $\|z_V b^* b\| \geq \nu(\tilde{a})$ for some $\nu \in W$, and proceed in the same way. □

In terms of pure states, the condition in Proposition 3 follows if for any open $V \subset M$, there is $z_V \in \mathcal{Z}(V)$ and $\nu \in \mathcal{P}_\mathcal{A}$ such that

$$\begin{aligned}&\nu_0\left(z_V\tilde{a} + \sum_{s\neq t\in G_0} \alpha_s^{-1}\left(a_s^* a_t\right) z_V U_{s^{-1}t}\right)\\ &\geq \mu(z_V\tilde{a}) \quad \text{for all} \quad \mu \in \mathcal{P}_\mathcal{A}, \text{ with } \mu_{|\mathcal{Z}} \in V,\end{aligned}$$

and ν_0 an extension of ν to a state in $\mathcal{B}$. In fact it suffices to check this for $\mu \in \mathcal{P}_\mathcal{A}$ with $\mu(z_V) = 1$.

We arrive at condition (B1):

Condition (B1): For any non-empty open set $V \subset M$, finite set $G_0 \subset G$ and $a_g \in \mathcal{A}$, $g \in G_0$, there exists a central element $z_V \in \mathcal{Z}(V)$ satisfying

$$\|z_V a_e\| \leq \left\| z_V \sum_{g\in G_0} a_g U_g \right\|. \tag{12}$$

As a matter of terminology, if (12) or (11) hold for $b = \sum_{g\in G_0} a_g U_g \in \mathcal{B}_0$, we say that $(B1)$ holds for b, and if it holds on a subset $\mathcal{B}'_0$ dense in a C^*-subalgebra $\mathcal{B}' \subset \mathcal{B}$ we say that (B1) holds in $\mathcal{B}'$.

Of course, if (B0) holds, then (B1) holds for any $z_V \in \mathcal{Z}(V)$. On the other hand, the condition in Proposition 3 amounts to (B1) for the positive elements b^*b and $\tilde{a}$. We have then that for M-localizable algebras $\mathcal{A}$, as in (9), the global condition (B0) is equivalent to the M-local condition (B1).

Together with Proposition 2, we obtain another condition for the isomorphism of $\mathcal{B}$ with the crossed product algebra. Recall that condition (A1) means that (id, U) is a covariant representation of $(\mathcal{A}, G, \alpha)$, where α acts by automorphisms both of $\mathcal{A}$ and of the central subalgebra $\mathcal{Z}$.

Corollary 1 *Assume that $\mathcal{A}$ satisfies (C), condition (A1) is satisfied for* $(\mathcal{A}, G, \alpha)$ *and that the action α is amenable. Then*

$$(B1) \quad \Leftrightarrow \quad (B0) \quad \Leftrightarrow \quad \Phi = id \rtimes U : \mathcal{A} \rtimes_\alpha G \to \mathcal{B} \textit{ is an isomorphism.}$$

In the spirit of Remark 3, if we are given an arbitrary faithful representation $\pi : \mathcal{A} \to B(H_\pi)$, with H_π a Hilbert space, and a covariant representation (π, U), with $U : G \to B(H_\pi)$ a unitary representation, we can also easily deduce a M-local condition to ensure that $\mathcal{A} \rtimes_\alpha G \cong \mathcal{B}_{\pi,U} := \text{alg}(\pi(\mathcal{A}), U_G)$.

We will see in the next section that, in fact, for these results to hold, we only need to check (B1) for a subclass of open subsets of M.

Remark 4 We remark that conditions (B0) and (B1) are not, in general, related to the action α but only to the way $\mathcal{A}$ interacts with the unitaries U_g. In some situations, we may have unitaries U_g and U'_g defining the same action $\alpha : G \to \text{Aut}(\mathcal{A})$ such that the algebras $\mathcal{B} = \text{alg}(\mathcal{A}, U_G)$ and $\mathcal{B}' = \text{alg}(\mathcal{A}, U'_G)$ are not isomorphic, if (B0) is satisfied in $\mathcal{B}$ but not in $\mathcal{B}'$ (see for instance Example 12.11 in [5]). This cannot happen if G is amenable and the action is *topologically free*, by the isomorphism theorems in [18] (Theorem 3.3) and [5] (Corollary 12.16), as in this case we would have $\mathcal{B} \cong \mathcal{B}' \cong \mathcal{A} \rtimes_\alpha G$.

We note further that if the action is *not* topologically free, then the fact that conditions (B0) and (B1) hold may depend on the way $\mathcal{A}$ is represented as an algebra of bounded operators; see Examples 1 and 2 at the end of the next section.

4.2 *Fixed Points*

We show here how the structure of fixed points plays a role in ensuring the localized condition (B1). We keep the assumption that condition (A1) holds for our dynamical system $(\mathcal{A}, G, \alpha)$, in that the action α leaves the central subalgebra $\mathcal{Z} = C(M)$ invariant and (id, U) is a covariant representation of $(\mathcal{A}, G, \alpha)$ on some Hilbert space H, with U a unitary representation of G in $B(H)$, and $\mathcal{B} = \text{alg}(\mathcal{A}, U_G)$. We also consider the induced action of G on M given by $\beta_g : M \to M$ such that $z\left(\beta_g(m)\right) = \left(\alpha_g(z)\right)(m)$, $z \in \mathcal{Z}$, $m \in M$, $g \in G$.

Note that given an open set $V \subset M$ and finite $G_0 \subset G$, if $m_0 \in V$ is such that $\beta_g(m_0) \neq m_0$ for all $g \in G_0$, then there exists an open $\Delta \subset V$, $m_0 \in \Delta$, such that

$$\beta_g(m) \neq m \quad \text{for all} \quad m \in \Delta, g \in G_0, \quad \text{and} \quad \beta_g(\Delta) \cap \Delta = \emptyset. \tag{13}$$

In general, we have the following lemma.

Lemma 2 *Given a finite set $G_0 \subset G$ and a non-empty open set $V \subset M$, there exists a non-empty open set $\Delta \subset V$ such that for $g \in G_0$, either $\beta_g\big|_\Delta = id_\Delta$ or $\beta_g(\Delta) \cap \Delta = \emptyset$. In particular, we have $G_0 = \tilde{D} \cup D_0$ such that*

$$\beta_g\big|_\Delta = id_\Delta, \quad \textit{for } g \in \tilde{D}, \quad \beta_g(\Delta) \cap \Delta = \emptyset, \quad \textit{for } g \in D_0 = G_0 - \tilde{D}.$$

Moreover, if G_0 is closed for inverses, then D_0 and $\tilde{D}$ are closed for inverses.

Proof Induction on number of elements of G_0: if $G_0 = \{g\}$, then either $\beta_g\big|_V = id_V$, and $\tilde{D} := G_0$, or there exists $m \in V$ with $\beta_g(m) \neq m$, in which case we can take $\Delta \subset V$ open with $\beta_g(\Delta) \cap \Delta = \emptyset$ and $D_0 = G_0$.

Assuming now that $G_0 = \{g_k : 0 \leq k \leq n\}$, let $\Delta' \subset V$ be such that $G_0 \setminus \{g_n\} = \tilde{D}' \cup D_0'$ with $\beta_g\big|_{\Delta'} = id_{\Delta'}$, for $g \in \tilde{D}'$ and $\beta_g(\Delta') \cap \Delta' = \emptyset$, for $g \in D_0'$. As above, if $\beta_{g_n}\big|_{\Delta'} = id_{\Delta'}$, we take $\Delta = \Delta'$, $\tilde{D} = \tilde{D}' \cup \{g_n\}$, $D_0 = D_0'$. Otherwise, there exists $m \in \Delta'$ with $\beta_{g_n}(m) \neq m$, in which case we can take $\Delta \subset \Delta'$ open with $\beta_{g_n}(\Delta) \cap \Delta = \emptyset$ and $D_0 = D_0' \cup \{g_n\}$.

Assuming G_0 closed for inverses, then $\beta_g\big|_\Delta = id_\Delta \Leftrightarrow \beta_{g^{-1}}\big|_\Delta = id_\Delta$ and $\beta_g(\Delta) \cap \Delta = \emptyset \Leftrightarrow \beta_{g^{-1}}(\Delta) \cap \Delta = \emptyset$, so also D_0 and $\tilde{D}$ are closed for inverses. □

Our goal now is to see that to prove condition (B1), it suffices to consider open sets of M that are fixed by a finite subset of G.

Remark 5 For $z \in \mathcal{Z}$ and $g \in G$, we have $zU_g = U_g\alpha_g^{-1}(z) = U_g(z \circ \beta_{g^{-1}})$. Let $\Delta \subset M$ be open and $z_\Delta \in \mathcal{Z}(\Delta) \subset C(M)$, that is, z_Δ is a non-negative function, supported in Δ, with $\|z_\Delta\| = 1$. We will use the following.

(i) If $\beta_g\big|_\Delta = id_\Delta$, then $z_\Delta U_g = U_g z_\Delta$, that is, z_Δ and U_g commute. (This holds as long as $z_\Delta \in \mathcal{Z}$ is zero outside Δ.)
(ii) If $\beta_g(\Delta) \cap \Delta = \emptyset$, then $z_\Delta \circ \beta_{g^{-1}}$ is zero on Δ. In particular, if μ_0 is a state of $\mathcal{B}$ that restricts to a pure state in $\mathcal{A}$, so that $\mu_{0|\mathcal{Z}} = m_\mu \in \mathcal{Z}$, then $\mu_0(z_\Delta a U_g) = 0$: if $m_\mu \notin \Delta$ it follows from (10), if $m_\mu \in \Delta$ then

$$|\mu_0(z_\Delta a U_g)|^2 = |\mu_0(aU_g(z_\Delta \circ \beta_{g^{-1}}))|^2 \leq \mu(aU_gU_g^*a^*)(z_\Delta \circ \beta_{g^{-1}})^2(m_\mu) = 0.$$

Proposition 4 *Given $G_0 \subset G$ finite and a non-empty open set $V \subset M$, let $G_0' = \{s^{-1}t : s, t \in G_0\} = D_0 \cup \tilde{D}$, and $\emptyset \neq \Delta \subset V$ be an open set such that $\beta_g\big|_\Delta = id|_\Delta$ for $g = s^{-1}t \in \tilde{D}$, and $\beta_g(\Delta) \cap \Delta = \emptyset$, for $g = s^{-1}t \in D_0 = G_0' - \tilde{D}$, $s, t \in G_0$. Then for $b = \sum_{g \in G_0} a_g U_g$, $a_g \in \mathcal{A}$, and any $z_\Delta \in \mathcal{Z}(\Delta)$, we have that*

$$\|z_\Delta b^* b\| \geq \left\| z_\Delta \left(\tilde{a} + \sum_{s \neq t, s^{-1}t \in \tilde{D}} \alpha_s^{-1}\left(a_s^* a_t\right) U_{s^{-1}t} \right) \right\|,$$

with $\tilde{a} = \sum_{s \in G_0} \alpha_s^{-1}\left(a_s^* a_s\right)$.

Note that the existence of such an open set Δ and sets $D_0, \tilde{D} \subset G$ is a consequence of applying Lemma 2 to the finite set $G_0' = \{s^{-1}t : s, t \in G_0\}$.

Proof Let $\emptyset \neq V \subset M$ be open and let $b = \sum_{g \in G_0} a_g U_g$, where we assume without loss of generality that G_0 is closed for inverses, so also G_0' is closed for inverses. Write $G_0' = D_0 \cup \tilde{D}$ satisfying the conditions above, with $\Delta \subset M$ open.

Then we can write

$$b^* b = \sum_{s,t \in G_0} \alpha_s^{-1}\left(a_s^* a_t\right) U_{s^{-1}t}$$

$$= \tilde{a} + \sum_{s \neq t, s^{-1}t \in D_0} \alpha_s^{-1}\left(a_s^* a_t\right) U_{s^{-1}t} + \sum_{s \neq t, s^{-1}t \in \tilde{D}} \alpha_s^{-1}\left(a_s^* a_t\right) U_{s^{-1}t}.$$

Let

$$\tilde{b} := \tilde{a} + \sum_{s \neq t, s^{-1}t \in \tilde{D}} \alpha_s^{-1}\left(a_s^* a_t\right) U_{s^{-1}t}.$$

Note that $\tilde{b}$ is self-adjoint since $\tilde{D}$ is closed for inverses, and

$$\left(\alpha_s^{-1}\left(a_s^* a_t\right) U_{s^{-1}t}\right)^* = \alpha_t^{-1}\left(a_t^* a_s\right) U_{t^{-1}s}.$$

Consider the C^*-subalgebra $\mathcal{Z}_0(\Delta) \subset \mathcal{Z} = C(M)$ of functions that are zero outside Δ and let $z_\Delta \in \mathcal{Z}_0(\Delta)$. Since $\beta_g\big|_\Delta = id_\Delta$, for $g = s^{-1}t \in \tilde{D}$, we have that $z_\Delta U_g = U_g(z_\Delta \circ \beta_{g^{-1}}) = U_g z_\Delta$, from which follows that z_Δ and $\tilde{b}$ commute.

In particular, $z_\Delta \tilde{b}$ is also self-adjoint, hence normal, so there always exists a pure state $\nu_{\mathcal{B}} \in \mathcal{P}_{\mathcal{B}}$ such that $\|z_\Delta \tilde{b}\| = |\nu_{\mathcal{B}}(z_\Delta \tilde{b})|$. We see now that we can pick such a $\nu_{\mathcal{B}}$ such that $\nu_{\mathcal{B}}(z_\Delta b^* b) = \nu_{\mathcal{B}}(z_\Delta \tilde{b})$, which proves our claim since in this case

$$\left\|z_\Delta b^* b\right\| \geq |\nu_{\mathcal{B}}(z_\Delta b^* b)| = |\nu_{\mathcal{B}}(z_\Delta \tilde{b})| = \|z_\Delta \tilde{b}\|.$$

Consider then the C^*-subalgebras

$$C := \text{alg}\{\tilde{b}, \mathcal{Z}_0(\Delta), Id\} \subset \mathcal{B} \quad \text{and} \quad \tilde{C} := \text{alg}\{C, \mathcal{Z}\}.$$

Let $z_\Delta \in \mathcal{Z}(\Delta)$, that is, $z_\Delta \in \mathcal{Z}_0(\Delta)$ such that $\|z_\Delta\| = 1$, and assume $z_\Delta \tilde{b} \neq 0$ (otherwise there is nothing to prove). Since C is a commutative C^*-algebra, as $\tilde{b}$ commutes with any element of $Z_0(\Delta)$, there exists a pure state ν of C such that

$$\|z_\Delta \tilde{b}\| = |\nu(z_\Delta \tilde{b})|.$$

Moreover, $\nu(z_\Delta \tilde{b}) = \nu(z_\Delta)\nu(\tilde{b}) \neq 0$, so $\nu(z_\Delta) \neq 0$.

Since $C \subset \tilde{C}$ is a subalgebra and ν is a pure state of C, there is an extension of ν that is a pure state of $\tilde{C}$, $\nu_{\tilde{C}} \in \mathcal{P}_{\tilde{C}}$. Let $f \in \mathcal{Z}$ such that $f z_\Delta = 0$. Then since $z_\Delta \in \tilde{C}$ is a central element of this subalgebra, then

$$0 = \nu_{\tilde{C}}\left(z_\Delta f\right) = \nu_{\tilde{C}}\left(z_\Delta\right) \nu_{\tilde{C}}(f)$$

and since $\nu_{\tilde{C}}\left(z_\Delta\right) = \nu\left(z_\Delta\right) \neq 0$ then $\nu_{\tilde{C}}(f) = 0$. Thus we have that

$$\nu_{\tilde{C}}(f) = 0 \quad \text{for all } f \in \mathcal{Z} : f z_\Delta = 0.$$

Since $\tilde{C} \subset \mathcal{B}$ is a subalgebra, we can extend the pure state $\nu_{\tilde{C}}$ to a pure state $\nu_{\mathcal{B}} \in \mathcal{P}_{\mathcal{B}}$. Then if $s^{-1}t \in D_0$ we have that

$$z_\Delta \alpha_s^{-1}\left(a_s^* a_t\right) U_{s^{-1}t} = \alpha_s^{-1}\left(a_s^* a_t\right) U_{s^{-1}t}\left(z_\Delta \circ \beta_{t^{-1}s}\right)$$

and since $\beta_g(\Delta) \cap \Delta = \emptyset$, for $g = s^{-1}t \in D_0$, we have that

$$\left(z_\Delta \circ \beta_{t^{-1}s}\right)^2 z_\Delta = 0$$

and so

$$\nu_{\mathcal{B}}\left(\left(z_\Delta \circ \beta_{t^{-1}s}\right)^2\right) = \nu_{\tilde{C}}\left(\left(z_\Delta \circ \beta_{t^{-1}s}\right)^2\right) = 0$$

for $s \neq t$ and $s^{-1}t \in D_0$. By the Cauchy-Schwartz inequality for states, we have that:

$$\begin{aligned} \left|\nu_{\mathcal{B}}\left(z_\Delta \alpha_s^{-1}\left(a_s^* a_t\right) U_{s^{-1}t}\right)\right|^2 &= \left|\nu_{\mathcal{B}}\left(\alpha_s^{-1}\left(a_s^* a_t\right) U_{s^{-1}t}\left(z_\Delta \circ \beta_{t^{-1}s}\right)\right)\right|^2 \\ &\leq \nu_{\mathcal{B}}\left(\alpha_s^{-1}\left(a_s^* a_t\right) U_{s^{-1}t}\left(\alpha_s^{-1}\left(a_s^* a_t\right) U_{s^{-1}t}\right)^*\right) \nu_{\mathcal{B}}\left(\left(z_\Delta \circ \beta_{t^{-1}s}\right)^2\right) = 0. \end{aligned}$$

Thus we have that

$$\left|\nu_{\mathcal{B}}\left(z_\Delta b^* b\right)\right| = \left|\nu\left(z_\Delta \tilde{b}\right)\right| = \|z_\Delta \tilde{b}\|$$

and from this we conclude that $\|z_\Delta b^* b\| \geq \|z_\Delta \tilde{b}\|$. □

One first consequence applies when the action is topologically free, as in condition (A3) of Sect. 3. It follows from (13) that if (A3) holds, that is, if the action is topologically free, then for any open $V \subset M$, and $G_0 \subset G$ finite, we can always find $\Delta \subset V$ with $\beta_g(\Delta) \cap \Delta = \emptyset$ for all $g \in G_0$. Hence, in Lemma 2 we can always take $D_0 = G_0$ and $\tilde{D} = \emptyset$.

We obtain the following version of the result in [18] (Theorem 3.2) that establishes the isomorphism of $\mathcal{B}$ with the crossed product. Recall that we always assume condition (A1), so that in particular (id, U) is a covariant representation of (A, G, α), and that (C) holds if $\mathcal{A}$ is M-localizable as in (9).

Corollary 2 *Assume that (A1) is satisfied and that the action is topologically free, in that (A3) is satisfied, then (B1) holds in $\mathcal{B}$. If moreover (C) holds for $\mathcal{A}$ and the action α is amenable, then $id \rtimes U : A \rtimes_\alpha G \to \mathcal{B}$ is an isomorphism.*

Proof Let $b = \sum_{g\in G_0} a_g U_g$, $a_g \in \mathcal{A}$, G_0 finite, and $\emptyset \neq V \subset M$ be open. If $\mathcal{A}$ satisfies (C) and (A3) holds, we can take $\tilde{D} = \emptyset$ in Proposition 4, so we get for any $z_\Delta \in \mathcal{Z}(\Delta)$,

$$\|z_\Delta b^* b\| \geq \|z_\Delta \tilde{a}\|,$$

with $\tilde{a} = \sum_{s\in G_0} \alpha_s^{-1}\left(a_s^* a_s\right)$. We conclude that (B1) holds for b. It follows from Corollary 1 that if the action α is amenable, then $id \rtimes U$ is an isomorphism. □

Even if the action is not topologically free, we can still get a criterion for the isomorphism of $\mathcal{B}$ with the crossed product to hold. The following result comes directly from Proposition 4, as in this case, (11) holds.

Corollary 3 *Let $G_0 \subset G$ be finite, $a_g \in \mathcal{A}$, $g \in G_0$, and $\tilde{a} = \sum_{g\in G_0} \alpha_g^{-1}(a_g^* a_g)$. If for every non-empty open set $V \subset M$ such that $\beta_g\big|_V = id|_V$ for $g \in D$, with $D \subset \{s^{-1}t : s, t \in G_0\}$ arbitrary, there is a central element $z_V \in \mathcal{Z}(V)$ satisfying*

$$\left\| z_V \left(\tilde{a} + \sum_{s\neq t, s^{-1}t\in D} \alpha_s^{-1}\left(a_s^* a_t\right) U_{s^{-1}t} \right) \right\| \geq \|z_V \tilde{a}\|$$

then $(B1)$ holds for $b = \sum_{G_0} a_g U_g$.

We arrive at condition (B2):

Condition (B2): For any finite set $D \subset G$ and any non-empty open $V \subset M$ such that $\beta_g\big|_V = id|_V$ for all $g \in D$, and $a_g \in \mathcal{A}$, $g \in D$, there exists a central element $z_V \in \mathcal{Z}(V)$ satisfying

$$\|z_V a_e\| \leq \left\| z_V \sum_{g\in D} a_g U_g \right\|. \tag{14}$$

We can assume that a_e is positive and $\sum_{g\neq e\in D} a_g U_g$ is self-adjoint. If (14) holds for $b = \sum_{g\in G_0} a_g U_g \in \mathcal{B}_0$, we say that (B2) holds for b, and if it holds on a subset $\mathcal{B}_0'$ dense in a C^*-subalgebra $\mathcal{B}' \subset \mathcal{B}$ we say that (B2) holds in $\mathcal{B}'$.

In the notation of Corollary 3, applying (B2) to the element

$$\tilde{b} = \tilde{a} + \sum_{s\neq t, s^{-1}t\in D} \alpha_s^{-1}\left(a_s^* a_t\right) U_{s^{-1}t},$$

we obtain that (B1) holds for $b = \sum_{G_0} a_g U_g$. Of course, if (B1) holds then we can apply it on open sets V such that $\beta_g\big|_V = id|_V$ for all $g \in D$, to obtain (B2). Note that such sets exist only when the action is not topologically free.

It follows from the previous corollary that we can improve Corollary 1 to get conditions for the isomorphism of $\mathcal{B} = \text{alg}(\mathcal{A}, U_G)$ with the crossed product algebra. Recall that (C) means that $\mathcal{A}$ is M-localizable as in (9) and that condition (A1) is simply that (id, U) is a covariant representation of $(\mathcal{A}, G, \alpha)$, where α leaves the central subalgebra $\mathcal{Z}$ invariant.

Corollary 4 *Assume that $\mathcal{A}$ satisfies (C), that condition (A1) is satisfied for $(\mathcal{A}, G, \alpha)$ and that the action α is amenable. Then*

$$(B1) \quad \Leftrightarrow \quad (B2) \quad \Leftrightarrow \quad \Phi = id \rtimes U : \mathcal{A} \rtimes_\alpha G \to \mathcal{B} \textit{ is an isomorphism.}$$

In the setting of arbitrary faithful representations, as in Remark 3, if we are given a faithful representation $\pi : \mathcal{A} \to B(H_\pi)$, with H_π a Hilbert space, and a covariant representation (π, U) of $(\mathcal{A}, G, \alpha)$, with $U : G \to B(H_\pi)$ a unitary representation, we can replace (14) in (B2) by

$$\|z_V a_e\| \leq \left\| \pi(z_V) \sum_{g \in D} \pi(a_g) U_g \right\|,$$

and, assuming that $\mathcal{A}$ satisfies (C) and the action α is amenable, we can conclude that $\mathcal{A} \rtimes_\alpha G \cong \mathcal{B}_{\pi,U} := \text{alg}(\pi(\mathcal{A}), U_G)$.

We now apply our results to show that if topological freeness of the action is not satisfied for some system $(\mathcal{A}, G, \alpha)$, it is possible to construct covariant representations (π, U) for which (B2) is not satisfied and so the C^* algebra $\text{alg}(\mathcal{A}, U_G)$ is not isomorphic to the crossed product. (Note that for a commutative algebra, the existence of such a covariant representation follows from Theorem 2 in [7].)

Example 1 Let $\mathcal{A} = C(M)$ for some compact $M \subset \mathbb{R}^N$ such that M is the closure of an open set and M has smooth boundary, and G be an amenable discrete group. Let $\beta : G \times M \to M$ be an action by diffeomorphisms such that $\beta_e = id$ and $\beta_{gs} = \beta_g \beta_s$, $g, s \in G$. Assume that for some $g \in G$ where $g \neq e$ there exists a non-empty open set $V \subset M$ such that

$$\beta_g|_V = id|_V.$$

Let $\alpha : G \to \text{Aut}(\mathcal{A})$ be given by $\alpha_g(a) = a \circ \beta_g$, $g \in G$. We claim that there exists a covariant representation (π, U) of $(\mathcal{A}, G, \alpha)$ such that condition (B2) is not satisfied in $\mathcal{B} = \text{alg}(\pi(\mathcal{A}), U)$.

Indeed, consider the Hilbert space given by $H = L^2(M)$ with the Lebesgue measure. Consider the faithful covariant representation on $B(H)$ given by

$$[\pi(a)f](x) = a(x)f(x), \qquad [U_g f](x) = |\det d\beta_g|^{1/2} f(\beta_g(x)),$$

for $a \in \mathcal{A} = C(M)$ and $f \in L^2(M)$. Let $e \neq g \in G$ and $\emptyset \neq V \subset M$ be such that $\beta_g|_V = id|_V$, and let z_V be a continuous non-negative function on M with compact support in V with $\|z_V\| = 1$. Then since $\beta_g|_V = id|_V$, we have

$$[(\pi(z_V)U_g)f](x) = z_V(x)|\det d\beta_g|^{1/2} f(\beta_g(x)) = z_V(x)f(x) = [\pi(z_V)f](x).$$

Thus

$$0 = \|\pi(z_V) - \pi(z_V)U_g\| < \|\pi(z_V)\| = \|z_V\| = 1,$$

and so (B2), hence also (B0), is not satisfied. Since $\mathcal{A}$ is M-localizable, as it is commutative, and G is amenable, we conclude from Corollary 4, or directly from Proposition 2 (see also Remark 3), that $\mathcal{B}$ is not isomorphic to the crossed product $\mathcal{A} \rtimes_\alpha G$. Note that this does not contradict the isomorphism theorems in [5, 18] as the action α is not topologically free,

Although the above example shows that for such an algebra $\mathcal{A}$ and action α some covariant representations of (A, G, α) may not satisfy condition (B0), however it is possible to construct concrete representations which satisfy it.

Example 2 Consider the C^*-algebra $\mathcal{A} = C(M)$ for some compact $M \subset \mathbb{R}^N$ such that M is the closure of an open set and M has smooth boundary. Let $G = \mathbb{Z}$, which is amenable, and consider the action $\beta : \mathbb{Z} \times M \to M$ given by $\beta_n = \varphi^n$, $n \in \mathbb{Z}$, with φ a diffeomorphism of M that fixes a non-empty open set of M. As in Example 1, consider the action of $\mathbb{Z}$ on $C(M)$ given by $\alpha_n(a) = a \circ \beta_n$, $a \in \mathcal{A}$, $n \in \mathbb{Z}$. Clearly, α is not topologically free.

We will construct a covariant representation of $(\mathcal{A}, \mathbb{Z}, \alpha)$ that satisfies condition (B2). Consider the Hilbert space given by $H = L^2(M \times S^1)$, where M has the usual Lebesgue measure and S^1 is endowed with the normalized Lebesgue measure, so that it has measure 1. Consider the faithful representation $\pi : \mathcal{A} \to B(H)$ and the unitary representation $U : \mathbb{Z} \to B(H)$ given by

$$[\pi(a)f](x,t) = a(x)f(x,t), \qquad [U_n f](x,t) = e^{int}|\det d\varphi^n|^{1/2} f(\varphi^n(x), t).$$

We thus obtain

$$U_n \pi(a) U_n^* = \pi(\alpha_n(a)),$$

and so (π, U) is a covariant representation of $(\mathcal{A}, \mathbb{Z}, \alpha)$.

Let $G_0 \subset \mathbb{Z}$ be finite and $\emptyset \neq V \subset M$ be any open set such that $\varphi^n|_V = id|_V$ for all $n \in G_0$. Let $z_V : M \to \mathbb{R}$ be a continuous non-negative function with support in V and $f \in L^2(M \times S^1)$ be a function constant in the variable $t \in S^1$ such that $\|f\|_{L^2(M\times S^1)} = 1$. Since $\varphi^n|_V = id|_V$, we have that

$$z_V(x)|\det d\varphi^n|^{1/2} f(\varphi^n(x)) = z_V(x) f(x),$$

thus for any $a_n \in \mathcal{A}$, $n \in G_0$,

$$\begin{aligned}
\left\|\sum_{n\in G_0} \pi(z_V)\pi(a_n)U_n\right\|^2_{B(L^2(M\times S^1))} &\geq \left\|\sum_{n\in G_0} [\pi(z_V)\pi(a_n)U_n](f)\right\|^2_{L^2(M\times S^1)} \\
&= \int_M \int_{S^1} \left|\sum_{n\in G_0} z_V(x)e^{int} a_n(x)|\det d\varphi^n|^{1/2} f(\varphi^n(x))\right|^2 dtdx \\
&= \int_M \int_{S^1} \left|\sum_{n\in G_0} z_V(x)a_n(x)e^{int} f(x)\right|^2 dtdx \\
&= \int_M \sum_{n\in G_0} |z_V(x)|^2 |a_n(x)|^2 |f(x)|^2 dx.
\end{aligned}$$

Let $x_0 \in V$ be the maximum of the function $|z_V a_0|$. Now take a sequence $f_k : M \to \mathbb{R}$ such that $\|f_k\|_{L^2(M\times S^1)} = 1$ and f_k^2 is such that for any function $h \in C(M)$, we have $\int_M h(x) f_k^2(x)dx \to h(x_0)$, as $k \to \infty$ (for instance $f_k(x) = \chi_{B_{1/k}(x_0)\cap M}(x)/\sqrt{m(B_{1/k}(x_0) \cap M)}$, with m the Lebesgue measure and $B_{1/k}(x_0)$ the ball centered in x_0 with radius $1/k$). Thus we obtain that

$$\begin{aligned}
\lim_{k\to\infty} \left\|\sum_{n\in G_0} [\pi(z_V)\pi(a_n)U_n](f_k)\right\|^2_{L^2(M\times S^1)} &= \sum_{n\in G_0} |z_V(x_0)|^2 |a_n(x_0)|^2 \\
&\geq |z_V(x_0)a_0(x_0)|^2.
\end{aligned}$$

We conclude that

$$\left\|\sum_{n\in G_0} \pi(z_V)\pi(a_n)U_n\right\|_{B(L^2(M\times S^1))} \geq |z_V(x_0)a_0(x_0)| = \|z_V a_0\|,$$

which proves condition (B2) in $\mathcal{B} := \mathrm{alg}(\pi(\mathcal{A}), U_\mathbb{Z})$. It then follows from Corollary 4 that the integrated representation $\pi \rtimes U$ gives an isomorphism of $\mathcal{A} \rtimes_\alpha \mathbb{Z}$ with $\mathcal{B} = \mathrm{alg}(\pi(\mathcal{A}), U_\mathbb{Z})$.

In this example, $\mathbb{Z}$ may be replaced by a generic discrete, amenable, group G, replacing S^1 by the compact space $\hat{G}$, which is the Pontryagin dual of G.

4.3 Subalgebras and the Commutative Case

We consider here the equivalent conditions (B1) and (B2) on subalgebras of $\mathcal{B} = \text{alg}(\mathcal{A}, U_G)$. We start with considering the class of operators with scalar coefficients, which, as we shall see, will be relevant in the commutative case.

Let $\tilde{\mathcal{B}} := \text{alg}(U_G) \subset \mathcal{B}$ be the C^*-subalgebra generated by U_G, which is given by the closure of

$$\tilde{\mathcal{B}}_0 := \tilde{\mathcal{B}} \cap \mathcal{B}_0 = \left\{ \sum_{g \in G_0} (c_g I) U_g : G_0 \text{ finite}, \ c_g \in \mathbb{C} \right\}.$$

Let $b_0 = \sum_{g \in G_0} (c_g I) U_g$ with $c_g \in \mathbb{C}$, then we can write

$$b_0^* b_0 = \tilde{a}_0 I + \sum_{s \neq t \in G_0} \overline{c_s} c_t U_{s^{-1}t}, \quad \text{with} \quad \tilde{a}_0 := \sum_{s \in G_0} |c_s|^2 \in \mathbb{C}.$$

Assuming $\tilde{a}_0 > 0$, condition (B1) for b_0 can be written in the following way: for any non-empty open set $V \subset M$, there exists $z_V \in \mathcal{Z}(V)$ such that

$$\begin{aligned} \left\| z_V \left(\tilde{a}_0 + \sum_{s \neq t \in G_0} \overline{c_s} c_t U_{s^{-1}t} \right) \right\| &= \tilde{a}_0 \left\| z_V \left(I + \sum_{s \neq t \in G_0} \overline{c_s} c_t / \tilde{a}_0 U_{s^{-1}t} \right) \right\| \\ &\geq \| z_V \tilde{a}_0 \| = \tilde{a}_0, \end{aligned}$$

that is, condition (B1) holds in $\tilde{\mathcal{B}} \subset \mathcal{B}$ if, and only if, for any finite $G_0 \subset G$ and $c_g \in \mathbb{C}$, $g \in G_0$ and open $V \subset M$, there exists $z_V \in \mathcal{Z}(V)$ such that

$$\left\| z_V \left(I + \sum_{g \in G_0} c_g U_g \right) \right\| \geq \| z_V \| = 1.$$

We can write condition (B2) in a similar way. It follows from Corollary 3 that (B1) being satisfied in $\tilde{\mathcal{B}}$ is equivalent to (B2) being satisfied in $\tilde{\mathcal{B}}$.

Assume now $\mathcal{A}$ is commutative. The point of considering the subalgebra $\tilde{\mathcal{B}}$ is that in this case any $b \in \mathcal{B}_0$ can be approximated by some $b_0 \in \tilde{\mathcal{B}}_0 = \tilde{\mathcal{B}} \cap \mathcal{B}_0$ on open sets.

Proposition 5 *Assume $\mathcal{A} = C(M)$ is commutative, M a compact Hausdorff space. Let $b = \sum_{g \in G_0} a_g U_g$, with $a_g \in \mathcal{A}$, and $\tilde{a} = \sum_{g \in G_0} \alpha_g^{-1}(a_g^* a_g)$. Given $\varepsilon > 0$, for any non-empty open $V \subset M$, and $m \in V$, there exists $b_0 \in \tilde{\mathcal{B}}_0$ and a non-empty open $\Delta \subset V$, $m \in \Delta$, such that*

$$\|z_\Delta(b^*b - b_0^*b_0)\| < \varepsilon \quad \textit{and} \quad \|z_\Delta(\tilde{a} - \tilde{a}_0)\| < \varepsilon$$

for any $z_\Delta \in \mathcal{Z}(\Delta)$ with $z_\Delta(m) = 1$.

Proof Let $\emptyset \neq V \subset M$ be open and fix $m \in V$. We have

$$b^*b = \sum_{s,t \in G_0} \alpha_s^{-1}\left(a_s^* a_t\right) U_{s^{-1}t}.$$

By continuity of the functions a_s, we can take a neighborhood $m \in \Delta \subset V$ small enough such that

$$\left|\alpha_s^{-1}\left(a_s^* a_t\right)(x) - \alpha_s^{-1}\left(a_s^* a_t\right)(m)\right| < \varepsilon \quad \text{for all} \quad x \in \Delta, \quad s, t \in G_0.$$

Let $c_{s,t} = \alpha_s^{-1}\left(a_s^* a_t\right)(m) \in \mathbb{C}$, then for any $z_\Delta \in \mathcal{Z}(\Delta)$, assume also $z_\Delta(m) = 1$,

$$\left\|z_\Delta\left(\alpha_s^{-1}\left(a_s^* a_t\right) - c_{s,t} I\right)\right\| = \sup_{x \in \Delta} z_\Delta(x)\left|\alpha_s^{-1}\left(a_s^* a_t\right)(x) - c_{s,t}\right| \leq \varepsilon.$$

Let $b_0 = \sum_{g \in G_0} a_g(m) U_g$ then $b_0^* b_0 = \sum_{s,t \in G_0} c_{s,t} U_{s^{-1}t}$, and

$$\|z_\Delta(b^*b - b_0^*b_0)\| < K\varepsilon, \quad \|z_\Delta(\tilde{a} - \tilde{a}_0)\| < K'\varepsilon$$

for some $K, K' \in \mathbb{N}$ so the claim follows. □

Recall that we assume throughout that we have a covariant representation (id, U) of $(\mathcal{A}, G, \alpha)$ with α an action that leaves a central subalgebra $\mathcal{Z}$ invariant, that is, that condition (A1) holds.

Theorem 2 *Let $\mathcal{A} = C(M)$ be commutative and assume (A1) is satisfied. If (B2) holds in $\tilde{\mathcal{B}}$, that is, if for any finite $G_0 \subset G$ and $c_g \in \mathbb{C}$, $g \in G_0$ and any non-empty open V satisfying $\beta_g|_V = id|_V$ for all $g \in G_0$, there exists $z_V \in \mathcal{Z}(V)$ such that*

$$\left\|z_V\left(I + \sum_{g \in G_0} c_g U_g\right)\right\| \geq \|z_V\| = 1, \tag{15}$$

then (B2) holds in $\mathcal{B}$. Assuming the action α is amenable, then $id \rtimes U : \mathcal{A} \rtimes_\alpha G \to \mathcal{B}$ is an isomorphism.

Proof Let $b = \sum_{g\in G_0} a_g U_g \in \mathcal{B}_0$, $a_g \in \mathcal{A}$, be arbitrary and $\tilde{a}$ be as before, then from the previous proposition, given $\varepsilon > 0$, for any open $V \subset M$ and $m \in V$, there exists $b_0 \in \mathcal{B}_0 \cap \tilde{\mathcal{B}}$ and an open $\Delta \subset V$ such that

$$\|z_\Delta(b_0^* b_0)\| < \|z_\Delta(b^* b)\| + \varepsilon \quad \text{and} \quad \|z_\Delta(\tilde{a}_0)\| > \|z_\Delta(\tilde{a})\| - \varepsilon$$

for any $z_\Delta \in \mathcal{Z}(\Delta)$. Assuming condition (15) gives that for any such open V with $\beta_g|_V = id|_V$, there is such a z_Δ satisfying $\|z_\Delta b_0^* b_0\| \geq \|z_\Delta \tilde{a}_0\|$ hence

$$\|z_\Delta b^* b\| \geq \|z_\Delta \tilde{a}\| - 2\varepsilon$$

so that (B1) holds in b, hence (B2) holds in $\mathcal{B}$. Since $\mathcal{A}$ is commutative, it satisfies (C), hence from Corollary 1, (B0) holds and $id \rtimes U : \mathcal{A} \rtimes_\alpha G \to \mathcal{B}$ is an isomorphism. □

Hence, in the commutative case, we have that condition (B1), respectively, condition (B2), in $\mathcal{B} = \text{alg}(\mathcal{A}, U_G)$ is equivalent to condition (B1), respectively, condition (B2), in $\tilde{\mathcal{B}} = \text{alg}(U_G)$. Note that this condition depends on M-localization and on the unitary group U_G. We will see now that under an extra condition, the equivalent conditions (B1) and (B2) are guaranteed just from conditions on U_G.

Lemma 3 *Let $\tilde{A} \subset B(H)$ and $\tilde{B} \subset B(H)$ be commutative unital C^*-subalgebras that commute with each other, that is, for all $a \in \tilde{A}$ and $b \in \tilde{B}$ we have $ab = ba$. Let $\tilde{A} \cong C(X)$ and $\tilde{B} \cong C(Y)$ and $C = alg(\tilde{A}, \tilde{B}) \cong C(Z)$ and consider the induced maps by inclusion:*

$$\pi_X : Z \to X, \quad \pi_Y : Z \to Y.$$

If for all $0 \neq a \in \tilde{A}$ and $0 \neq b \in \tilde{B}$ we have

$$ab \neq 0, \tag{16}$$

then for every neighborhood $x \in V \subset X$ we have that $\pi_Y(\pi_X^{-1}(V)) = Y$.

Proof Suppose by contradiction that there exists $V \subset X$ open, $x \in V$, such that $\pi_Y(\pi_X^{-1}(V)) \neq Y$. Since X is normal (being compact and Hausdorff), we consider an open set $x \in U \subset V$ such that $\overline{U} \subset V$. By Uryshon's lemma, there exists $\rho \in \tilde{A} \cong C(X)$ such that:

$$\rho(x) = 1, \quad \text{and} \quad \rho|_{X-\overline{U}} = 0.$$

We have that $K = \pi_Y(\pi_X^{-1}(\overline{U})) \subset Y$ is compact, thus $Y - K$ is a non-empty open set. We consider $y \in Y - K$, and by Uryshon's lemma, a function $g \in \tilde{B} \cong C(Y)$ such that

$$g(y) = 1, \quad \text{and} \quad g|_K = 0.$$

Then we have that $\rho g = 0$: if $z \in Z$ and $\pi_X(z) \notin \overline{U}$ then $\rho(z) = 0$, on the other hand, if $\pi_X(z) \in \overline{U}$, then $\pi_Y(z) \in K$ and so $g(z) = 0$. We obtain that $\rho(z)g(z) = 0$ for every $z \in Z$, thus $\rho g = 0$. This is in contradiction with (16), since $\rho \neq 0$ and $g \neq 0$. We have then that $\pi_Y(\pi_X^{-1}(V)) = Y$ for all open sets $x \in V$ of X, which concludes the proof. □

We now have the following sufficient conditions for the isomorphism with the crossed product in the commutative case. Note that condition (17) below is in fact (B0) for $\tilde{\mathcal{B}} = \mathrm{alg}(U_G)$, so that it is equivalent to $\tilde{\mathcal{B}} \cong \mathbb{C} \rtimes_\alpha G = C^*(G)$, the group algebra. It can be regarded as a strong form of linear independence of U_g, $g \in G$.

Theorem 3 *Assume (A1) is satisfied and that $\mathcal{A}$ is commutative. Let $\tilde{\mathcal{B}} := \mathrm{alg}\{U_g : g \in G\} \subset B(H)$, such that for all finite $G_0 \subset G$, where $c_g \in \mathbb{C}$, we have*

$$\left\| \sum_{g \in G_0} c_g U_g \right\| \geq |c_e|. \tag{17}$$

Assume that for all $0 \neq a \in \mathcal{A}$ and $0 \neq b \in \tilde{\mathcal{B}}$ we have

$$ab \neq 0, \tag{18}$$

then $(B2)$ holds in $\mathcal{B} = \mathrm{alg}(\mathcal{A}, U_G)$. If the action α is amenable, then $id \rtimes U : \mathcal{A} \rtimes_\alpha G \to \mathcal{B}$ is an isomorphism.

Proof By Theorem 2, it suffices to prove that given a finite set $G_0 \subset G$ and an open set $V \subset M$ such that the action $\beta_g|_V = id|_V$ is trivial for all $g \in G_0$, and a function $z_V \in \mathcal{Z}(V) \subset \mathcal{A}$ we have

$$\left\| z_V \sum_{g \in G_0} c_g U_g \right\| \geq \|c_e z_V\| = |c_e|. \tag{19}$$

Moreover, we only need to prove (19) for positive elements. Consider then $u = \sum_{g \in G_0} c_g U_g \in \tilde{\mathcal{B}}$ a positive element and $z_V \in \mathcal{Z}(V)$ such that $\beta_g|_V = id|_V$ for all $g \in G_0$.

From Remark 5, u and z_V commute. We now consider $\overline{A} = \mathrm{alg}\{z_V, Id\}$ and $\overline{B} = \mathrm{alg}\{u, Id\}$. We have that $\overline{A}$ is a commutative C^* algebra, since it is a sub-algebra of a commutative C^* algebra and $\overline{B}$ is a commutative C^* algebra since it is generated by a positive element u. Thus we have that $\overline{A} \cong C(X)$ and $\overline{B} \cong C(Y)$. Since z_V commutes with u we have that $C := \mathrm{alg}\{\overline{A}, \overline{B}\} \cong C(Z)$ is commutative, and by condition (18), since $\overline{A} \subset \mathcal{A}$ and $\overline{B} \subset \tilde{\mathcal{B}}$ for all $0 \neq a \in \overline{A}$ and $b \in \overline{B}$ we have $ab \neq 0$. Thus we can apply Lemma 3. We consider $x \in V$ such that

$$z_V(x) = 1.$$

We also consider $y \in Y$ such that

$$u(y) = \|u\| \geq |c_e|$$

by Eq. (17). For $\varepsilon > 0$, consider the open set $V = \{x' \in X : z_V(x') > 1 - \varepsilon\}$. Thus by Lemma 3, there exists $w \in Z$ such that $\pi_X(w) \in V$ and $\pi_Y(w) = y$. Then, with $x' = \pi_X(w)$,

$$(z_V u)(w) = z_V(x')u(y) \geq (1 - \varepsilon)|c_e|.$$

But then, since w identifies with a pure state in C, we conclude that

$$\|z_V u\| \geq (z_V u)(w) \geq (1 - \varepsilon)|c_e| \quad \text{for all} \quad \varepsilon > 0,$$

which yields what we wanted to prove. □

Put in a general framework where $\mathcal{A}$ is an arbitrary unital C^*-algebra, not necessarily commutative, the results in this section can be regarded as density results in the following way.

Let $\mathcal{A}'$ be a C^*-subalgebra of $\mathcal{A}$, containing $1_{\mathcal{A}}$, where α also acts by $\mathrm{Aut}(\mathcal{A}')$ and let $\mathcal{B}' := \mathrm{alg}(\mathcal{A}', U_G) \subset \mathcal{B}$, with $\mathcal{B}_0'$ the set of elements of the form $\sum_{g \in G_0} c_g U_g$, G_0 finite, $c_g \in \mathcal{A}'$. The conditions in Proposition 5 can be easily formulated in this setting:

For $b = \sum_{g \in G_0} a_g U_g \in \mathcal{B}_0$ and $\tilde{a} = \sum_{g \in G_0} \alpha_g^{-1}(a_g^* a_g) \in \mathcal{A}$, given $\varepsilon > 0$, and a non-empty open set $V \subset M$, $m \in V$, there exists $b_0 = \sum_{g \in G_0} c_g U_g \in \mathcal{B}_0'$ and a non-empty open $\Delta \subset V$, such that

$$\|z_\Delta(b^* b - b_0^* b_0)\| < \varepsilon \quad \text{and} \quad \|z_\Delta(\tilde{a} - \tilde{a}_0)\| < \varepsilon \tag{20}$$

for any $z_\Delta \in \mathcal{Z}(\Delta)$ with $z_\Delta(m) = 1$, where $\tilde{a}_0 = \sum_{g \in G_0} \alpha_g^{-1}(c_g^* c_g) \in \mathcal{A}'$.

If condition (20) is satisfied for any $b \in \mathcal{B}_0$, we say that $\mathcal{B}'$ and $\mathcal{A}'$ are M-*locally dense* in $\mathcal{B}$ and $\mathcal{A}$, respectively. We have shown that when $\mathcal{A}$ is commutative, then $\mathcal{A}' = \mathbb{C}$ and $\mathcal{B}' = \mathrm{alg}(U_G)$ are locally dense in $\mathcal{A}$ and $\mathcal{B}$, even though they are not dense with the strong operator topology.

Of course, in this situation, the proof of Theorem 2 stands exactly in the same way, assuming now that (B1) holds in $\mathcal{B}'$. (Note that if $\mathcal{A}'$ contains $\mathcal{Z}$, then this comes down to $\mathcal{B}' \cong \mathcal{A}' \rtimes_\alpha G$.)

Proposition 6 *Assume (A1) is satisfied. Let $\mathcal{A}'$ be a C^*-subalgebra of $\mathcal{A}$ containing the identity, where α also acts by* $\mathrm{Aut}(\mathcal{A}')$, *and let $\mathcal{B}' := \mathrm{alg}(\mathcal{A}', U_G) \subset \mathcal{B}$, such that $\mathcal{B}'$ and $\mathcal{A}'$ are M-locally dense subalgebras of $\mathcal{B}$ and $\mathcal{A}$, respectively. If (B2) holds in $\mathcal{B}'$, then (B2) holds in $\mathcal{B}$. If (C) holds and the action α is amenable, then $id \rtimes U : \mathcal{A} \rtimes_\alpha G \to \mathcal{B}$ is an isomorphism.*

As for Theorem 3, we see from the proof that, even in the non-commutative case, if (B0) holds in $\mathcal{B}'$, so in particular, assuming the action is amenable, $\mathcal{B}' \cong \mathcal{A}' \rtimes_\alpha G$, and if $ab \neq 0$ for $a \in \mathcal{A}$, $b \in \mathcal{B}'$, then (B1) also holds in $\mathcal{B}'$. We then obtain:

Theorem 4 *Assume (A1) is satisfied and that the action α is amenable. Let $\mathcal{A}'$ be a C^*-subalgebra of $\mathcal{A}$ containing the identity, where α also acts ameanably by* $\mathrm{Aut}(\mathcal{A}')$, *and let $\mathcal{B}' := \mathrm{alg}(\mathcal{A}', U_G) \subset \mathcal{B}$. Assume that $\mathcal{B}'$ and $\mathcal{A}'$ are M-locally dense subalgebras of $\mathcal{B}$ and $\mathcal{A}$, respectively, and that (B0) holds in $\mathcal{B}'$, in particular, $\mathcal{B}' \cong \mathcal{A}' \rtimes_\alpha G$. Assume that $ab \neq 0$ for $0 \neq a \in \mathcal{A}$, $0 \neq b \in \mathcal{B}'$. Then (B2) holds in $\mathcal{B}$, and if (C) holds for $\mathcal{A}$, then $id \rtimes U : \mathcal{A} \rtimes_\alpha G \to \mathcal{B}$ is an isomorphism.*

Proof Similarly to the proof of Theorem 3, from Corollary 4 and the previous proposition, it suffices to check condition (B2) on $\mathcal{B}'$. Given a finite set $G_0 \subset G$ and an open set $V \subset M$ such that the action $\beta_g|_V = id|_V$ is trivial for all $g \in G_0$, and a function $z_V \in \mathcal{Z}(V) \subset \mathcal{A}$, we want to show that, for any positive element $u = \sum_{g \in G_0} c_g U_g \in \mathcal{B}'_0$, $c_g \in \mathcal{A}'$,

$$\left\| z_V \sum_{g \in G_0} c_g U_g \right\| \geq \| z_V c_e \|.$$

Again from Remark 5, u and z_V commute, so that we can show, using Lemma 3 and that, by assumption, $\|u\| \geq \|c_e\|$, that

$$\|z_V u\| \geq (1-\varepsilon)\|c_e\|, \quad \text{for all } \varepsilon > 0,$$

which shows that $\|z_V u\| \geq \|c_e\| \geq \|z_V c_e\|$. In particular (B2) holds in $\mathcal{B}'$ and the result follows. □

5 Back to the Local Trajectories Method

We consider here, as in Sect. 3, the local trajectories representations and give alternative conditions to establish the local trajectory method, Theorem 1, in the algebra $\mathcal{B} = \mathrm{alg}(\mathcal{A}, U_G)$. We assume, as always, that condition (A1) is satisfied, that is, that (id, U) is a covariant representation of $(\mathcal{A}, G, \alpha)$ preserving a central subalgebra $\mathcal{Z} = C(M)$. We will replace condition (A2) that the group G is amenable by the more general notion of amenability of the action α.

Let Ω be the orbit space of the induced action β on M. Recall that for $\omega \in \Omega$, π_ω is a representation on $B(\ell^2(G, H_\omega))$ such that for $a \in \mathcal{A}$, $\xi \in \ell^2(G, H_\omega)$, $g, s \in G$,

$$[\pi_\omega(a)\xi](s) = \pi'_\omega(\alpha_s^{-1}(a))(\xi(s)), \quad [\pi_\omega(U_g)\xi](s) = \lambda_g(\xi)(s) = \xi\left(g^{-1}s\right),$$

where $\pi'_\omega : \mathcal{A} \to B(H_\omega)$ is $\pi'_\omega = \phi_\omega \circ \rho_\omega$ with $\rho_\omega = \rho_{m_\omega} : \mathcal{A} \to \mathcal{A}/J_{m_\omega}$ the quotient map, and ϕ_ω an isometry.

Then π_ω is defined in $\mathcal{A} \rtimes_\alpha G$ as the regular representation induced by π'_ω. As for $\mathcal{B} = \text{alg}(\mathcal{A}, U_G)$, we assume that elements $b = \sum_{g \in G_0} a_g U_g$ uniquely determine the coefficients a_g, $g \in G_0$, so that π_ω is also well-defined in $\mathcal{B}_0$. Typically, we are interested in the case when $\mathcal{B} \cong \mathcal{A} \rtimes_\alpha G$ so this condition is guaranteed.

Throughout this section, we let $\pi = \oplus_{\omega \in \Omega} \pi_\omega$. As a matter of terminology, we say that the local trajectories method works on $\mathcal{B}$ if π is well-defined and faithful in $\mathcal{B}$. Recall from Proposition 1 that π is always faithful in $\mathcal{A}$.

We have noted in Remark 1 that for the local trajectories method to work on $\mathcal{B}$ it suffices to give conditions such that the maps

$$id \rtimes_\alpha U : \mathcal{A} \rtimes_\alpha G \to \mathcal{B}, \quad id \rtimes_{\alpha'} \pi(U) : \pi(\mathcal{A}) \rtimes_{\alpha'} G \to \pi(\mathcal{B}) \tag{21}$$

are isomorphisms, where $\alpha'_g(\pi(a)) := \pi(U_g)\pi(a)\pi(U_g^*) = \pi(U_g a U_g^*)$. Note that $\pi(\mathcal{B}) \cong \text{alg}(\pi(\mathcal{A}), \pi(U_G))$.

The goal of this section is to apply the criteria obtained in Sect. 4 to study in particular the second map in (21), where we have regular representations on spaces of the form $B(\ell^2(G, H))$.

We consider first the commutative case where we can use the results in Sect. 4.3. Let $\tilde{\mathcal{B}}_0 = \tilde{\mathcal{B}} \cap \mathcal{B}_0$ and $\tilde{\mathcal{B}} = \text{alg}(U_G)$ and consider operators of the form

$$b_0 = I + \sum_{e \neq s \in G_0} c_s U_s \in \tilde{\mathcal{B}}_0, \quad \text{with } \pi_\omega(b_0) = I + \sum_{e \neq s \in G_0} c_s \lambda_s, c_s \in \mathbb{C}.$$

Then for $\xi \in \ell^2(G, H_\omega)$ and $g \in G$, we have

$$[\pi_\omega(b_0)(\xi)](g) = \xi(g) + \sum_{g \neq s \in G_0} c_s \xi(g^{-1}s).$$

We have that the localized condition (B1) always holds in $\pi(\tilde{\mathcal{B}}_0)$. For that we make use of the following lemma.

Lemma 4 *Let $\rho \in \mathcal{Z} \cong C(M)$ such that $0 \leq \rho \leq 1$ and $\|\rho\| = 1$. Then there exist $\omega \in \Omega$ and $g \in G$ such that*

$$\alpha_g^{-1}(\rho) = 1_{\mathcal{A}} + J_{m_\omega}. \tag{22}$$

In particular, $\pi'_\omega(\alpha_g^{-1}(\rho)) = I \in B(H_\omega)$ and $[\pi_\omega(\rho)\xi](g) = \xi(g)$, for any $\xi \in \ell^2(G, H_\omega)$. Conversely, given $\omega \in \Omega$ and $g \in G$, there is $\rho \in \mathcal{Z} \cong C(M)$ such that $0 \leq \rho \leq 1$, and $\|\rho\| = 1$ satisfying (22).

Proof Let $m \in M$ be such that $\rho(m) = 1 = \max_{x \in M} \rho(x)$. Take the orbit $\omega \in \Omega$ such that $m \in \omega$, so $m = \beta_g^{-1}(m_\omega)$ for $m_\omega \in \omega$ and $g \in G$. Note that $\alpha_g^{-1}(\rho) = \rho \circ$

β_g^{-1} will have its maximum in m_ω since $\beta_g^{-1}(m_\omega) = m$ and ρ attains its maximum at m. This gives that $\alpha_g^{-1}(\rho)(m_\omega) = 1$, hence $\alpha_g^{-1}(\rho) - 1_{\mathcal{A}} \in J_{m_\omega}$ and

$$\pi'_\omega\left(\alpha_g^{-1}(\rho)\right) = \pi'_{m_\omega}\left(\alpha_g^{-1}(\rho)\right) = \pi'_\omega(1) = I \in B(H_\omega),$$

from which follows $\pi_\omega(\rho)\eta(g) = \pi'_\omega\left(\alpha_g^{-1}(\rho)\right)\eta(g) = \eta(g)$, for any $\eta \in \ell^2(G, H_\omega)$. For the converse, let $\omega \in \Omega$ and $g \in G$ and pick any $\rho \in \mathcal{Z}$ such that $0 \leq \rho \leq 1$ and $\rho(m) = 1$, with $m = \beta_g^{-1}(m_\omega)$. □

Proposition 7 *Assume condition (A1). Let $\pi = \oplus_{\omega\in\Omega}\pi_\omega$ be the local trajectories representation, assumed well defined in $\mathcal{B}_0$. Then for any $V \subset M$ open and $c_s \in \mathbb{C}$, $s \in G_0 \subset G$ finite, and for all $z_V \in \mathcal{Z}(V)$ we have*

$$\left\| z_V \pi\left(I + \sum_{e\neq s\in G} c_s U_s\right)\right\| \geq 1.$$

In particular, (B1) holds in $\pi(\tilde{\mathcal{B}}_0) \subset \pi(\mathcal{B})$.

Proof Let $z_V \in \mathcal{Z}(V)$, then since π is an isomorphism on $\mathcal{Z}$, write $z_V = \pi(\rho_V)$, for $\rho_V \in \mathcal{Z}$. Then $\|\rho_V\| = \|\pi(\rho_V)\| = 1$ and since π is a $*$-homomorphism and $z_V \geq 0$, also $\rho_V \geq 0$.

Let $g \in G$ and $\omega \in \Omega$ be as in the previous lemma applied to ρ_V. Then we have

$$(\pi_\omega(\rho_V)\xi)(g) = \pi'_\omega\left(\alpha_g^{-1}(\rho_V)\right)(\xi(g)) = \xi(g),$$

for any $\xi \in \ell^2(G, H_\omega)$. Now take $\xi_g \in \ell^2(G, H_\omega)$ such that $\xi_g(s) = 0$ for $s \neq g$ and $\xi_g(g) = u$, with $u \in H_\omega$, $\|u\| = 1$, arbitrary. Let $b_0 = I + \sum_{e\neq s\in G_0} c_s U_s \in \tilde{\mathcal{B}}_0$. It follows that

$$\left(\pi_\omega(\rho_V b_0)\xi_g\right)(g) = \pi_\omega(\rho_V)\left(\xi_g(g) + \sum_{e\neq s\in G} c_s\xi_g\left(s^{-1}g\right)\right) = \xi_g(g) = u.$$

We have then

$$\begin{aligned}\|\pi_\omega(\rho_V b_0)\| &= \sup_{\|\xi\|=1} \|\pi_\omega(\rho_V b_0)\xi\|_{\ell^2} \geq \|\pi_\omega(\rho_V b_0)\xi_g\|_{\ell^2}\\ &\geq \|(\pi_\omega(\rho_V b_0)\xi_g)(g)\| = \|u\| = 1.\end{aligned}$$

Since we wrote $z_V = \pi(\rho_V)$, we conclude that

$$\left\| z_V \pi \left(I + \sum_{e \neq s \in G} c_s U_s \right) \right\| = \sup_{\omega \in \Omega} \left\| \pi_\omega(\rho_V) \pi_\omega \left(I + \sum_{e \neq s \in G} c_s U_s \right) \right\| \geq 1.$$

In particular, (B1) holds in $\pi(\tilde{\mathcal{B}}_0)$. □

Assume now that $\mathcal{A}$ is commutative; then both $\mathcal{A}$ and $\pi(\mathcal{A})$ satisfy the M-localization condition (C) as in (9). Moreover, since $\tilde{\mathcal{B}}_0 = \mathcal{B}_0 \cap \tilde{\mathcal{B}}$ is M-locally dense in $\mathcal{B}$, by Proposition 5, also $\pi(\tilde{\mathcal{B}}_0) = \pi(\mathcal{B}_0) \cap \mathrm{alg}(\pi(U_G))$ is M-locally dense in $\pi(\mathcal{B})$, so we have from Theorem 2 that if (B2), or equivalently, (B1), holds in $\pi(\tilde{\mathcal{B}}_0)$ then it also holds in $\pi(\mathcal{B})$. In this case, using localization, it follows from Corollary 4 that:

Proposition 8 *Assume condition (A1) and that the action α is amenable. Let $\mathcal{A}$ be commutative with the local trajectories representation $\pi = \oplus_{\omega \in \Omega} \pi_\omega$ well defined in $\mathcal{B}_0$. Then (B1) holds in $\pi(\mathcal{B})$ and*

$$id \rtimes \pi(U) : \pi(\mathcal{A}) \rtimes_{\alpha'} G \to \pi(\mathcal{B}) = \mathrm{alg}(\pi(\mathcal{A}), \lambda_G)$$

is an isomorphism.

Following Remark 1, we conclude that, in the commutative case, for the local trajectories method to work, it suffices that $id \rtimes_\alpha U : \mathcal{A} \rtimes_\alpha G \to \mathcal{B}$ is an isomorphism. From Theorem 2 we obtain then the following sufficient condition.

Theorem 5 *Let $\mathcal{A}$ be commutative. Assume condition (A1) and that the action α is amenable. If (B2) holds in the subalgebra $\tilde{\mathcal{B}}_0$, that is, if for any finite $G_0 \subset G$ and $c_g \in \mathbb{C}$, $g \in G_0$ and any non-empty open set V satisfying $\beta_g|_V = id|_V$ for all $g \in G_0$, there exists $z_V \in \mathcal{Z}(V)$ such that*

$$\left\| z_V \left(I + \sum_{g \in G_0} c_g U_g \right) \right\| \geq \|z_V\| = 1,$$

then $\{\pi_\omega\}_{\omega \in \Omega}$ is a faithful family of representations of $\mathcal{B}$, that is, the local trajectories method works on $\mathcal{B}$.

We now show that in fact Proposition 8 still holds even if $\mathcal{A}$ is not commutative, relying essentially on properties of regular representations.

For elements $b = \sum_{s \in G_0} a_s U_s \in \mathcal{B}_0$, we have

$$(\pi_\omega(b)\xi)(g) = \sum_{s \in G_0} \left(\pi'_\omega(\alpha_g^{-1}(a_s)) \right) (\xi(s^{-1}g))$$

and with $\xi \in \ell^2(G, H_\omega)$ such that $\xi(t) = 0$ for $t \neq g$ and $\xi(g) = u$, $u \in H$, $\|u\| = 1$,

$$(\pi_\omega(b)\xi)(g) = (\pi_\omega(a_e)\xi)(g) = \left(\pi'_\omega(\alpha_g^{-1}(a_e)\right)(\xi(g))$$

Introducing operators $j_g : H \to \ell^2(G, H)$ and $j_g^* : \ell^2(G, H) \to H$, $g \in G$, such that, for $h \in H$, $\xi \in \ell^2(G, H)$,

$$j_g(h)(g) = h, \quad j_g(h)(t) = 0, t \neq g, \qquad \text{and} \qquad j_g^*(\xi) = \xi(g), \tag{23}$$

we have that j_g is an isometry and $\|j_g\| = \|j_g^*\| = 1$, with $j_g^* j_g = I$. We can then write the equality above as

$$j_g^* \pi_\omega(b) j_g = \pi'_\omega(\alpha_g^{-1}(a_e)) \text{ in } B(H_\omega). \tag{24}$$

We make use of the following lemma (that holds in general for regular representations, similarly to the result just after).

Lemma 5 *For each $\omega \in \Omega$, $a \in \mathcal{A}$,*

$$\|\pi_\omega(a)\| = \sup_{g \in G} \|\pi'_\omega(\alpha_g^{-1}(a))\|_{B(H_\omega)}.$$

Proof For each $g \in G$, we can write, as in (24), $j_g^* \pi_\omega(a) j_g = \pi'_\omega(\alpha_g^{-1}(a))$. Hence, since $\|j_g\| = \|j_g^*\| = 1$,

$$\|\pi'_\omega(\alpha_g^{-1}(a))\| = \|j_g^* \pi_\omega(a) j_g\| \leq \|\pi_\omega(a)\|.$$

It follows that

$$\sup_{g \in G} \|\pi'_\omega(\alpha_g^{-1}(a))\|_{B(H_\omega)} \leq \|\pi_\omega(a)\|.$$

For the reverse, let $\xi \in \ell^2(G, H_\omega)$ with $\|\xi\|^2 = \sum_{g \in G} \|\xi(g)\|^2_{H_w}\| = 1$. Then we have that:

$$\begin{aligned}
\|\pi_w(a)\xi\|^2_{\ell^2(G, H_w)} &= \sum_{g \in G} \|\pi'_\omega(\alpha_g^{-1}(a))\xi(g)\|^2_{H_\omega} \\
&\leq \sum_{g \in G} \|\pi'_\omega(\alpha_g^{-1}(a))\|_{B(H_\omega)} \|\xi(g)\|^2_{H_w} \\
&\leq \sup_{g \in G} \|\pi'_\omega(\alpha_g^{-1}(a))\|_{B(H_\omega)} \sum_{g \in G} \|\xi(g)\|^2_{H_\omega} \\
&\leq \sup_{g \in G} \|\pi'_\omega(\alpha_g^{-1}(a))\|_{B(H_\omega)}.
\end{aligned}$$

Taking the supremum over $\xi \in \ell^2(G, H_\omega)$ such that $\|\xi\|_{\ell^2(G,H_\omega)} = 1$ we obtain:

$$\|\pi_w(a)\|_{\ell^2(G,H_\omega)} \le \sup_{g\in G} \|\pi'_\omega(\alpha_g^{-1}(a))\|_{B(H_\omega)},$$

which concludes the proof. □

A first consequence is the following version of (B0) in $\pi_\omega(\mathcal{B})$.

Proposition 9 *Assume condition (A1) and that the local trajectories representation* $\pi = \bigoplus_{\omega\in\Omega}\pi_\omega$ *is well defined in* $\mathcal{B}_0$. *Then, for any finite set* $G_0 \subset G$ *and* $b = \sum_{s\in G_0} a_s U_s$, *and for any orbit* $\omega \in \Omega$, *we have*

$$\|\pi_\omega(b)\| \ge \|\pi_\omega(a_e)\|.$$

Proof With the notation as in (23), we have for each $\omega \in \Omega$ and all $g \in G$, $j_g^*\pi_\omega(b)j_g = \pi'_\omega(\alpha_g^{-1}(a_e))$, hence

$$\|\pi'_\omega(\alpha_g^{-1}(a_e))\| = \|j_g^*\pi_\omega(b)j_g\| \le \|\pi_\omega(b)\| \quad \text{for all} \quad g \in G.$$

Since, from Lemma 5, $\|\pi_\omega(a_e)\| = \sup_{g\in G}\|\pi'_\omega(\alpha_g^{-1}(a_e))\|$, the result follows. □

We now show that the M-local condition (B1) holds in $\pi(\mathcal{B})$ so the second isomorphism in (21) always holds, in case our algebra $\mathcal{A}$ is M-localizable, that is, satisfies (C).

Proposition 10 *Assume condition (A1) and that the action* α *is amenable. Let* $\pi = \bigoplus_{\omega\in\Omega}\pi_\omega$ *be the local trajectories representation, assumed well defined in* $\mathcal{B}_0$. *Then for any* $V \subset M$ *open and* $a_s \in \mathcal{A}$, $s \in G_0 \subset G$ *finite, and for all* $z_V \in \mathcal{Z}(V)$ *we have*

$$\left\| z_V\pi\left(a_e + \sum_{e\neq s\in G_0} a_s U_s\right)\right\| \ge \|z_V\pi\,(a_e)\|\,.$$

In particular, (B1) holds in $\pi(\mathcal{B})$, *and if (C) holds for* $\mathcal{A}$, *then*

$$id \rtimes \pi(U) : \pi(\mathcal{A}) \rtimes_{\alpha'} G \to \pi(\mathcal{B})$$

is an isomorphism.

Proof Similarly to the proof of Proposition 7, writing $z_V = \pi(\rho_V)$, it suffices to prove that for any open $V \subset M$, $\rho_V \in \mathcal{Z}(V)$ and $b = \sum_{s\in G_0} a_s U_s \in \mathcal{B}_0$, we have $\|\pi(\rho_V b)\| \ge \|\pi(\rho_V a_e)\|$. This is straightforward from the previous proposition, since we have $\|\pi_\omega(\rho_V b)\| \ge \|\pi_\omega(\rho_V a_e)\|$ for all $\omega \in \Omega$. In particular, (B1) holds in $\pi(\mathcal{B})$, and the isomorphism follows from $\pi(\mathcal{A})$ also satisfying (C) and Corollary 1. □

It then follows again from Remark 1 that, if $\mathcal{A}$ is M-localizable, then the local trajectories method works on $\mathcal{B}$, as long as $id \rtimes_\alpha U : \mathcal{A} \rtimes_\alpha G \to \mathcal{B}$ is an isomorphism. We then obtain the following M-localized version of the local trajectories method, Theorem 1. (Note that if the action is topologically free, then there are no open sets of fixed points, so the condition is trivially satisfied.)

Theorem 6 *Assume condition (A1) and that the action α is amenable. If (C) holds for $\mathcal{A}$ and the M-local condition (B2) holds in $\mathcal{B}$, that is, if for every finite set $G_0 \subset G$ and non-empty open set $V \subset M$ such that $\beta_g|_V = id|_V$ for all $g \in D$, and for $a_g \in \mathcal{A}$, there exists $z_V \in \mathcal{Z}(V)$ such that*

$$\left\| z_V \sum_{g \in G_0} a_g U_g \right\| \geq \|z_V a_e\|,$$

then $\{\pi_\omega\}_{\omega \in \Omega}$ is a faithful family of representations of $\mathcal{B}$, that is, the local trajectories method works on $\mathcal{B}$.

Along the lines of the last results of the previous section, Theorems 3 and 4, we also conclude:

Theorem 7 *Let $\mathcal{A}$ be commutative. Assume condition (A1) and that the action α is amenable. Let $\tilde{\mathcal{B}} := \mathrm{alg}\{U_g : g \in G\} \subset B(H)$, and assume that for all finite $G_0 \subset G$, where $c_g \in \mathbb{C}$, we have*

$$\left\| \sum_{g \in G_0} c_g U_g \right\| \geq |c_e|,$$

that is, $\tilde{\mathcal{B}} \cong \mathbb{C} \rtimes_\alpha G = C^*(G)$, the group algebra. Assume that for all $0 \neq a \in \mathcal{A}$ and $0 \neq b \in \tilde{\mathcal{B}}$ we have

$$ab \neq 0.$$

Then $\mathcal{A} \rtimes_\alpha G \cong \mathcal{B}$ and $\{\pi_\omega\}_{\omega \in \Omega}$ is a faithful family of representations of $\mathcal{B}$, that is, the local trajectories method works on $\mathcal{B}$.

In general, if we have an isomorphism with the crossed product on M-locally dense subalgebras, as in (20), then:

Theorem 8 *Assume (A1) is satisfied and that the action α is amenable. Let $\mathcal{A}'$ be a C^*-subalgebra of $\mathcal{A}$ containing the identity, where α also acts ameanably by $\mathrm{Aut}(\mathcal{A}')$, and let $\mathcal{B}' := \mathrm{alg}(\mathcal{A}', U_G) \subset \mathcal{B}$. Assume that $\mathcal{B}'$ and $\mathcal{A}'$ are M-locally dense subalgebras of $\mathcal{B}$ and $\mathcal{A}$, respectively, and that $(B0)$ holds in $\mathcal{B}'$, in particular, $\mathcal{B}' \cong \mathcal{A}' \rtimes_\alpha G$.*

Assume that $ab \neq 0$ for $0 \neq a \in \mathcal{A}$, $0 \neq b \in \mathcal{B}'$. Then if (C) holds for $\mathcal{A}$, we also have $\mathcal{B} \cong \mathcal{A} \rtimes_\alpha G$ and $\{\pi_\omega\}_{\omega\in\Omega}$ is a faithful family of representations of $\mathcal{B}$, that is, the local trajectories method works on $\mathcal{B}$.

Acknowledgments The authors would like to thank an anonymous referee for carefully reading the manuscript and contributing with various comments and suggestions that helped to improve the paper.

Funding The three authors were supported by FCT–Fundação para a Ciência e Tecnologia, Portugal, under project UIDB/04721/2020. Center of Mathematical Analysis, Geometry and Dynamical Systems, under the FCT-Portugal project UIDB/04459/2020, with DOI identifier 10-54499/UIDP/04459/2020. (C.Carvalho). Research Foundation-Flanders (FWO) via the Odysseus II programme no. G0DBZ23N. (M.Dias).

References

1. Anantharaman-Delaroche, C.: Amenability and exactness for dynamical systems and their C^*-algebras. Trans. Am. Math. Soc. **354**(10), 4153–4178 (2002)
2. Anderson, J.: Extensions, restrictions, and representations of states on C^*-algebras. Trans. Am. Math. Soc. **249**(2), 303–329 (1979)
3. Antonevich, A.B.: On two methods of studying the invertibility of operators in C^*-algebras induced by dynamical systems. Math. USSR-Sb. **52**(1), 1–20 (1985)
4. Antonevich, A.B.: Linear Functional Equations. Operator Approach. Operator Theory: Advances and Applications, vol. 83. Birkhäuser Verlag, Basel (1996)
5. Antonevich, A.B., Lebedev, A.V.: Functional-Differential Equations. I. C^*-Theory. Pitman Monographs and Surveys in Pure and Applied Mathematics, vol. 70. Longman Scientific & Technical, Harlow (1994)
6. Archbold, R.J.: Extensions of states of C^*-algebras. J. London Math. Soc. (2) **21**(2), 351–354 (1980)
7. Archbold, R.J., Spielberg, J.S.: Topologically free actions and ideals in discrete C^*-dynamical systems. Proc. Edinb. Math. Soc. (2) **37**(1), 119–124 (1994)
8. Bastos, M.A., Fernandes, C.A., Karlovich, Y.I.: Spectral measures in C^*-algebras of singular integral operators with shifts. J. Funct. Anal. **242**(1), 86–126 (2007)
9. Bastos, M.A., Fernandes, C.A., Karlovich, Y.I.: A C^*-algebra of singular integral operators with shifts and piecewise quasicontinuous coefficients. In: Operator Theory, Operator Algebras, and Matrix Theory. Operator Theory: Advances and Applications, vol. 267, pp. 25–64. Birkhäuser/Springer, Cham (2018)
10. Bastos, M.A., Fernandes, C.A., Karlovich, Y.I.: Invertibility criteria in C^*-algebras of functional operators with shifts and PQC coefficients. Integral Equ. Oper. Theory **91**(3), Paper No. 19, 25 (2019)
11. Böttcher, A., Karlovich, Y.I., Spitkovsky, I.M.: The C^*-algebra of singular integral operators with semi-almost periodic coefficients. J. Funct. Anal. **204**(2), 445–484 (2003)
12. Brown, N.P., Ozawa, N.: C^*-Algebras and Finite-Dimensional Approximations. Graduate Studies in Mathematics, vol. 88. American Mathematical Society, Providence (2008)
13. Chu, C.-H., Kusuda, M.: On factor states of C^*-algebras and their extensions. Proc. Am. Math. Soc. **124**(1), 207–215 (1996)

14. Dixmier, J.: C^*-Algebras. North-Holland Mathematical Library, vol. 15. North-Holland Publishing Co., Amsterdam/New York/Oxford (1977). Translated from the French by Francis Jellett
15. Kadison, R.V., Ringrose, J.R.: Fundamentals of the Theory of Operator Algebras, vol. I. Graduate Studies in Mathematics, vol. 15. American Mathematical Society, Providence (1997). Elementary theory, Reprint of the 1983 original
16. Karlovich, Y.I.: The local trajectory method of studying invertibility in C^*-algebras of operators with discrete groups of shifts. Soviet Math. Dokl. **37**(2), 407–412 (1988)
17. Karlovich, Y.I.: C^*-algebras of operators of convolution type with discrete groups of shifts and oscillating coefficients. Soviet Math. Dokl. **38**(2), 301–307 (1989)
18. Karlovich, Y.I.: A Local-Trajectory Method and Isomorphism Theorems for Nonlocal C^*-algebras. Operator Theory: Advances and Applications, vol. 170, pp. 137–166. Birkhäuser, Basel (2007)
19. Palma, R.: Produtos cruzados C^*. invertibilidade numa álgebra de operadores funcionais. Master's Thesis, Instituto Superior Técnico, University of Lisbon, 2008
20. Pedersen, G.K.: C^*-Algebras and Their Automorphism Groups. Pure and Applied Mathematics (Amsterdam). Academic Press, London (2018). Second edition of [MR0548006], Edited and with a preface by Søren Eilers and Dorte Olesen
21. Shvydkoy, R., Latushkin, Y.: Operator algebras and the Fredholm spectrum of advective equations of linear hydrodynamics. J. Funct. Anal. **257**(10), 3309–3328 (2009)
22. Williams, D.P.: Crossed Products of C^*-Algebras. Mathematical Surveys and Monographs, vol. 134. American Mathematical Society, Providence (2007)

Factorisation of Symmetric Matrices and Applications in Gravitational Theories

M. Cristina Câmara and Gabriel Lopes Cardoso

To Yuri Karlovich on his 75th birthday

Abstract We consider the canonical Wiener-Hopf factorisation of 2×2 symmetric matrices $\mathcal{M}$ with respect to a contour Γ. For the case where the quotient q of the two diagonal elements of $\mathcal{M}$ is a rational function, we show that due to the symmetric nature of the matrix $\mathcal{M}$, the second column in each of the two matrix factors that arise in the factorisation is determined in terms of the first column in each of these matrix factors, by multiplication by a rational matrix, and we give a method for determining the second columns of these factors. We illustrate our method with two examples in the context of a Riemann-Hilbert approach to obtaining solutions to the Einstein field equations.

1 Introduction

Let Γ be a simple closed contour in the complex plane encircling the origin and denote by $\mathbb{D}_\Gamma^+$ and $\mathbb{D}_\Gamma^-$ the interior and the exterior of Γ, respectively.

If $\mathcal{M}$ is an $n \times n$ matrix function whose elements are in $L^\infty(\Gamma)$, i.e., essentially bounded functions on Γ, a bounded Wiener-Hopf (WH for short) factorisation of $\mathcal{M}$ with respect to (w.r.t.) Γ is a representation

$$\mathcal{M}(\tau) = \mathcal{M}_-(\tau)\, D(\tau)\, \mathcal{M}_+(\tau), \qquad \tau \in \Gamma, \tag{1}$$

where $\mathcal{M}_\pm$ and their inverses $\mathcal{M}_\pm^{-1}$ are analytic and bounded in $\mathbb{D}_\Gamma^\pm$ (we say that their elements are in $H_\pm^\infty$, respectively), and $D(\tau) = \mathrm{diag}(\tau^{k_1}, \tau^{k_2}, \ldots, \tau^{k_n})$ with

M. C. Câmara (✉) · G. L. Cardoso
Center for Mathematical Analysis, Geometry and Dynamical Systems, Department of Mathematics, Instituto Superior Técnico, Universidade de Lisboa, Lisboa, Portugal
e-mail: cristina.camara@tecnico.ulisboa.pt; gabriel.lopes.cardoso@tecnico.ulisboa.pt

A. Böttcher et al. (eds.), *Achievements and Challenges in the Field of Convolution Operators*, Operator Theory: Advances and Applications 306,
https://doi.org/10.1007/978-3-031-80486-1_6

$k_1 \leq k_2 \leq \cdots \leq k_n$, $k_i \in \mathbb{Z}$ for $i = 1, 2, \ldots, n$. If $k_i = 0$ for all $i = 1, 2, \ldots, n$, then we say that

$$\mathcal{M}(\tau) = \mathcal{M}_-(\tau)\,\mathcal{M}_+(\tau), \quad \tau \in \Gamma,$$

is a canonical bounded WH factorisation. In what follows, we will omit the term 'bounded'.

Denoting by C^μ the algebra of all Hölder continuous functions with exponent $\mu \in]0, 1[$ defined on Γ [14], if $\mathcal{M}$ is invertible in $(C^\mu)^{n\times n}$, i.e. $\mathcal{M} \in (C^\mu)^{n\times n}$ and $\det \mathcal{M} \neq 0$ on Γ, then $\mathcal{M}$ admits a factorisation of the form (1) with $\mathcal{M}_\pm \in (C^\mu_\pm)^{n\times n}$, where $C^\mu_\pm = C^\mu \cap H^\infty_\pm$ [14]. If the factorisation is canonical, then it is unique if we impose a normalising condition on one of the factors; here we will look for canonical WH factorisations with the factor $\mathcal{M}_+$ normalised to the identity at 0, in which case it will be denoted by X,

$$\mathcal{M} = \mathcal{M}_- X \quad \text{on} \quad \Gamma, \quad \text{with} \quad X(0) = \mathbb{I}. \tag{2}$$

We will be particularly interested in 2×2 matrix functions $\mathcal{M}$, having in view applications of WH factorisation to solving certain gravitational field equations. For these applications, $\mathcal{M}$ must be a symmetric 2×2 matrix function

$$\mathcal{M} = \begin{bmatrix} a & b \\ b & d \end{bmatrix}$$

with $a/d \in \mathcal{R}$, where $\mathcal{R}$ denotes the space of all rational functions without poles on Γ, i.e.

$$\frac{a}{d} = \frac{p_1}{p_2} =: q, \tag{3}$$

where p_1 and p_2 are polynomials and p_2 does not vanish on Γ, and $\mathcal{M}$ must admit a canonical WH factorisation.

One may then naturally ask how all these conditions, in particular that $\mathcal{M}$ is symmetric, are reflected in the form of the factors $\mathcal{M}_\pm$ of a canonical WH factorisation.

We study this question in Sect. 2 by applying and extending an approach which was first presented in [8], although in a different form, and we show that there indeed exists a certain relation between the two columns in each of the factors $\mathcal{M}_-$ and X^{-1} (related by $\mathcal{M}X^{-1} = \mathcal{M}_-$), which is determined by the structure of the original matrix $\mathcal{M}$ in terms of its symmetry and the quotient q between its diagonal elements, in such a way that the second column can be obtained from the first by multiplication by a certain rational matrix. This, on the one hand, brings out a connection between the form of the matrix $\mathcal{M}$ that is to be factorised and a certain structure of the factors $\mathcal{M}_\pm$, which we present in Theorem 1; on the other hand, it may also allow for a simpler determination of one of the columns in the factors and avoid repetition of similar calculations, as we illustrate in the examples in Sect. 3.

The question of how to determine the factors of a (not necessarily bounded) canonical WH factorisation of a 2×2 matrix function $\mathcal{M}$, $\mathcal{M} = \mathcal{M}_-\mathcal{M}_+$, from a solution to the Riemann-Hilbert problem $\mathcal{M}\phi_+ = \phi_-$, where $\phi_+ \in (H_+^\infty)^2$ and $\phi_- \in (H_-^\infty)^2$ can be taken as the first columns in $\mathcal{M}_+^{-1}$ and $\mathcal{M}_-$, respectively, has been previously studied using different approaches. For example, using the corona theorem in the context of the real line, a class of matrix functions was considered in [6] for which a solution to $\mathcal{M}\phi_+ = \phi_-$ was given and, assuming $\phi_\pm$ to be corona pairs in $\mathbb{C}^\pm$ [9, 15], explicit formulae for the factors in terms of these solutions were obtained. In [2] explicit formulae were also given, for general 2×2 matrices with determinant 1, in terms of a solution to $\mathcal{M}\phi_+ = \phi_-$, provided that $\phi_\pm$ are corona pairs, requiring however knowledge of such a solution as well as the solutions of two associated corona problems with data ϕ_+ and ϕ_-.

This also naturally leads to the question of how to determine the first column in X^{-1} and in $\mathcal{M}_-$. Regarding this question, we focus here on rational matrices possessing a canonical WH factorisation, which are of great importance when considering applications in gravitational theories. Obtaining the first columns is equivalent to determining the (unique) solution to

$$\mathcal{M}\phi_+ = \phi_- \quad \text{on} \quad \Gamma, \quad \text{with} \quad \phi_\pm = (\phi_{1\pm}, \phi_{2\pm}) \in \left(C_\pm^\mu\right)^2, \quad \phi_+(0) = (1, 0).$$

We show that it is also possible to simplify the calculations for the first column, reducing the problem of analyticity of the solution to that of the first component $\phi_{1\pm}$ (in $\mathbb{D}_\Gamma^\pm$).

We illustrate these results in Sect. 3 by applying them to solving the Einstein field equations by a Riemann-Hilbert approach based on [1, 3]. It is well known that Wiener-Hopf factorisation is very important in the study of singular integral equations, convolution equations and in many applications in Mathematics, Physics and Engineering [10, 14]. Determining explicit solutions to the Einstein field equations by means of a Riemann-Hilbert (RH) approach is one of the recent applications of WH factorisation theory.

2 Canonical Factorisation and Structure of the Factors

In what follows, for any algebra A, let $\mathcal{G} A$ denote the group of invertible elements in A.

Let

$$\mathcal{M}(\tau) = \begin{bmatrix} a(\tau) & b(\tau) \\ b(\tau) & d(\tau) \end{bmatrix}$$

with a, b, d analytic in a neighbourhood of Γ, where Γ is a simple closed contour, be such that $\mathcal{M}$ admits a canonical WH factorisation w.r.t. Γ of the form (2),

$$\mathcal{M} = \mathcal{M}_- X \quad \text{on} \quad \Gamma, \quad \text{with} \quad X(0) = \mathbb{I}.$$

Then, $\delta = \det \mathcal{M}$ also admits a canonical WH factorisation w.r.t. Γ,

$$\delta = \delta_- \, \delta_+ \quad \text{with} \quad \delta_\pm \in \mathcal{G}C^\mu_\pm, \quad \delta_+(0) = 1. \tag{4}$$

Let

$$\frac{a}{d} = q \in \mathcal{R}, \quad q = \frac{p_1}{p_2}, \tag{5}$$

where p_1 and p_2 are polynomials without common zeroes and such that q is bounded at ∞.

The symmetric structure of $\mathcal{M}$, which will be reflected in the form of the factors X^{-1} and $\mathcal{M}_-$, can be characterised by the following relation.

Proposition 1 *Let q be defined by* (5) *and let*

$$Q_1 = \text{diag}\,(1, -q), \quad Q_2 = \text{diag}\,(q, -1). \tag{6}$$

Then

$$\mathcal{M} Q_1 \mathcal{M} = \delta \, Q_2 \quad \textit{with} \quad \delta = \det \mathcal{M} = ad - b^2. \tag{7}$$

Remark 1 Note that, conversely, if $\tilde{\mathcal{M}}$ is a 2×2 matrix

$$\tilde{\mathcal{M}} = \begin{bmatrix} a & b \\ c & d \end{bmatrix} \quad \text{with} \quad a, d, ad - bc \neq 0$$

and

$$\tilde{\mathcal{M}}^T Q_1 \tilde{\mathcal{M}} = (\det \tilde{\mathcal{M}}) \, Q_2$$

with $Q_1 = \text{diag}\,(1, -q)$, $Q_2 = \text{diag}\,(q, -1)$ for some q, then we have that $\tilde{\mathcal{M}}^T = \tilde{\mathcal{M}}$ and $q = a/d$. So the relation (7) does indeed characterise the structure of $\mathcal{M}$.

The first columns of X^{-1} and $\mathcal{M}_-$, denoted f_+ and f_- respectively, are the unique solution to

$$\mathcal{M} f_+ = f_- \quad \text{with} \quad f_\pm \in \left(C^\mu_\pm\right)^2, \quad f_+(0) = \begin{bmatrix} 1 \\ 0 \end{bmatrix}; \tag{8}$$

the second columns of X^{-1} and $\mathcal{M}_-$, denoted s_+ and s_-, respectively, constitute the unique solution to

$$\mathcal{M}s_+ = s_- \quad \text{with} \quad s_\pm \in \left(C_\pm^\mu\right)^2, \quad s_+(0) = \begin{bmatrix} 0 \\ 1 \end{bmatrix}. \tag{9}$$

We therefore have that

$$\mathcal{M}\left[f_+ \;\; s_+\right] = \left[f_- \;\; s_-\right], \tag{10}$$

where

$$X^{-1} = \left[f_+ \;\; s_+\right], \quad \mathcal{M}_- = \left[f_- \;\; s_-\right], \quad X(0) = \mathbb{I}.$$

To study the relation between $f_\pm$ and $s_\pm$ in this case, we will use the following results, which can be easily verified.

Proposition 2 *If A is a 2×2 matrix, $A = [f \;\; s]$, and*

$$J = \begin{bmatrix} 0 & -1 \\ 1 & 0 \end{bmatrix},$$

then

(i) $AJA^T = (\det A)\, J$,
(ii) $\det A = s^T J f$,
(iii) $A^{-1} = -(\det A)^{-1} J A^T J$ *if* $\det A \neq 0$.

Also note that, for any vector f with two components,

$$f^T J f = 0. \tag{11}$$

Proposition 3 *Let $\mathcal{M}f_+ = f_-$ with $\mathcal{M}$ satisfying (7). Then*

$$\mathcal{M}(JQ_2 f_+) = JQ_1 f_-. \tag{12}$$

Proof We have that $Q_2 = \delta^{-1}\mathcal{M}Q_1\mathcal{M}$, hence

$$\mathcal{M}(JQ_2 f_+) = \delta^{-1}\mathcal{M}J\,(\mathcal{M}Q_1\mathcal{M})\, f_+ = \left(\delta^{-1}\mathcal{M}J\mathcal{M}\right) Q_1\,(\mathcal{M}f_+) = JQ_1 f_-,$$

where we used Proposition 2 (i). □

Thus we have

$$\mathcal{M}\left[f_+ \;\; JQ_2 f_+\right] = \left[f_- \;\; JQ_1 f_-\right]$$

and, applying determinants and using (4), we obtain

$$\delta_+ \det\left[f_+ \quad J Q_2 f_+\right] = \delta_-^{-1} \det\left[f_- \quad J Q_1 f_-\right] = r_1, \tag{13}$$

where $r_1 \in \mathcal{R}$, with r_1 bounded at ∞, since the left hand side of the first equality is meromorphic in $\mathbb{D}_\Gamma^+$ and the right hand side is meromorphic in $\mathbb{D}_\Gamma^-$ and bounded at ∞. Moreover, from (6) we see that the poles of r_1 must be those of q (see (13)), so we may assume that

$$r_1 = \frac{\tilde{p}_1}{p_2}, \tag{14}$$

where $\tilde{p}_1$ is a polynomial of degree not greater than $\deg(p_2)$. We may also assume that r_1 does not have zeroes on Γ, since, given the analyticity of a, b, d in a neighbourhood of Γ, the latter can be deformed if necessary.

Now, from (13), we have

$$\delta_+ \det\left[f_+ \quad r_1^{-1} J Q_2 f_+\right] = \delta_-^{-1} \det\left[f_- \quad r_1^{-1} J Q_1 f_-\right] = 1. \tag{15}$$

On the other hand, applying determinants to (10), we obtain

$$\delta_+ \det\left[f_+ \quad s_+\right] = \delta_-^{-1} \det\left[f_- \quad s_-\right] \tag{16}$$

and, since the left hand side of (16) is in C_+^μ, while the right hand side is in C_-^μ, we conclude that both sides are equal to a constant $k = 1$, taking into account that the left hand side must equal 1 at $\tau = 0$. Thus we have

$$\delta_+ \det\left[f_+ \quad s_+\right] = 1 = \delta_-^{-1} \det\left[f_- \quad s_-\right]. \tag{17}$$

Then, combining (17) with (15) gives

$$\delta_+ \det\left[f_+ \quad s_+ - r_1^{-1} J Q_2 f_+\right] = \delta_-^{-1} \det\left[f_- \quad s_- - r_1^{-1} J Q_1 f_-\right] = 0.$$

It follows that

$$s_+ - r_1^{-1} J Q_2 f_+ = \lambda_1 f_+, \tag{18}$$

$$s_- - r_1^{-1} J Q_1 f_- = \lambda_2 f_-, \tag{19}$$

where λ_1 and λ_2 are functions of τ which, due to (8), (9) and (12) must satisfy

$$\lambda_1 = \lambda_2 =: \lambda.$$

Proposition 4 *With the notation above, $\lambda \in \mathcal{R}$ and its poles are the zeroes of r_1.*

Proof From (11) and (18), with $\lambda_1 = \lambda$, we have that

$$0 = s_+^T J s_+ = \lambda s_+^T J f_+ + r_1^{-1} s_+^T J J Q_2 f_+ = \lambda \det\begin{bmatrix} f_+ & s_+ \end{bmatrix} - r_1^{-1} s_+^T Q_2 f_+$$
$$= \lambda\, \delta_+^{-1} - r_1^{-1} s_+^T Q_2 f_+,$$

and hence

$$\lambda = \delta_+ r_1^{-1} s_+^T Q_2 f_+ \in H_+^\infty + \mathcal{R}. \tag{20}$$

On the other hand, from (11) and (19) with $\lambda_2 = \lambda$, we get in an analogous way

$$\lambda = \delta_-^{-1} r_1^{-1} s_-^T Q_1 f_- \in H_-^\infty + \mathcal{R}. \tag{21}$$

It follows from (20) and (21) that $\lambda \in \mathcal{R}$. Since r_1 and q have the same denominator, we see that the poles of Q_1 and Q_2 are cancelled and therefore the poles of λ are those of r_1^{-1}, i.e., the zeroes of r_1. □

As a consequence of the previous results, we obtain the following relation between the two columns of X^{-1} and $\mathcal{M}_-$. In what follows we use the notation $f_\pm = (f_{1\pm}, f_{2\pm})$ and $s_\pm = (s_{1\pm}, s_{2\pm})$.

Theorem 1 *With the same notation as in Proposition 3, there exists $r_2 \in \mathcal{R}$, bounded at ∞, with the same poles as q, such that*

$$s_+ = r_1^{-1} \left(r_2 \mathbb{I} + J Q_2\right) f_+, \tag{22}$$
$$s_- = r_1^{-1} \left(r_2 \mathbb{I} + J Q_1\right) f_-. \tag{23}$$

Proof From (18) and (19) we have

$$s_+ = r_1^{-1} \left(\lambda r_1 \mathbb{I} + J Q_2\right) f_+, \tag{24}$$
$$s_- = r_1^{-1} \left(\lambda r_1 \mathbb{I} + J Q_1\right) f_-, \tag{25}$$

and using (11) and (17), we obtain

$$\lambda r_1 = \delta_-^{-1} s_-^T Q_1 f_- = \delta_+ s_+^T Q_2 f_+, \tag{26}$$

so the result holds with $r_2 = \lambda r_1$ and r_2 is bounded at ∞, since q is bounded there. □

We note that (22) is equivalent to

$$\begin{cases} s_{1+} = r_1^{-1} \left(r_2 f_{1+} + f_{2+}\right), \\ s_{2+} = r_1^{-1} \left(r_2 f_{2+} + q f_{1+}\right), \end{cases} \tag{27}$$

and that (23) is equivalent to

$$\begin{cases} s_{1-} = r_1^{-1}\,(r_2 f_{1-} + q f_{2-})\,, \\ s_{2-} = r_1^{-1}\,(r_2 f_{2-} + f_{1-})\,. \end{cases} \tag{28}$$

Now we address the question of determining r_2. It must be such that $s_\pm$, given by (27) and (28), are analytic in $\mathbb{D}_\Gamma^\pm$, $s_\pm \in (H_\pm^\infty)^2$, and $s_+(0) = \begin{bmatrix} 0 \\ 1 \end{bmatrix}$; in that case s_+ and s_- are the second columns of X^{-1} and $\mathcal{M}_-$, respectively, as formulated next.

Proposition 5 *If* $r_2 \in \mathcal{R}$ *is such that* s_+ *and* s_-*, defined by the right hand side of* (27) *and* (28)*, respectively, are functions in* $(H_\mp^\infty)^2$*, with* $s_+(0) = \begin{bmatrix} 0 \\ 1 \end{bmatrix}$*, then* s_+ *and* s_- *are the second columns of* X^{-1} *and* $\mathcal{M}_-$*, respectively. In that case,* r_2 *is unique.*

Proof We have that

$$\begin{aligned} \mathcal{M}s_+ &= \mathcal{M}\left(r_1^{-1}\,(r_2\mathbb{I} + JQ_2)\,f_+\right) = r_1^{-1}\,(\mathcal{M}r_2 f_+ + \mathcal{M}JQ_2 f_+) \\ &= r_1^{-1}\,(r_2 f_- + JQ_1 f_-) = s_-, \end{aligned}$$

and since $s_\pm \in (H_\pm^\infty)^2$ with $s_+(0) = \begin{bmatrix} 0 \\ 1 \end{bmatrix}$, we have the second columns of X^{-1} and $\mathcal{M}_-$. Moreover, from (26) we have $r_2 = \delta_+\, s_+^T Q_2 f_+$, and the uniqueness of r_2 follows from the uniqueness of f_+ and s_+. □

Remark 2 Using (26), we note that r_2 can also be expressed as $r_2 = \delta_-^{-1} s_-^T Q_1 f_-$.

Since r_2 has the form

$$r_2 = \frac{R_2}{p_2},$$

where p_2 is the denominator of q as in (3) and R_2 is a polynomial of degree not greater than $\deg(p_2)$ (since r_2 is bounded at ∞), we only have to determine the coefficients of R_2, which must be such that the zeroes of $(r_2\mathbb{I} + JQ_2)f_+$ cancel the zeroes of r_1 in $\mathbb{D}_\Gamma^+$ and the zeroes of $(r_2\mathbb{I} + JQ_1)f_-$ cancel the zeroes of r_1 in $\mathbb{D}_\Gamma^-$.

We now present a systematic method to obtain those coefficients in the case where r_1 has simple or double zeroes; the method can however be generalised, following the same reasoning, for higher order of zeroes.

Let each zero of r_1 be denoted by z_i^+ if it belongs to $\mathbb{D}_\Gamma^+$, and by z_i^- if it belongs to $\mathbb{D}_\Gamma^-$. Imposing that $(r_2\mathbb{I} + JQ_2)f_+$ vanishes at a zero z_i^+ of r_1 means that, for $f_+ = (f_{1+}, f_{2+})$, and r_1 given by (14), we must have

$$\begin{cases} R_2(z_i^+)\, f_{1+}(z_i^+) + p_2(z_i^+)\, f_{2+}(z_i^+) = 0, \\ R_2(z_i^+)\, f_{2+}(z_i^+) + p_1(z_i^+)\, f_{1+}(z_i^+) = 0. \end{cases} \tag{29}$$

Note that, since r_1 is given by (13), we have that

$$r_1 = \delta_+ \frac{p_1 f_{1+}^2 - p_2 f_{2+}^2}{p_2},$$

so

$$p_1(z_i^+)\, f_{1+}^2(z_i^+) = p_2(z_i^+)\, f_{2+}^2(z_i^+), \tag{30}$$

and it follows that, since f_{1+} and f_{2+} cannot vanish simultaneously,

$$f_{2+}(z_i^+) = 0 \Longrightarrow p_1(z_i^+) = 0,$$
$$f_{1+}(z_i^+) = 0 \Longrightarrow p_2(z_i^+) = 0.$$

We will show that (29) reduces to just one equation, for which we consider three cases:

(i) if $f_{2+}(z_i^+) = 0$, the first equation in (29) is equivalent to

$$R_2(z_i^+) = 0, \tag{31}$$

while the second equation is satisfied for any $R_2(z_i^+)$;

(ii) if $f_{1+}(z_i^+) = 0$, the second equation in (29) is equivalent to (31), while the first equation is satisfied for any $R_2(z_i^+)$;

(iii) if $f_{1+}(z_i^+), f_{2+}(z_i^+) \neq 0$, then multiplying the second equation in (29) by $f_{1+}(z_i^+)$ we get

$$\begin{aligned} & R_2(z_i^+)\, f_{2+}(z_i^+)\, f_{1+}(z_i^+) + p_1(z_i^+)\, f_{1+}^2(z_i^+) = 0 \\ \Longleftrightarrow\ & R_2(z_i^+)\, f_{2+}(z_i^+)\, f_{1+}(z_i^+) + p_2(z_i^+)\, f_{2+}^2(z_i^+) = 0 \\ \Longleftrightarrow\ & f_{2+}(z_i^+)\left[R_2(z_i^+)\, f_{1+}(z_i^!) + p_2(z_i^+)\, f_{2+}(z_i^+)\right] = 0, \end{aligned} \tag{32}$$

which is equivalent to the first equation in (29), i.e.

$$R_2(z_i^+)\, f_{1+}(z_i^+) + p_2(z_i^+)\, f_{2+}(z_i^+) = 0. \tag{33}$$

Analogously, for $f_- = (f_{1-}, f_{2-})$ and a zero z_i^- of r_1, we get

$$R_2(z_i^-) = 0, \quad \text{if} \quad f_{1-}(z_i^-) = 0 \quad \text{or} \quad f_{2-}(z_i^-) = 0, \tag{34}$$

$$R_2(z_i^-)\, f_{1-}(z_i^-) + p_1(z_i^-)\, f_{2-}(z_i^-) = 0, \quad \text{if} \quad f_{1-}(z_i^-), f_{2-}(z_i^-) \neq 0. \tag{35}$$

If all the zeroes of r_1 are simple zeroes, then (31)/(33) and (34)/(35) provide a system of equations allowing to determine all but one coefficient of R_2; the remaining coefficient is obtained from the normalising condition $s_{1+}(0) = 0$.

Remark 3 Note that the normalising condition $s_{1+}(0) = 0$ implies that $s_{2+}(0) = 1$, as follows. First we note that

$$r_1 = \delta_+ \left(q \, f_{1+}^2 - f_{2+}^2 \right). \tag{36}$$

Using (27) and (36) we have

$$\begin{aligned} s_{2+} f_{1+} &= r_1^{-1} \left(r_2 \, f_{1+} f_{2+} + q \, f_{1+}^2 \right) \\ &= r_1^{-1} r_2 \, f_{1+} f_{2+} + r_1^{-1} \left(\delta_+^{-1} r_1 + f_{2+}^2 \right) \\ &= r_1^{-1} \left(r_2 \, f_{1+} + f_{2+} \right) f_{2+} + \delta_+^{-1} \\ &= s_{1+} f_{2+} + \delta_+^{-1}. \end{aligned}$$

Imposing $s_{1+}(0) = 0$ and using $\delta_+(0) = 1$, this yields $s_{2+}(0) = 1$.

To extend the method presented above to the case where r_1 has double zeroes, let us assume that $\tilde{p}_1$ has a double zero at $z_i^+ \in \mathbb{D}_\Gamma^+$. Using the above results, we see that the conditions that we have to impose on r_2 are the same as above and moreover

$$\begin{cases} R_2'(z_i^+) \, f_{1+}(z_i^+) + R_2(z_i^+) \, f_{1+}'(z_i^+) + p_2'(z_i^+) \, f_{2+}(z_i^+) + p_2(z_i^+) \, f_{2+}'(z_i^+) = 0, \\ R_2'(z_i^+) \, f_{2+}(z_i^+) + R_2(z_i^+) \, f_{2+}'(z_i^+) + p_1'(z_i^+) \, f_{1+}(z_i^+) + p_1(z_i^+) \, f_{1+}'(z_i^+) = 0. \end{cases} \tag{37}$$

To show that this pair of conditions can be reduced to one equivalent condition, we consider once again three cases. First, however, note that saying that $\tilde{p}_1$ has a double zero at z_i^+ means that (cf. (30))

$$\begin{aligned} & p_1'(z_i^+) \, f_{1+}^2(z_i^+) + 2 p_1(z_i^+) \, f_{1+}'(z_i^+) \, f_{1+}(z_i^+) \\ & - p_2'(z_i^+) \, f_{2+}^2(z_i^+) - 2 p_2(z_i^+) \, f_{2+}'(z_i^+) \, f_{2+}(z_i^+) = 0, \end{aligned}$$

i.e.

$$\begin{aligned} & p_1'(z_i^+) \, f_{1+}^2(z_i^+) + p_1(z_i^+) \, f_{1+}'(z_i^+) \, f_{1+}(z_i^+) \\ & = p_2'(z_i^+) \, f_{2+}^2(z_i^+) - p_1(z_i^+) \, f_{1+}'(z_i^+) \, f_{1+}(z_i^+) + 2 p_2(z_i^+) \, f_{2+}'(z_i^+) \, f_{2+}(z_i^+). \end{aligned} \tag{38}$$

With this in mind,

(i) if $f_{2+}(z_i^+) = 0$, from (31) and the first equation in (37) we get

$$\begin{cases} R_2(z_i^+) = 0, \\ R_2'(z_i^+)\, f_{1+}(z_i^+) + p_2(z_i^+)\, f_{2+}'(z_i^+) = 0, \end{cases} \tag{39}$$

while the second equation in (37) is satisfied for any $R_2(z_i^+)$ because $f_{2+}(z_i^+) = p_1(z_i^+) = R_2(z_i^+) = 0$ and (38) implies that $p_1'(z_i^+) = 0$;

(ii) if $f_{1+}(z_i^+) = 0$, the equations analogously reduce to

$$\begin{cases} R_2(z_i^+) = 0, \\ R_2'(z_i^+)\, f_{2+}(z_i^+) + p_1(z_i^+)\, f_{1+}'(z_i^+) = 0; \end{cases}$$

(iii) if $f_{1+}(z_i^+), f_{2+}(z_i^+) \neq 0$, then multiplying the second equation in (37) by $f_{1+}(z_i^+)$ we see that it is equivalent to

$$\begin{aligned} &R_2'(z_i^+)\, f_{2+}(z_i^+)\, f_{1+}(z_i^+) + p_2(z_i^+)\, f_{2+}(z_i^+)\, f_{2+}'(z_i^+) \\ &+ p_2'(z_i^+)\, f_{2+}^2(z_i^+) - p_1(z_i^+)\, f_{1+}(z_i^+)\, f_{1+}'(z_i^+) = 0, \end{aligned} \tag{40}$$

where we used (33) and (38). Now, from (33) and (30) we have that

$$R_2(z_i^+)\, f_{2+}(z_i^+) + p_1(z_i^+)\, f_{1+}(z_i^+) = 0,$$

and substituting in (40) we obtain

$$\begin{aligned} &f_{2+}(z_i^+)\big[R_2'(z_i^+)\, f_{1+}(z_i^+) + p_2(z_i^+)\, f_{2+}'(z_i^+) \\ &\quad + p_2'(z_i^+)\, f_{2+}(z_i^+) + R_2(z_i^+)\, f_{1+}'(z_i^+)\big] = 0, \end{aligned}$$

which is equivalent to the first condition in (37). Therefore we find that imposing a double zero for $\tilde{p}_1$ at z_i^+ is equivalent to imposing, in this case,

$$\begin{cases} R_2(z_i^+)\, f_{1+}(z_i^+) + p_2(z_i^+)\, f_{2+}(z_i^+) = 0, \\ R_2'(z_i^+)\, f_{1+}(z_i^+) + R_2(z_i^+)\, f_{1+}'(z_i^+) + (p_2\, f_{2+})'\,(z_i^+) = 0. \end{cases}$$

Analogously, for $f_- = (f_{1-}, f_{2-})$ and a double zero z_i^- of r_1 in $\mathbb{D}_\Gamma^-$, we obtain the conditions:

(i) if $f_{1-}(z_i^-) = 0$,

$$\begin{cases} R_2(z_i^-) = 0, \\ R_2'(z_i^-)\, f_{2-}(z_i^-) + p_2(z_i^-)\, f_{1-}'(z_i^-) = 0; \end{cases}$$

(ii) if $f_{2-}(z_i^-) = 0$,

$$\begin{cases} R_2(z_i^-) = 0, \\ R_2'(z_i^-)\, f_{1-}(z_i^-) + p_1(z_i^-)\, f_{2-}'(z_i^-) = 0; \end{cases}$$

(iii) if $f_{1-}(z_i^-), f_{2-}(z_i^-) \neq 0$,

$$\begin{cases} R_2(z_i^-)\, f_{1-}(z_i^-) + p_1(z_i^-)\, f_{2-}(z_i^-) = 0, \\ R_2'(z_i^+)\, f_{1-}(z_i^-) + R_2(z_i^-)\, f_{1-}'(z_i^-) + (p_1\, f_{2-})'\,(z_i^-) = 0. \end{cases} \tag{41}$$

Proceeding analogously for every double zero of r_1, and as in the previous step for every single zero of r_1, and adding the normalising condition $s_{1+}(0) = 0$, we obtain a linear system providing the coefficients of the numerator of r_2.

Remark 4 Note that this linear system is what we would have obtained had we only imposed that one of the components of s_+ (respectively s_-) is analytic in $\mathbb{D}_\Gamma^+$ (respectively $\mathbb{D}_\Gamma^-$) and satisfies a certain normalising condition; the analyticity and normalising condition for the other component follows from there.

We end this section with a proposition that extends this result to general 2×2 matrices $\mathcal{M}$, provided they are rational. It relates the two components ϕ_{1+} and ϕ_{2+} of any solution to the Riemann-Hilbert problem of the form

$$\mathcal{M}\phi_+ = \phi_- \quad \text{on} \quad \Gamma, \quad \text{with} \quad \phi_\pm = \begin{bmatrix} \phi_{1+} \\ \phi_{2+} \end{bmatrix} \in (H_\pm^\infty)^2, \tag{42}$$

such as (8) and (9), in the case that $\mathcal{M}$ is a rational 2×2 (not necessarily symmetric) matrix function

$$\mathcal{M} = \begin{bmatrix} q_{11} & q_{12} \\ q_{21} & q_{22} \end{bmatrix}.$$

From (42) we have that

$$\begin{aligned} q_{11}\phi_{1+} + q_{12}\phi_{2+} &= \phi_{1-} = T_1 \in \mathcal{R}, \\ q_{21}\phi_{1+} + q_{22}\phi_{2+} &= \phi_{2-} = T_2 \in \mathcal{R}, \end{aligned} \tag{43}$$

where T_1 and T_2 are rational functions bounded at ∞, whose denominators are defined by the poles of q_{ij} in $\mathbb{D}_\Gamma^+$ $(i, j = 1, 2)$. We can therefore reduce the problem to the following case with polynomial coefficients,

$$\begin{aligned} p_{11}\phi_{1+} + p_{12}\phi_{2+} &= P_1, \\ p_{21}\phi_{1+} + p_{22}\phi_{2+} &= P_2, \end{aligned}$$

where p_{ij}, P_i $(i, j = 1, 2)$ are polynomials. We have the following, which is a slight generalisation of Lemma 3.9 in [5].

Proposition 6 *Let p_{ij}, P_i $(i, j = 1, 2)$ be polynomials such that $p_{11}\, p_{22} - p_{12}\, p_{21}$ does not vanish on Γ. Assume moreover that p_{1i}, p_{2i} do not have common zeroes in $\mathbb{D}_\Gamma^+$ and consider the solution*

$$\phi = \begin{bmatrix} \phi_1 \\ \phi_2 \end{bmatrix}$$

of the system

$$\left.\begin{aligned} p_{11}\,\phi_1 + p_{12}\,\phi_2 = P_1 \\ p_{21}\,\phi_1 + p_{22}\,\phi_2 = P_2 \end{aligned}\right\} \quad on \quad \Gamma. \tag{44}$$

Then, if ϕ_1 is analytic in $\mathbb{D}_\Gamma^+$, ϕ_2 is also analytic in $\mathbb{D}_\Gamma^+$, and vice versa.

Proof By Cramer's rule we have that

$$\phi_1 = \frac{P_1\, p_{22} - P_2\, p_{12}}{p_{11}\, p_{22} - p_{12}\, p_{21}}.$$

On the other hand, from (44) we have

$$\phi_2 = \frac{P_1 - p_{11}\,\phi_1}{p_{12}} = \frac{P_2 - p_{21}\,\phi_1}{p_{22}}. \tag{45}$$

If ϕ_1 is analytic in $\mathbb{D}_\Gamma^+$ and p_{12}, p_{22} do not have common zeroes in $\mathbb{D}_\Gamma^+$, then in the neighbourhood of any zero of p_{12} in $\mathbb{D}_\Gamma^+$ we see from the second equality in (45) that $\frac{P_2 - p_{21}\,\phi_1}{p_{22}}$ must be analytic. Thus ϕ_2 is analytic in $\mathbb{D}_\Gamma^+$.

Conversely, if ϕ_2 is analytic in $\mathbb{D}_\Gamma^+$, it follows by an analogous argument that ϕ_1 is also analytic in $\mathbb{D}_\Gamma^+$. □

Remark 5 To obtain a solution to (8) when $\mathcal{M}$ is a rational matrix, one has to solve (43) for the unknown scalar rational functions T_1 and T_2, which must be determined such that both ϕ_{1+} and ϕ_{2+} belong to H_+^∞. The result given above shows that it suffices to determine T_1 and T_2 such that ϕ_{1+} is analytic in $\mathbb{D}_\Gamma^+$.

Rational matrices are of great importance in gravitational theories, yielding solutions to the Einstein field equations, such as the famous Schwarzschild solution and the non-extremal Kerr black hole, via WH factorisation, as explained in the next section. There are different methods to study their WH factorisation. Classical methods, such as those presented in [4, 12] (see also [7]), while providing conditions for existence and estimates for the partial indices of that factorisation, are, in general, computationally extremely difficult to apply in order to obtain explicit enough formulae for the factors allowing one to determine from these the explicit form of the space-time metric. This problem is aggravated in the case of monodromy matrices, considered in the next section, by the fact that the elements of the matrices depend not only on the complex variable τ with respect to which they are rational, but also on several parameters in a very non-trivial way. This is true even in the case of apparently simple 2×2 monodromy matrices, such as the non-extremal Kerr monodromy matrix [5]. So different classes of rational matrices may need different factorisation techniques. The method presented here, which consists in obtaining first one of the columns of X^{-1} and $\mathcal{M}_-$, using Proposition 6, and then looking for a scalar rational function r_2 providing the second column by Theorem 1 and Proposition 2, is in general computationally simpler and can moreover be applied to non-rational 2×2 matrices. Note that, for matrices of the form considered in this paper, the main computational difficulties, when determining a second column using Theorem 1, are connected with the degree of the denominator of q (which is the maximum degree of the numerator of r_1, see (14)) and not with the number of poles of the elements of the matrix $\mathcal{M}$.

3 The Einstein Field Equations and the Monodromy Matrix

The Einstein field equations, a system of 10 nonlinear second order PDE's in 4 variables for the space-time metric g, relate the geometry of space-time to the distribution of matter and energy, described by the stress-energy-momentum tensor T, within it. In the following, we will assume the absence of a cosmological constant term in the field equations. When $T = 0$ in the region under consideration, the field equations are also referred to as the vacuum field equations.

Obtaining exact solutions to the field equations is, in general, a non-trivial task. Exact solutions, which play an important role in Physics and Mathematics, can however be obtained under simplifying assumptions, such as symmetry conditions. We focus on the subspace of solutions of the field equations possessing two commuting isometries, so that the theory can be reduced to two dimensions using a well-known 2-step procedure [13], and the problem of solving the field equations is reduced to a system of nonlinear second order PDE's depending on two coordinates, which we denote by ρ and v, called Weyl coordinates, with $\rho > 0$. We identify these solutions with matrix functions $M(\rho, v)$ of class C^2, which satisfy the field equations

$$d\,(\rho \star A) = 0, \quad A = M^{-1} dM, \tag{46}$$

where $\star$ is the Hodge star operator, $\det M = 1$ and $M = M^{\natural}$ [13]. Here $\natural$ denotes a certain involution called generalised transposition. When $T = 0$, i.e. when dealing with the vacuum field equations, M is a 2×2 matrix and $\natural$ denotes matrix transposition.

Determining explicit solutions to the field equations (46) by means of a Riemann-Hilbert (RH) approach is one of the most recent applications of WH factorisation theory. Here the factorisation is considered with respect to an admissible contour, by which we mean a closed simple contour in the complex plane, encircling the origin and invariant under the involution $\tau \underset{i_\lambda}{\mapsto} -\frac{\lambda}{\tau}$, where $\lambda = \pm 1$ depending on the physics context. This RH approach crucially involves introducing a complex parameter τ, called the spectral parameter, which is allowed to vary on an algebraic curve, called the spectral curve, given by the relation

$$\omega = v + \frac{\lambda}{2}\rho\,\frac{\lambda - \tau^2}{\tau}, \qquad \tau \in \mathbb{C}\backslash\{0\}. \tag{47}$$

Given an $n \times n$ matrix $\mathcal{M}(\omega)$ with $\det \mathcal{M}(\omega) = 1$ and $\mathcal{M}(\omega) = \mathcal{M}^{\natural}(\omega)$, we consider then

$$\mathcal{M}_{\rho,v}(\tau) = \mathcal{M}(\omega)\big|_{\omega = v + \frac{\lambda}{2}\rho\frac{\lambda-\tau^2}{\tau}} = \mathcal{M}(v + \frac{\lambda}{2}\rho\,\frac{\lambda - \tau^2}{\tau}), \tag{48}$$

which we call the monodromy matrix.

We can now state the main theorem of [1], where it was shown that, under very general assumptions, the canonical WH factorisation of $\mathcal{M}_{\rho,v}(\tau)$ w.r.t. an admissible contour Γ in the τ-complex plane, normalised at 0, determines a solution to the field equations (46).

Theorem 2 ([1, Theorem 6.1]) *Let the following assumptions hold:*

(1) There exists an open set S such that, for every $(\rho_0, v_0) \in S$, one can find a simple closed contour Γ in the τ-plane, encircling the origin and invariant under $\tau \underset{i_\lambda}{\mapsto} -\frac{\lambda}{\tau}$, such that: for all (ρ, v) in a neighbourhood of (ρ_0, v_0), the matrix $\mathcal{M}_{\rho,v}(\tau)$ given by (48)*, as well as its inverse, is analytic in a region O in the τ-plane containing Γ. We require O to be invariant under i_λ, and such that $\mathcal{M}^{\natural}_{\rho,v}(\tau) = \mathcal{M}_{\rho,v}(\tau)$ on O;*

(2) for any (ρ, v) in a neighbourhood of (ρ_0, v_0), $\mathcal{M}_{\rho,v}(\tau)$ admits a canonical Wiener-Hopf factorisation with respect to Γ,

$$\mathcal{M}_{\rho,v}(\tau) = \mathcal{M}^{-}_{\rho,v}(\tau)\, X(\tau, \rho, v) \quad on \quad \Gamma,$$

where the "plus" factor X is normalised by

$$X(0, \rho, v) = \mathbb{I}_{n \times n}$$

for all (ρ, v) in a region $S \subset \mathbb{R}^+ \times \mathbb{R}$;

(3) the matrix function $X(\tau, \rho, v)$, for each $\tau \in \mathbb{D}^+_\Gamma \cup O$, and

$$M(\rho, v) := \lim_{\tau \to \infty} \mathcal{M}^{-}_{\rho,v}(\tau) \tag{49}$$

are of class C^2 *(w.r.t.* (ρ, v)*) and* $\frac{\partial X}{\partial \rho}$ *and* $\frac{\partial X}{\partial v}$ *are analytic as functions of* τ *in* $\mathbb{D}_\Gamma^+ \cup O$.

Then $M(\rho, v)$ *defined by* (49) *is a solution to the field equations* (46).

Note that these assumptions, which may at first seem complicated, allow in fact for a broad range of applicability of the results and they are easily seen to be satisfied in all known physically significant cases. Also note that different choices of the contour Γ lead to different factorisations of the same monodromy matrix, and therefore to different solutions to the field equations originating from the same monodromy matrix [1].

To uniformise the notation, we will write $\mathcal{M}_-(\tau)$ instead of $\mathcal{M}^-_{\rho,v}(\tau)$.

We now apply the method discussed in Sect. 2 to two specific 2×2 symmetric monodromy matrices $\mathcal{M}$ that possess a canonical WH factorisation. Their canonical factorisation yields stationary non-static solutions of the vacuum field equations. These solutions, which depend on parameters that introduce rotation in the space-time geometry, are seen to be deformations of static solutions belonging to the known class of $AIII$-metrics [1, 11]. Note that deforming a solution of the vacuum field equations in such a way as to obtain another solution is a non-trivial task when employing usual PDE methods; transformations that achieve this in the PDE context are sometimes called Kinnersley transformations [16, page 4].

3.1 An Example with Non-vanishing Component f_{2+}

Let

$$\mathcal{M}(\omega) = \begin{bmatrix} \frac{c^2}{\omega} + s^2\,\omega & cs\left(\frac{1}{\omega} + \omega\right) \\ cs\left(\frac{1}{\omega} + \omega\right) & c^2\,\omega + \frac{s^2}{\omega} \end{bmatrix} = \begin{bmatrix} \tilde{a}(\omega) & \tilde{b}(\omega) \\ \tilde{b}(\omega) & \tilde{d}(\omega) \end{bmatrix},$$

where $c, s \in \mathbb{C}$ with

$$c^2 - s^2 = 1.$$

Here we take $c, s \neq 0$.

We have

$$\tilde{q}(\omega) = \frac{\tilde{a}(\omega)}{\tilde{d}(\omega)} = \frac{c^2 + s^2\,\omega^2}{c^2\,\omega^2 + s^2} \tag{50}$$

and

$$q(\tau) = \tilde{q}\left(v + \frac{\rho}{2}\frac{1+\tau^2}{\tau}\right),$$

obtained from (50) by substituting the relation given in (47) with $\lambda = -1$,

$$w = v + \frac{\rho}{2} \frac{1+\tau^2}{\tau}.$$

Note that

$$w = \frac{\rho}{2} \frac{(\tau - \tau_0)}{\tau} (\tau - \tilde{\tau}_0),$$

where

$$\tau_0 = \frac{-v + \sqrt{v^2 - \rho^2}}{\rho}, \quad \tilde{\tau}_0 = \frac{1}{\tau_0} = \frac{-v - \sqrt{v^2 - \rho^2}}{\rho}.$$

We choose an admissible contour Γ such that τ_0 lies inside the contour, in which case $\tilde{\tau}_0$ lies in $\mathbb{D}_\Gamma^-$ [1].

The following useful relations will be used in determining a canonical factorisation of $\mathcal{M}_{\rho,v}(\tau)$ defined by

$$\mathcal{M}_{\rho,v}(\tau) = \mathcal{M}\left(v + \frac{\rho}{2} \frac{1+\tau^2}{\tau}\right)$$

(which will simply be denoted by $\mathcal{M}(\tau)$ in the following):

$$\left(c^2 A^{-1} \pm s^2 A\right)\left(s^2 A^{-1} \pm c^2 A\right) - c^2 s^2 \left(A \pm A^{-1}\right)^2 = \pm 1 \tag{51}$$

for any $A \in \mathbb{C}$. In particular, one obtains from (51) that

$$\det \mathcal{M}(\tau) = 1.$$

To establish *the existence of a canonical factorisation* for $\mathcal{M}$, we first solve the Riemann-Hilbert problem

$$\mathcal{M}\phi_+ = \phi_- \quad \text{on} \quad \Gamma, \quad \text{with} \quad \phi_+ \in C_+^\mu, \quad \phi_- \in C_{-,0}^\mu, \tag{52}$$

where $C_{-,0}^\mu$ consists of the functions in C_-^μ vanishing at ∞. Note that

$$w = m_- m_+,$$

where

$$m_- = -\frac{\rho\,\tilde{\tau}_0}{2} \frac{(\tau - \tau_0)}{\tau}, \quad m_+ = \frac{\tilde{\tau}_0 - \tau}{\tilde{\tau}_0}, \quad \tau_0 \in \mathbb{D}_\Gamma^+, \quad \tilde{\tau}_0 \in \mathbb{D}_\Gamma^-. \tag{53}$$

From (52) we have that

$$\begin{cases} \left(c^2\, m_-^{-1} m_+^{-1} + s^2\, m_- m_+\right) \phi_{1+} + cs \left(m_-^{-1} m_+^{-1} + m_- m_+\right) \phi_{2+} = \phi_{1-} = R_{1,0}, \\ cs \left(m_-^{-1} m_+^{-1} + m_- m_+\right) \phi_{1+} + \left(c^2\, m_- m_+ + s^2\, m_-^{-1} m_+^{-1}\right) \phi_{2+} = \phi_{2-} = R_{2,0}, \end{cases} \tag{54}$$

where $R_{1,0}$ and $R_{2,0}$ are rational functions vanishing at ∞ and with the same poles as m_-, m_-^{-1}, i.e. with simple poles at $\tau = 0, \tau_0$. Using Cramer's rule, we have from (54) that

$$\begin{aligned} \phi_{1+} &= \begin{vmatrix} R_{1,0} & cs \left(m_-^{-1} m_+^{-1} + m_- m_+\right) \\ R_{2,0} & c^2\, m_- m_+ + s^2\, m_-^{-1} m_+^{-1} \end{vmatrix} \\ &= \left(R_{1,0}\, c^2 - R_{2,0}\, cs\right) m_- m_+ + \left(R_{1,0}\, s^2 - R_{2,0}\, cs\right) m_-^{-1} m_+^{-1}, \end{aligned} \tag{55}$$

and we look for functions $R_{1,0}$, $R_{2,0}$ of the form

$$R_{1,0} = \frac{A_1 \tau + A_0}{\tau(\tau - \tau_0)}, \qquad R_{2,0} = \frac{B_1 \tau + B_0}{\tau(\tau - \tau_0)},$$

such that ϕ_{1+} given by (55) is analytic in $\mathbb{D}_\Gamma^+$, i.e. with a double zero at 0 for the numerator of the first term in (55) and a double zero at τ_0 for the second term, taking (53) into account. This implies that $A_0 = B_0 = A_1 = B_1 = 0$, so (52) admits only the zero solution $\phi_+ = \phi_- = 0$ and we conclude that $\mathcal{M}$ has a canonical factorisation.

To *obtain the canonical factorisation*

$$\mathcal{M}(\tau) = \mathcal{M}_-(\tau)\, X(\tau) \quad \text{on} \quad \Gamma, \quad \text{with} \quad X(0) = \mathbb{I},$$

we determine the two columns of X^{-1} and $\mathcal{M}_-$ separately. The *first columns* of X^{-1} and $\mathcal{M}_-$ are given by the (unique) solution to

$$\mathcal{M}\, f_+ = f_- \quad \text{on} \quad \Gamma, \quad \text{with} \quad f_\pm \in (C_\pm^\mu)^2, \quad f_+(0) = (1, 0).$$

Denoting $f_\pm = (f_{1\pm}, f_{2\pm})$, we get a system as in (54), with $R_{1,0}$ and $R_{2,0}$ replaced by R_1 and R_2, respectively, where R_1 and R_2 are rational functions bounded at ∞ and with simple poles at 0 and τ_0, yielding

$$\begin{aligned} f_{1+} &= \begin{vmatrix} R_1 & cs \left(m_-^{-1} m_+^{-1} + m_- m_+\right) \\ R_2 & c^2\, m_- m_+ + s^2\, m_-^{-1} m_+^{-1} \end{vmatrix} \\ &= \left(R_1\, c^2 - R_2\, cs\right) m_- m_+ + \left(R_1\, s^2 - R_2\, cs\right) m_-^{-1} m_+^{-1}, \end{aligned} \tag{56}$$

$$f_{2+} = \begin{vmatrix} c^2\, m_-^{-1} m_+^{-1} + s^2\, m_- m_+ & R_1 \\ cs\left(m_-^{-1} m_+^{-1} + m_- m_+\right) & R_2 \end{vmatrix}$$
$$= \left(R_2\, s^2 - R_1\, cs\right) m_- m_+ + \left(R_2\, c^2 - R_1\, cs\right) m_-^{-1} m_+^{-1}. \tag{57}$$

Taking the form of (56) into account, we look for $\alpha_1, \alpha_2 \in \mathbb{C}$ such that

$$R_1\, c^2 - R_2\, cs = \alpha_1\, m_-^{-1},$$
$$R_1\, s^2 - R_2\, cs = \alpha_2\, m_-,$$

which is equivalent to having

$$R_1 = \alpha_1\, m_-^{-1} - \alpha_2\, m_-,$$
$$R_2 = \frac{R_1\, c^2 - \alpha_1\, m_-^{-1}}{cs} = \frac{R_1\, s^2 - \alpha_2\, m_-}{cs} = \frac{\alpha_1\, m_-^{-1}\, s^2 - \alpha_2\, m_-\, c^2}{cs}. \tag{58}$$

It is easy to see that, for R_1 and R_2 given by (58), f_{1+} and f_{2+} given by (56) and (57) are analytic in $\mathbb{D}_\Gamma^+$ with

$$f_{1+} = \alpha_1\, m_+ + \alpha_2\, m_+^{-1},$$
$$f_{2+} = -\alpha_2\, \frac{c}{s}\, m_+^{-1} - \alpha_1\, \frac{s}{c}\, m_+,$$

and from the normalising conditions $f_{1+}(0) = 1$, $f_{2+}(0) = 0$, we get $\alpha_1 = c^2$, $\alpha_2 = -s^2$, so that

$$f_+ = \begin{bmatrix} c^2\, m_+ - s^2\, m_+^{-1} \\ sc\left(m_+^{-1} - m_+\right) \end{bmatrix}, \quad f_- = \begin{bmatrix} c^2\, m_-^{-1} + s^2\, m_- \\ sc\left(m_-^{-1} + m_-\right) \end{bmatrix}.$$

The *second columns in* X^{-1} *and* $\mathcal{M}_-$ can be determined analogously with different normalising conditions. However we will obtain them here using the results of Sect. 2, namely Theorem 1. Noting that, by (51),

$$f_{1+}\left(c^2\, m_+^{-1} - s^2\, m_+\right) = 1 - f_{2+}^2,$$

we have that

$$r_1 = q\, f_{1+}^2 - f_{2+}^2 = \left(q\, f_{1+} + c^2\, m_+^{-1} - s^2\, m_+\right) f_{1+} - 1,$$

and hence

$$-\left[(q\,f_{1+} + c^2\,m_+^{-1} - s^2\,m_+)f_{2+}\right] f_{1+} + f_{2+} = r_1\,(-f_{2+})\,. \tag{59}$$

Comparing (59) with the equation for s_{1+} given in (27), we take

$$r_2 = -(q\,f_{1+} + c^2\,m_+^{-1} - s^2\,m_+)f_{2+}$$

and $s_{1+} = -f_{2+}$. Using (27) and (28), this uniquely determines s_+ and s_- to be

$$s_+ = \begin{bmatrix} sc\left(m_+ - m_+^{-1}\right) \\ c^2\,m_+^{-1} - s^2\,m_+ \end{bmatrix}, \quad s_- = \begin{bmatrix} sc\left(m_-^{-1} + m_-\right) \\ s^2\,m_-^{-1} + c^2\,m_- \end{bmatrix}.$$

In particular we see that $s_{1+} = -f_{2+}$ and $s_{1-} = f_{2-}$.

The matrix $M(\rho, v)$ is obtained from the matrix $\mathcal{M}_-$,

$$\mathcal{M}_- = \begin{bmatrix} c^2\,m_-^{-1} + s^2\,m_- & sc\left(m_-^{-1} + m_-\right) \\ sc\left(m_-^{-1} + m_-\right) & s^2\,m_-^{-1} + c^2\,m_- \end{bmatrix},$$

by using (49) and reads,

$$M(\rho, v) = \begin{bmatrix} c^2\,f^{-1} + s^2\,f & sc\left(f^{-1} + f\right) \\ sc\left(f^{-1} + f\right) & s^2\,f^{-1} + c^2\,f \end{bmatrix},$$

$$f(\rho, v) = -\frac{\rho\,\tilde{\tau}_0}{2} = \frac{1}{2}\left(v + \sqrt{v^2 - \rho^2}\right), \tag{60}$$

where we restrict to the region $v > \rho$ to ensure that we obtain a solution that is real.

Given a matrix $M(\rho, v)$, it encodes a solution to the vacuum field equations, i.e. a space-time metric [3]. Namely, denoting the matrix elements of the matrix M by

$$M = \begin{bmatrix} \Delta + \frac{\tilde{B}^2}{\Delta} & \frac{\tilde{B}}{\Delta} \\ \frac{\tilde{B}}{\Delta} & \frac{1}{\Delta} \end{bmatrix}, \tag{61}$$

and defining the function B by the relation $\rho \star d\tilde{B} = \Delta^2\,dB$, the associated space-time metric is given by

$$ds_4^2 = -\lambda\,\Delta\,(dt + B\,d\phi)^2 + \Delta^{-1}\left(e^{\psi}\,(d\rho^2 + \lambda\,dv^2) + \rho^2 d\phi^2\right), \tag{62}$$

where $\psi(\rho, v)$ is a function obtained from $M(\rho, v)$ by integration [1, 13]. For the example (60) we have $\lambda = -1$ as well as $\star d\rho = dv, \star dv = d\rho$. Restricting to region $v > \rho$, we obtain the space-time metric (62) with

$$\Delta = \frac{\frac{1}{2}\left(v + \sqrt{v^2 - \rho^2}\right)}{s^2 + c^2 \frac{1}{4}\left(v + \sqrt{v^2 - \rho^2}\right)^2},$$

$$B = 2s\,c\,\rho\tau_0 = -2cs\left(v - \sqrt{v^2 - \rho^2}\right),$$

$$e^{\psi} = \frac{v + \sqrt{v^2 - \rho^2}}{2\sqrt{v^2 - \rho^2}}.$$

We note that $B \neq 0$ provided $s \neq 0$; in this case the metric is stationary and non-static. When $s = 0$, the metric is static and describes a known metric that belongs to the class of $AIII$-metrics [1, 11]. Thus, the stationary metric (with $s \neq 0$) can be viewed as a deformation of a static metric (with $s = 0$).

3.2 An Example with Vanishing Component f_{2+}

Now we consider

$$\mathcal{M}(\omega) = \begin{bmatrix} \frac{1}{\omega} & \frac{\epsilon}{\omega} \\ \frac{\epsilon}{\omega} & \omega + \frac{\epsilon^2}{\omega} \end{bmatrix} = \begin{bmatrix} \tilde{a}(\omega) & \tilde{b}(\omega) \\ \tilde{b}(\omega) & \tilde{d}(\omega) \end{bmatrix}, \qquad \det \mathcal{M}(\omega) = 1,$$

where $\epsilon \in \mathbb{C}$. Here we take $\epsilon \neq 0$.

We have

$$\tilde{q}(\omega) = \frac{\tilde{a}(\omega)}{\tilde{d}(\omega)} = \frac{1}{\omega^2 + \epsilon^2} \tag{63}$$

and

$$q(\tau) = \tilde{q}\left(v + \frac{\rho}{2}\frac{1 - \tau^2}{\tau}\right), \tag{64}$$

obtained from (63) by substituting the relation given in (47) with $\lambda = 1$,

$$w = v + \frac{\rho}{2}\frac{1 - \tau^2}{\tau}.$$

Note that

$$w = -\frac{\rho}{2}\,\frac{(\tau - \tau_0)}{\tau}\,(\tau - \tilde{\tau}_0),$$

where

$$\tau_0 = \frac{v - \sqrt{v^2 + \rho^2}}{\rho}, \quad \tilde{\tau}_0 = -\frac{1}{\tau_0} = \frac{v + \sqrt{v^2 + \rho^2}}{\rho}.$$

We choose an admissible contour Γ such that τ_0 lies inside the contour, in which case $\tilde{\tau}_0$ lies in $\mathbb{D}_\Gamma^-$ [1]. We define

$$m_- = \frac{\rho}{2}\,\tilde{\tau}_0\,\frac{(\tau - \tau_0)}{\tau}, \quad m_+ = \frac{\tilde{\tau}_0 - \tau}{\tilde{\tau}_0}, \quad \tau_0 \in \mathbb{D}_\Gamma^+, \quad \tilde{\tau}_0 \in \mathbb{D}_\Gamma^-.$$

and note that

$$w = m_- m_+.$$

It can be shown, as in the previous example, that the monodromy matrix

$$\mathcal{M}_{\rho,v}(\tau) = \mathcal{M}\left(v + \frac{\rho}{2}\,\frac{1 - \tau^2}{\tau}\right)$$

possesses a canonical WH factorisation (2) w.r.t. Γ, with the factors X^{-1} and $\mathcal{M}_-$ possessing the following first columns (denoted by f_+ and f_-, respectively),

$$f_+ = \begin{bmatrix} f_{1+} \\ f_{2+} \end{bmatrix} = \begin{bmatrix} m_+ \\ 0 \end{bmatrix}, \quad f_- = \begin{bmatrix} f_{1-} \\ f_{2-} \end{bmatrix} = \begin{bmatrix} m_-^{-1} \\ \epsilon\, m_-^{-1} \end{bmatrix}.$$

Note that $f_{2+}(\tau) = 0$ and $f_{2-}(\tau) = \epsilon f_{1-}(\tau)$).

To determine the second columns of X^{-1} and $\mathcal{M}_-$, we use the method described in Sect. 2. The second columns will be denoted by s_+ and s_-, respectively.

First we express the rational function q in (64) as

$$q = \frac{p_1(\tau)}{p_2(\tau)},$$

where

$$p_1(\tau) = \tau^2, \quad p_2(\tau) = \frac{\rho^2}{4}(\tau - \tau_0)^2(\tau - \tilde{\tau}_0)^2 + \epsilon^2\tau^2. \tag{65}$$

The quantity r_1, given in (36), takes the form

$$r_1 = \frac{\tilde{p}_1}{p_2}$$

with

$$\tilde{p}_1 = \frac{\tau^2(\tau - \tilde{\tau}_0)^2}{\tilde{\tau}_0^2}$$

and p_2 given in (65). Both $\tilde{p}_1$ and p_2 are polynomials of degree 4. $\tilde{p}_1$ has two double zeroes, one located in the interior of Γ (at $\tau = 0$) and one located in the exterior of Γ (at $\tau = \tilde{\tau}_0$).

Next, we determine the rational function r_2, by demanding that the zeroes of $(r_2 + J\,Q_2)f_+$ cancel the zeroes of r_1 in $\mathbb{D}_\Gamma^+$ and the zeroes of $(r_2 + J\,Q_1)f_-$ cancel the zeroes of r_1 in $\mathbb{D}_\Gamma^-$. Using (39) and (41), we obtain (where we use that $f_{2+}(\tau) = 0$ and $f_{2-}(\tau) = \epsilon f_{1-}(\tau)$),

$$\begin{cases} R_2(0) & = 0, \\ R_2'(0) & = 0, \\ R_2(\tilde{\tau}_0) + \epsilon p_1(\tilde{\tau}_0) & = 0, \\ R_2'(\tilde{\tau}_0) + \epsilon p_1'(\tilde{\tau}_0) & = 0. \end{cases} \tag{66}$$

Now we use that R_2 is a polynomial of degree not greater than $\deg(p_2) = 4$, i.e.

$$R_2(\tau) = A\tau^4 + B\tau^3 + C\tau^2 + D\tau + E,$$

where the constants A, B, C, D, E are determined by the four conditions (66) as well as by the normalising condition $s_{1+}(0) = 0$ (cf. Remark 3). Imposing the first two conditions in (66) gives $D = E = 0$, while imposing the two remaining conditions in (66) gives

$$\begin{aligned} B &= -2A\,\tilde{\tau}_0, \\ C &= -\epsilon + A\,\tilde{\tau}_0^2. \end{aligned} \tag{67}$$

Next we use the normalising condition $s_{1+}(0) = 0$. Using the expression for s_{1+} given in (27) we obtain $C = 0$, which when combined with (67) gives

$$A = \frac{\epsilon}{\tilde{\tau}_0^2}, \quad B = -2\frac{\epsilon}{\tilde{\tau}_0}.$$

The polynomial R_2 is therefore determined to be

$$R_2 = \epsilon\, \tau^3 \left(\frac{\tau}{\tilde{\tau}_0^2} - \frac{2}{\tilde{\tau}_0} \right).$$

Having determined r_2 using only knowledge about the first column vectors f_+ and f_-, the second column vectors s_+ and s_- are uniquely determined using (27) and (28),

$$s_+ = \begin{bmatrix} \epsilon \left(m_+ - m_+^{-1} \right) \\ m_+^{-1} \end{bmatrix}, \quad s_- = \begin{bmatrix} \epsilon\, m_-^{-1} \\ m_- + \epsilon^2\, m_-^{-1} \end{bmatrix}.$$

The matrix $M(\rho, v)$ is obtained from the matrix $\mathcal{M}_-$,

$$\mathcal{M}_- = \begin{bmatrix} m_-^{-1} & \epsilon\, m_-^{-1} \\ \epsilon\, m_-^{-1} & m_- + \epsilon^2\, m_-^{-1} \end{bmatrix},$$

by using (49) and reads,

$$M(\rho, v) = \begin{bmatrix} f^{-1} & \epsilon\, f^{-1} \\ \epsilon\, f^{-1} & f + \epsilon^2\, f^{-1} \end{bmatrix}, \quad f(\rho, v) = \frac{\rho\, \tilde{\tau}_0}{2} = \frac{v + \sqrt{v^2 + \rho^2}}{2}.$$

Using (61) and (62) with $\lambda = 1$ as well as $\star d\rho = -dv, \star dv = d\rho$, the associated space-time metric is given by

$$ds_4^2 = -\Delta\, (dt + B\, d\phi)^2 + \Delta^{-1} \left(e^{\psi}\, (d\rho^2 + dv^2) + \rho^2 d\phi^2 \right),$$

with

$$\Delta = \frac{\frac{1}{2} \left(v + \sqrt{v^2 + \rho^2} \right)}{\epsilon^2 + \frac{1}{4} \left(v + \sqrt{v^2 + \rho^2} \right)^2},$$

$$B = 2\epsilon\, \rho \tau_0 = 2\epsilon \left(v - \sqrt{v^2 + \rho^2} \right),$$

$$e^{\psi} = \frac{v + \sqrt{v^2 + \rho^2}}{2\sqrt{v^2 + \rho^2}}.$$

We note that $B \neq 0$ provided $\epsilon \neq 0$; as in the previous example, the metric is stationary. When $\epsilon = 0$, the metric is static and describes a known metric that belongs to the class of $AIII$-metrics [1, 11] but is different from the $AIII$-metric mentioned in the previous subsection.

Funding Work funded by FCT/Portugal through project UIDB/04459/2020 with DOI identifier 10-54499/UIDP/04459/2020.

References

1. Aniceto, P., Câmara, M.C., Cardoso, G.L., Rosselló, M.: Weyl metrics and Wiener-Hopf factorization. J. High Energy Phys. **05**, 124 (2020)
2. Bastos, M.A., Karlovich, Y.I., dos Santos, A.F.: Oscillatory Riemann-Hilbert problems and the corona theorem. J. Funct. Anal. **197**, 347–397 (2003)
3. Breitenlohner, P., Maison, D.: On the Geroch group. Ann. Inst. H. Poincaré Phys. Théor. **46**, 215–246 (1987)
4. Clancey, K.F., Gohberg, I.: Factorisation of Matrix Functions and Singular Integral Operators. Operator Theory: Advances and Applications, vol. 3. Birkhäuser Verlag, Basel (1981)
5. Câmara, M.C., Cardoso, G.L.: Riemann-Hilbert problems, Toeplitz operators and ergosurfaces. J. High Energy Phys. **06**, 27 (2024)
6. Câmara, M.C., dos Santos, A.F.: Wiener-Hopf factorization for a class of oscillatory symbols. Integral Equ. Oper. Theory **36**, 409–432 (2000)
7. Câmara, M.C., Lebre, A.B., Speck, F.O.: Meromorphic factorization, partial index estimates and elastodynamic diffraction problems. Math. Nachr. **157**, 291 (1992)
8. Câmara, M.C., dos Santos, A.F., Carpentier M.P.: Explicit Wiener-Hopf factorization and nonlinear Riemann-Hilbert problems. Proc. R. Soc. Edinb. Sect. A **132**, 45–74 (2002)
9. Câmara, M.C., Diogo, C., Rodman, L.: Fredholmness of Toeplitz operators and corona problems. J. Funct. Anal. **259**, 1273–1299 (2010)
10. Duduchava, R.: Integral Equations in Convolution with Discontinuous Presymbols, Singular Integral Equations with Fixed Singularities, and Their Applications to Some Problems of Mechanics, vol. 1. Aufl, B.G. Teubner, Leipzig (1979)
11. Griffiths, J.B., Podolsky, J.: Exact Space-Times in Einstein's General Relativity. Cambridge Monographs on Mathematical Physics. Cambridge University Press, Cambridge (2009)
12. Litvinchuk, G., Spitkovsky, I.: Factorisation of Measurable Matrix Functions. Operator Theory: Advances and Applications, vol. 25. Birkhäuser Verlag, Basel (1987). Translated from Russian by B. Luderer, with a foreword by B. Silbermann
13. Lü, H., Perry, M.J., Pope, C.N.: Infinite-dimensional symmetries of two-dimensional coset models. Nucl. Phys. B **806**, 656–683 (2009)
14. Mikhlin, S.G., Prössdorf, S.: Singular Integral Operators. Springer-Verlag, Berlin (1986)
15. Sarason, D.: Doubly shift-invariant spaces in H^2. J. Oper. Theory **16**, 75–97 (1986)
16. Richterek, L., Horský, J.: Weyl metrics and the generating conjecture. Czechoslovak J. Phys. **54**, 1451–1459 (2004)

Invertibility of Toeplitz Plus Hankel Operators on l^p-Spaces

Victor Didenko and Bernd Silbermann

Dedicated to Yuri Karlovich on the occasion of his 75-th birthday

Abstract The invertibility of Toeplitz plus Hankel operators $T(a) + H(b)$, $a, b \in L^\infty$ acting on l^p-spaces is studied. If the generating functions a and b satisfy the equation

$$a(t)a(1/t) = b(t)b(1/t),$$

various sufficient conditions for the invertibility and one-sided invertibility of the operators $T(a)+H(b)$ are obtained and the corresponding inverses are constructed. Necessary conditions of one-sided invertibility are also discussed. Besides, we suggest a generalization of the above condition for the functions a and b, which allows to extend the approach used to a substantially wider class of Toeplitz plus Hankel operators.

1 Introduction

Let $\mathbb{R}$ and $\mathbb{C}$ be respectively the sets of all real and all complex numbers, $\mathbb{T}$ the counterclockwise oriented unit circle in the complex plane $\mathbb{C}$, and $L^p = L^p(\mathbb{T})$ the complex Banach space of all Lebesgue measurable functions f on $\mathbb{T}$ equipped with the norm

V. Didenko (✉)
Institute for Advanced Study, BNU-HKBU United International College, Zhuhai, China
e-mail: diviol@gmail.com

B. Silbermann
Technische Universität Chemnitz, Fakultät für Mathematik, Chemnitz, Germany
e-mail: silbermn@mathematik.tu-chemnitz.de

A. Böttcher et al. (eds.), *Achievements and Challenges in the Field of Convolution Operators*, Operator Theory: Advances and Applications 306,
https://doi.org/10.1007/978-3-031-80486-1_7

$$\|f\|_p := \begin{cases} \left(\dfrac{1}{2\pi}\displaystyle\int_{\mathbb{T}} |f(t)|^p\,|dt|\right)^{1/p} < \infty, & \text{if} \quad 1 \le p < \infty, \\ \operatorname*{ess\,sup}_{t\in\mathbb{T}} |f(t)| < \infty, & \text{if} \quad p = \infty. \end{cases}$$

Toeplitz plus Hankel operators and Wiener-Hopf plus Hankel operators arise in statistical mechanics, random matrix theory, scattering theory [1, 4, 16, 18–23]. The invertibility of such operators plays an important role in the asymptotics of Toeplitz plus Hankel determinants and approximation methods for related operator equations [3]. However, unlike the scalar Toeplitz operators, Fredholmness of Toeplitz plus Hankel and Wiener-Hopf plus Hankel operators does not imply their one-sided invertibility. Therefore, finding effective invertibility conditions and constructing the corresponding inverses face essential difficulties. In particular, although it is known that a Toeplitz plus Hankel operator $T(a) + H(b)$ is invertible if and only if the so-called partial indices in the so-called special antisymmetric factorization of the matrix

$$V(a,b) = \begin{pmatrix} b\widetilde{a}^{-1} & a - b\widetilde{b}\widetilde{a}^{-1} \\ \widetilde{a}^{-1} & -\widetilde{a}^{-1}\widetilde{b} \end{pmatrix} \tag{1}$$

are equal to zero [14, 15], there are no methods allowing to evaluate these indices for general symbols a and b. Here and throughout the following we denote by $\tilde{g}$ the function defined by $\tilde{g}(t) = g(1/t)$. On the other hand, if (a, b) is a matching pair—i.e. if this duo satisfies the so-called matching condition

$$a(t)a(1/t) = b(t)b(1/t), \quad t \in \mathbb{T}, \tag{2}$$

the invertibility, one-sided invertibility, generalized invertibility of the operators $T(a) + H(b)$ acting on classical Hardy spaces can be successfully treated and the corresponding inverses can be constructed—cf. [2, 6, 8, 10, 11]. It is worth noting that the approach employed in this situation is not related to the antisymmetric factorization of the matrix (1) but exploits the Wiener-Hopf factorization of auxiliary scalar functions associated with the symbols a and b. Moreover, in most cases, the corresponding inverses can be expressed in explicit form.

Passing to Toeplitz plus Hankel operators acting on the sequence spaces l^p, we note that the methods of [6, 8, 10, 11] do not always work even for Toeplitz operators. This happens because the invertibility conditions obtained there are based on Wiener-Hopf factorization of scalar functions in L^p-spaces. However, for the operators acting on l^p-spaces, $p \neq 2$, such a factorization can be exploited only for operators with symbols which are l^p-multipliers along with their Wiener-Hopf factors. Nevertheless, this factorization can still be of use when studying Toeplitz plus Hankel operators on the spaces of sequences but additional conditions are needed. The corresponding approach was developed in [9] and some results of that work are used here when considering the invertibility of the operator $T(a) + H(b)$

on l^p-spaces and constructing the corresponding inverses. Along with the one-sided invertibility and invertibility of Toeplitz plus Hankel operators, we also consider a few situations where these operators are only generalized invertible and determine their generalized inverses. Note that an operator B is called generalized inverse for the operator A if

$$ABA = A.$$

This work is organized as follows. In Sect. 2, we recall some properties of Toeplitz operators acting on l^p-spaces which are needed in what follows. In Sect. 3, sufficient conditions for the one-sided invertibility, invertibility, and generalized invertibility of Toeplitz plus Hankel operators are presented and the corresponding inverses are constructed. The invertibility condition given in this section are based on the invertibility of auxiliary Toeplitz operators, called the subordinated operators, and are not directly associated with the properties of a Wiener-Hopf factorization of their generating functions—cf. Theorem 1. The formulas for the inverses of Toeplitz plus Hankel operators contain the inverses of the subordinated Toeplitz operators acting on l^p-spaces, $p \neq 2$. Therefore, we also recall recent results related to this issue. Section 4 deals with more delicate invertibility conditions for a special class of Toeplitz plus Hankel operators. The main feature of this class is that the corresponding subordinated operators, generated by the same matching pair on the space l^p and on the Hardy space H^q, $1/p + 1/q = 1$, should have the same indices. The later requirement allows to use the Wiener-Hopf factorization in case of Toeplitz operators acting on l^p-spaces. Consequently, we obtain necessary conditions for one-sided invertibility of Toeplitz plus Hankel operators and new sufficient conditions different from the ones in Sect. 3. In particular, we note that Toeplitz plus Hankel operators can be invertible even if the subordinated operators are only one-sided invertible from the same or from different sides. Finally, in conclusion we note a generalization of the condition (2) which allows to apply the approach used to a substantially larger class of Toeplitz plus Hankel operators.

2 Operators on l^p-Spaces

Let $\mathbb{Z}$ and $\mathbb{Z}_+$ respectively denote the sets of all integers and all non-negative integers. Consequently, $\widetilde{l}^p = \widetilde{l}^p(\mathbb{Z})$ and $l^p = l^p(\mathbb{Z}_+)$, $1 \leq p < \infty$ are the Banach spaces of all sequences $\xi = (\xi_n)_{n\in\mathbb{Z}}$ and $\xi = (\xi_n)_{n\in\mathbb{Z}_+}$ of complex numbers such that

$$\|\xi\|_p = \left(\sum_n |\xi_n|^p\right)^{1/p} < \infty,$$

where the summation is over $\mathbb{Z}$ or $\mathbb{Z}_+$. In what follows, the space l^p is regarded as a natural subspace of $\widetilde{l^p}$, and by P we denote the canonical projection from $\widetilde{l^p}$ onto l^p. We also consider the spaces l_0 and c_0 of the sequences $(\xi_n)_{n\in\mathbb{Z}}$, the first of which consists of all finitely supported sequences and the second one of all sequences that tend to zero as $|n| \to \infty$.

Let $\mathcal{F}$: $L^1(\mathbb{T}) \to c_0$ refer to the Fourier transform

$$\mathcal{F}(f) = (\widehat{f_n})_{n\in\mathbb{Z}},$$

where

$$\widehat{f_n} = \frac{1}{2\pi}\int_0^{2\pi} f(e^{i\theta})e^{-in\theta}\,d\theta$$

are the Fourier coefficients of the function f. Consider now a function $a \in L^1(\mathbb{T})$ and define a linear operator on l_0 by

$$(L(a)\xi)_k := \sum_{m\in\mathbb{Z}} \widehat{a}_{k-m}\xi_m, \quad k \in \mathbb{Z}.$$

This operator is often called the Laurent operator generated by a.

We say that a is a multiplier on l^p if $L(a)\xi \in l^p$ for any $\xi \in l_0$ and

$$\|L(a)\| := \sup\left\{\|L(a)\xi\|_p : \xi \in l_0(\mathbb{Z}),\ \|\xi\|_p = 1\right\} < \infty. \tag{3}$$

If a satisfies the condition (3), then $L(a)$ extends to a bounded linear operator on the space l^p. This extension is again denoted by $L(a)$ and the set of all l^p-multipliers is referred to as M^p. Equipped with the norm $\|a\|_{M^p} := \|L(a)\|$, the set M^p becomes a commutative Banach algebra.

It is worth noting that M^1 is the Wiener algebra $W = W(\mathbb{T})$ of all functions having absolutely convergent Fourier series, that $M^2 = L^\infty$, and that every function $a \in L^\infty$ with bounded total variation $\operatorname{var}(a)$ belongs to any algebra M^p, $p \in (1,\infty)$. In particular, the sets of all trigonometric polynomials and all piecewise constant functions belong to all M^p, $p \in (1,\infty)$, and their closures, respectively denoted by C_p and PC_p, are subalgebras of M^p, as well. For other properties of multipliers, the reader can consult [5].

3 Invertibility: Direct Approach

Let J be the operator defined by

$$J\xi = J((\xi_n)_{n\in\mathbb{Z}}) := (\xi_{-n-1})_{n\in\mathbb{Z}}, \quad (\xi_n)_{n\in\mathbb{Z}} \in \widetilde{l^p},$$

and let $Q = I - P$, where I denotes the identity operator. The Toeplitz operator $T(a)$ and the Hankel operator $H(b)$ generated by functions a and b are defined on l^p (considered as a subspace of $\widetilde{l^p}$) by

$$T(a) := PL(a), \quad H(b) := PL(a)QJ.$$

If $a, b \in M^p$, then $T(a), H(b) : l^p \to l^p$ are bounded linear operators and if $(\xi_n) \in l^p$, then

$$T(a) : (\xi_j)_{j\in\mathbb{Z}_+} \to \left(\sum_{k\in\mathbb{Z}_+} \widehat{a}_{j-k}\xi_k\right)_{j\in\mathbb{Z}_+},$$

$$H(b) : (\xi_j)_{j\in\mathbb{Z}_+} \to \left(\sum_{k\in\mathbb{Z}_+} \widehat{b}_{j+k+1}\xi_k\right)_{j\in\mathbb{Z}_+}.$$

As was already mentioned, we are going to study the invertibility of Toeplitz plus Hankel operators $T(a) + H(b)$ the generating functions of which are connected in a special way. More exactly, from now on we assume that $a, b \in M^p$ and satisfy the matching condition (2). In addition, assuming that a and b belong to the group GM^p of invertible elements in M^p, we consider the functions

$$c := ab^{-1}(= \widetilde{b}\widetilde{a}^{-1}), \quad d := a\widetilde{b}^{-1}(= b\widetilde{a}^{-1}),$$

and note that the duo (c, d) is also a matching pair such that

$$c\widetilde{c} = d\widetilde{d} = 1.$$

This matching pair (c, d) and the Toeplitz operators $T(c)$, $T(d)$, respectively called the subordinated pair and the subordinated operators, play an outstanding role in the study of Toeplitz plus Hankel operators $T(a) + H(b)$. In particular, the operators $T(a) \pm H(b) : l^p \to l^p$, $p > 1$ are both Fredholm if and only if so are the operators $T(c)$ and $T(d)$. In what follows, any function g with the property $g\widetilde{g} = 1$ is called the matching function. Moreover, let us agree that for $a, b \in M^p$ all the operators mentioned in the corresponding statement are considered on the space l^p with the same index p, unless the other is specified.

Theorem 1 *Assume that $a, b \in M^p$, $p > 1$ constitute a matching pair. Then:*

1. *If the operators $T(c)$ and $T(d)$ are left-invertible, then the operator $T(a)+H(b)$ is also left-invertible and one of its left-inverses has the form*

$$B = T_l^{-1}(c)\left(H(\widetilde{a}^{-1}) + T(\widetilde{a}^{-1})T_l^{-1}(d)(I - H(d))\right), \tag{4}$$

where $T_l^{-1}(c)$ and $T_l^{-1}(d)$ are any left-inverses of the operators $T(c)$ and $T(d)$, respectively.

2. *If the operators $T(c)$ and $T(d)$ are right-invertible, then the operator $T(a) + H(b)$ is also right-invertible and one of its right-inverses has the form*

$$B = \left((I - H(\widetilde{c}))T_r^{-1}(c)T(\widetilde{a}^{-1}) + H(a^{-1})\right) T_r^{-1}(d), \tag{5}$$

where $T_r^{-1}(c)$ and $T_r^{-1}(d)$ are any right-inverses of the operators $T(c)$ and $T(d)$, respectively.

3. *If $T(c)$ and $T(d)$ are respectively right- and left-invertible operators, then the operator $T(a) + H(b)$ is generalized-invertible and one of its generalized-inverses has the form*

$$\begin{aligned} B = &-H(\widetilde{c}) \left(\mathbf{A}(I - H(d)) - \mathbf{B}H(\widetilde{a}^{-1})\right) \\ &+ H(a^{-1})\mathbf{D}(I - H(d)) + T(a^{-1}), \end{aligned} \tag{6}$$

where

$$\mathbf{A} = T_r^{-1}(c)T(\widetilde{a}^{-1})T_l^{-1}(d), \quad \mathbf{B} = -T_r^{-1}(c), \quad \mathbf{D} = T_l^{-1}(d). \tag{7}$$

Proof The proofs of the representations (4) and (5) are based on the well known Widom identities

$$\begin{aligned} T(\varphi\psi) &= T(\varphi)T(\psi) + H(\varphi)H(\widetilde{\psi}), \\ H(\varphi\psi) &= T(\varphi)H(\psi) + H(\varphi)T(\widetilde{\psi}). \end{aligned} \tag{8}$$

Here we only show the representation (4). The other one can be verified analogously. Recalling that $c = ab^{-1} = \widetilde{b}\widetilde{a}^{-1}$ and $d = a\widetilde{b}^{-1} = b\widetilde{a}^{-1}$, we first rearrange the products $H(d)T(a)$ and $H(d)H(b)$. Taking into account the relations (8), we write

$$\begin{aligned} H(d)T(a) &= H(d\widetilde{a}) - T(d)H(\widetilde{a}) = H(b) - T(d)H(\widetilde{a}), \\ H(d)H(b) &= T(d\widetilde{b}) - T(d)T(\widetilde{b}) = T(a) - T(d)T(\widetilde{b}). \end{aligned}$$

These representations yield

$$\begin{aligned} &T(\widetilde{a}^{-1})T_l^{-1}(d)\,(I - H(d))\,(T(a) + H(b)) \\ &= T(\widetilde{a}^{-1})T_l^{-1}(d)\left(T(a) + H(b) - \left(T(a) + H(b) - T(d)H(\widetilde{a}) - T(d)T(\widetilde{b})\right)\right) \\ &= T(\widetilde{a}^{-1})T_l^{-1}(d)\left(T(d)T(\widetilde{b}) + T(d)H(\widetilde{a})\right) \\ &= T(\widetilde{a}^{-1})T(\widetilde{b}) + T(\widetilde{a}^{-1})H(\widetilde{a}) \end{aligned}$$

$$= \left(T(\widetilde{a}^{-1}\widetilde{b}) - H(\widetilde{a}^{-1})H(b)\right) + \left(H(\widetilde{a}^{-1}\widetilde{a}) - H(\widetilde{a}^{-1})T(a)\right)$$
$$= \left(T(c) - H(\widetilde{a}^{-1})H(b)\right) + \left(H(1) - H(\widetilde{a}^{-1})T(a)\right)$$
$$= T(c) - H(\widetilde{a}^{-1})(T(a) + H(b)).$$

Therefore,

$$\left(H(\widetilde{a}^{-1}) + T(\widetilde{a}^{-1})T_l^{-1}(d)(I - H(d))\right)(T(a) + H(b)) = T(c)$$

and the representation (4) follows.

The generalized invertibility of $T(a) + H(b)$, $a, b \in L^\infty$ and the representation (6)–(7) have been established in [7] for Toeplitz plus Hankel operators acting on the classical Hardy spaces H^p. The corresponding proof is also valid for Toeplitz plus Hankel operators considered on the spaces of sequences. □

Corollary 1 *If both operators $T(c)$ and $T(b)$ are invertible, then the operator $T(a) + H(b)$ is also invertible and its inverse can be obtained from any of the representations* (4)–(6) *via replacing one-sided inverses by the corresponding inverses $T^{-1}(c)$ and $T^{-1}(d)$.*

Remark 1 According to the above corollary, we can have tree differently looking representations for the operator $(T(a) + H(b))^{-1}$. They can be transformed into each other by using the Widom identities (8).

Remark 2 If both operators $T(c)$ and $T(d)$ are invertible from the same side, then the formula (6) can also be used for constructing the left- or right-inverse of the operator $T(a) + H(b)$. However, the operators **A**, **B**, and **D** in (7) have to be modified—viz. the operator $T_r^{-1}(c)$ should be replaced by $T_l^{-1}(c)$ or the operator $T_l^{-1}(d)$ should be replaced by $T_r^{-1}(d)$. Nevertheless, any of the representations (4) and (5) looks much simpler than (6).

Theorem 1 shows that the inverses of Toeplitz plus Hankel operators are expressed via the inverses of Toeplitz operators. For Toeplitz operators on classical Hardy spaces, the inverses can be constructed by using the Wiener-Hopf factorization of the corresponding generating functions. In l^p-spaces this scheme does not always work. Therefore, here we recall some results from [9] when the Wiener-Hopf factorization of generating functions yields an appropriate factorization of Toeplitz operators and efficiently construct their one-sided inverses.

Let $H^p = H^p(\mathbb{T})$ and $\overline{H}^p = \overline{H^p(\mathbb{T})}$ denote the Hardy spaces of all functions $f \in L^p$ the Fourier coefficients $\widehat{f}$ of which vanish for all $n < 0$ and all $n > 0$, respectively. On the spaces L^p, $1 < p < \infty$, we consider the operators

$$\mathbf{J}f(t) := t^{-1}f(t^{-1}), \quad \mathbf{P}f(t) := \sum_{n \in \mathbb{Z}_+} \widehat{f}_n t^n,$$

and the operator $\mathbf{Q} = \mathbf{I} - \mathbf{P}$, where $\mathbf{I}$ denotes the identity operator. Consequently, if $a \in L^\infty$, then the Toeplitz and Hankel operators on the space H^p, $1 < p < \infty$ are respectively defined by

$$\mathbf{T}(a)f := \mathbf{P}af, \quad \mathbf{H}(a)f := \mathbf{P}a\mathbf{Q}\mathbf{J}f.$$

The representations (4)–(6) for various types of inverses of Toeplitz plus Hankel operators acting on l^p-spaces are based on the one-sided inverses of Toeplitz operators $T(c)$ and $T(b)$. For Toeplitz operators $\mathbf{T}(g)$, $g \in L^\infty$ considered on the classical Hardy spaces H^p, their one-sided inverses can be efficiently constructed. More exactly, assume that $\mathbf{T}(g)$ is Fredholm and $\operatorname{ind}\mathbf{T}(g) = n$, where

$$\operatorname{ind}\mathbf{T}(g) = \dim\ker\mathbf{T}(g) - \dim\ker\mathbf{T}^*(g)$$

is the index of the operator $\mathbf{T}(g)$. If $n \geq 0$, the operator $\mathbf{T}(g)$ is right invertible and one of its right inverses has the form

$$\mathbf{T}_r^{-1}(g) = \mathbf{T}^{-1}(t^n g)\mathbf{T}(t^n).$$

On the other hand, if $n \leq 0$, then the operator $\mathbf{T}(g)$ is left invertible and one of its left inverses has the form

$$\mathbf{T}_l^{-1}(g) = \mathbf{T}(t^n)\mathbf{T}^{-1}(t^n g).$$

The problem is now reduced to the construction of the inverse operator $\mathbf{T}^{-1}(t^n g)$. This can be done by using a factorization of the function g. Recall that a function $g \in L^\infty$ admits a generalized Wiener-Hopf factorization in H^p if it can be represented in the form

$$g(t) = g_-(t)\, t^{-n}\, g_+(t), \quad n \in \mathbb{Z}, \tag{9}$$

where $g_+ \in H^q$, $g_+^{-1} \in H^p$, $g_- \in \overline{H^p}$, $g_-^{-1} \in \overline{H^q}$, and the linear operator $g_+^{-1}\mathbf{P}g_-^{-1}\mathbf{I}$ defined on the set span $\{t^k : k \in \mathbb{Z}_+\}$ can be boundedly extended onto the whole space H^p. In what follows, such a representation is simply referred to as a Wiener-Hopf factorization. It should be noted that the above factorization strongly depends on the space H^p, but if p and the value of g_- at $z = \infty$ are fixed, the representation (9) is unique. Accordingly, in this work we always assume that $g_-(\infty) = 1$.

Wiener-Hopf factorization is closely connected to the Fredholmness of Toeplitz operators acting on the Hardy spaces H^p. Thus the operator $\mathbf{T}(g) : H^p \to H^p$ is Fredholm if and only if g admits the Wiener-Hopf factorization (9), cf. [5]. In particular, the representation (9) yields that $\operatorname{ind}\mathbf{T}(g) = n$ and

$$\mathbf{T}^{-1}(t^n g) = \mathbf{T}(g_+^{-1})\mathbf{T}(g_-^{-1}). \tag{10}$$

Consequently, if ind $\mathbf{T}(g) = 0$, the Toeplitz operator $\mathbf{T}(g) : H^p \to H^p$ is invertible and its inverse can be explicitly written in the form (10). On the other hand, for invertible Toeplitz operators $T(g)$ acting on sequence spaces l^p, the Wiener-Hopf factorization (9) of the generating function $g \in M^p$ does not automatically produce an explicit expression of the inverse operator $T^{-1}(g)$. Nevertheless, this representation can still be helpful, as the following theorem shows.

Theorem 2 (cf. [9]) *Assume that $g \in M^p$, the operator $T(g)\colon l^p \to l^p$ is invertible, and let $q := p/(p-1)$, i.e. $1/p + 1/q = 1$. If, in addition, the operator $\mathbf{T}(g)$ is invertible on the space H^q, then the function g admits the Wiener-Hopf factorization*

$$g = g_- g_+,$$

in H^q and the inverse for the operator $T(g)$ can be represented in the form

$$T^{-1}(g) := \mathcal{T}(g_+^{-1})\mathcal{T}(g_-^{-1}), \tag{11}$$

where

$$\mathcal{T}(g_+^{-1}) := (\widehat{g}^{-1}_{+,j-k})_{j,k=0}^{\infty}, \quad \mathcal{T}(g_-^{-1}) := (\widehat{g}^{-1}_{-,j-k})_{j,k=0}^{\infty},$$

and $(\widehat{g}^{-1}_{+,n})_{n\in\mathbb{Z}}$ and $(\widehat{g}^{-1}_{-,n})_{n\in\mathbb{Z}}$ are the sequences of the Fourier coefficients of the functions g_+^{-1} and g_-^{-1}, respectively.

The representation (11) allows to construct one-sided inverses for Fredholm Toeplitz operators acting on sequence spaces l^p and, consequently, to derive efficient formulas for one-sided and generalized inverses of operators $T(a) + H(b)$ with matching symbols a and b. However, Theorem 2 uses an important condition—viz. the operators $T(g)$ and $\mathbf{T}(g)$ should be simultaneously invertible—viz. the first on the space l^p and the second on the Hardy space H^q, $1/p + 1/q = 1$. This condition is, in particular, satisfied for the operators with symbols from decomposing subalgebras of M^p such as the Wiener algebra $W = W(\mathbb{T})$ of functions with absolutely convergent Fourier series. It is also known that for functions $g \in PC_p$, the corresponding operators $T(g) : l^p \to l^p$ and $\mathbf{T}(g) : H^q \to H^q$, $1/p + 1/q = 1$ are simultaneously Fredholm and have the same indices—cf. [12, 13] and [17]. Therefore, if $g \in PC_p$ and one of the operators mentioned is invertible, then so is the other.

Thus the problem of simultaneous invertibility of the operator $T(g)$ on l^p and the operator $\mathbf{T}(g)$ on H^q under the condition $1/p + 1/q = 1$ could be of interest in both the theory of Toeplitz operators and Wiener-Hopf factorization.

4 Invertibility: Factorization Based Approach

Although Wiener-Hopf factorization is a convenient machinery for constructing the inverses of Toeplitz plus Hankel operators, the invertibility conditions in the previous section do not involve any factorization arguments. On the other hand, applying such factorizations to the corresponding operators on the Hardy spaces H^p allows to derive more subtle invertibility results. At the same time, for l^p-spaces the factorization (9) cannot be immediately used for obtaining required information about the operators of interest. Therefore, here we consider a special class of Toeplitz plus Hankel operators. However, before we proceed, let us agree on the notation. First, as we already did earlier, the operators on Hardy spaces H^p will be written in boldface in contrast to the ones acting on l^p-spaces. Besides, since we consider operators on various H^p- and l^p-spaces, the index p will be sometimes directly incorporated in the operator notation in order to show the space where the corresponding operator acts. For example, $T_p(g)$ means the Toeplitz operator $T(g)$ acting on the space l^p, and $\mathbf{H}_q(g)$ the Hankel operator acting on the space H^q. From now on, we also presume that every time when the indices p and q appear together, they satisfy the relation $1/p + 1/q = 1$.

Note that if (a, b) is a matching pair with the subordinated pair (c, d) and $T_p(c)$ and $T_p(d)$ are Fredholm, then the factorizations [6, Eqs.(3.1) and (3.7))] yield the Fredholmness of $T_p(a) + H_p(b)$. Besides, the one-sided invertibility of $T_p(c)$, $T_p(d)$ allows to obtain various results about the invertibility of $T_p(a) + H_p(b)$, cf. Theorem 1 above. Thus the Toeplitz operators $T_p(c)$ and $T_p(d)$ generated by the pair subordinated for (a, b) are closely related to the corresponding Toeplitz plus Hankel operators and have to be studied in more detail. First of all, we note that c and d are special matching functions—viz.

$$c\widetilde{c} = 1 = d\widetilde{d},$$

and recall the properties of Toeplitz operators with such generating functions. For the reader's convenience, we summarize them in the proposition below.

Proposition 1 (cf. [6, Corollary 5.3 & Proposition 3.4], [9, Theorem 7]) *Let $g \in GM^p$ be a matching function—i.e. $g\widetilde{g} = 1$. Then the following assertions hold:*

1. *If the Toeplitz operator $\mathbf{T}_q(g)$ is Fredholm and* $\operatorname{ind} \mathbf{T}_q(g) = r$, *then the function g admits a Wiener-Hopf factorization in H^q of the form*

$$g(t) = \sigma(g)\, g_+(t)\, t^{-r}\, \widetilde{g}_+^{-1}(t), \quad g_-(\infty) = 1, \tag{12}$$

where $g_+ \in H^p$, $g_+^{-1} \in H^q$, and $\sigma(g) = g_+(0) = \pm 1$.

2. *If the Toeplitz operator* $T_p(g)$ *is invertible from the right, then the operators* $P_g^{\pm}$ *defined by*

$$P_g^{\pm} := \frac{1}{2}(I \pm JQgP) \tag{13}$$

are complementary projections on the null space $\ker T_p(g)$.
3. *Let* $T_p(g)$ *and* $\mathbf{T}_q(g)$ *be right-invertible operators such that*

$$\operatorname{ind} \mathbf{T}_q(g) = \operatorname{ind} T_p(g) = r,$$

the function g_+ *and the term* $\sigma(g)$ *be defined as in* (12), *and* $(\widehat{g_{+,j}^{-1}})$ *be the sequence of the Fourier coefficients of the function* g_+^{-1}. *Then the following sets* $\mathcal{B}_\pm$ *of sequences form bases in the image spaces* $\operatorname{im} P_g^{\pm}$:

(i) *If* $r = 2l$, $l \in \mathbb{N}$, *then*

$$\mathcal{B}_\pm := \left\{ (\widehat{g_{+,j-(l-k-1)}^{-1}} \pm \sigma(g)\widehat{g_{+,j-(l+k)}^{-1}})_{j\in\mathbb{Z}_+} : k = 0, \cdots, l-1 \right\}. \tag{14}$$

(ii) *If* $r = 2l + 1$, $l \in \mathbb{Z}_+$, *then*

$$\mathcal{B}_\pm := \left\{ (\widehat{g_{+,j-(l+k)}^{-1}} \pm \sigma(g)\widehat{g_{+,j-(l-k)}^{-1}})_{j\in\mathbb{Z}_+} : k = 0, \cdots, l \right\} \setminus \{0\}. \tag{15}$$

Using the above results, we can establish sufficient and necessary conditions of invertibility and one-sided invertibility for a class of Toeplitz plus Hankel operators on l^p. More exactly, let $\mathfrak{M}_{TH}^p(\kappa_1, \kappa_2)$ denote the set of Toeplitz plus Hankel operators $T_p(a) + H_p(b)$ such that:

1. The duo $(a, b) \in GM^p \times GM^p$ is a matching pair with the subordinated pair (c, d).
2. The operators $T_p(c)$ and $\mathbf{T}_q(c)$ are Fredholm and

$$\operatorname{ind} T_p(c) = \operatorname{ind} \mathbf{T}_q(c) = \kappa_1.$$

3. The operators $T_p(d)$ and $\mathbf{T}_q(d)$ are Fredholm and

$$\operatorname{ind} T_p(d) = \operatorname{ind} \mathbf{T}_q(d) = \kappa_2.$$

4.1 Necessary Conditions

Let us start with necessary conditions for the invertibility of Toeplitz plus Hankel operators from $\mathfrak{M}^p_{TH}(\kappa_1, \kappa_2)$. Note that in this section we will not directly appeal to the operators acting on the Hardy spaces $H^q(\mathbb{T})$. Therefore, in order to simplify the notation, we will write all operators without the subscript, which indicated the space of the operator action.

Theorem 3 *Assume that $T(a) + H(b) \in \mathfrak{M}^p_{TH}(\kappa_1, \kappa_2)$. Then the following assertions hold:*

1. If $T(a) + H(b)$ is left-invertible and $T(c)$ is right-invertible, then

$$\kappa_1 \le 1, \quad \kappa_2 \le 1. \tag{16}$$

2. If $T(a) + H(b)$ is right-invertible and $T(d)$ is left-invertible, then

$$\kappa_1 \ge -1, \quad \kappa_2 \ge -1. \tag{17}$$

3. If $T(a) + H(b)$ is invertible and both operators $T(c)$ and $T(d)$ are either left- or right-invertible or if $T(c)$ is right-invertible and $T(d)$ is left-invertible, then

$$|\kappa_1| \le 1, \quad |\kappa_2| \le 1. \tag{18}$$

Proof We start with the assertion 1. If $T(c)$ is right-invertible, then according to [9, Lemma 5], the kernel of the Toeplitz plus Hankel operator $T(a) + H(b)$ has the form

$$\ker(T(a) + H(b)) = \varphi(\operatorname{im} P_d^+) \dotplus \operatorname{im} P_c^-, \tag{19}$$

where $\varphi : \ker T(d) \to \ker(T(a) + H(b))$ is the injective operator defined by

$$\varphi(s) := \frac{1}{2}\left(T_r^{-1}(c)T(\widetilde{a}^{-1})s - JQcPT_r^{-1}(c)T(\widetilde{a}^{-1})s + JQ\widetilde{a}^{-1}s\right), \tag{20}$$

and the projections $P^\pm$ are introduced in (13). Note that the relations (14), (15) show that if $\operatorname{ind} T(g) = 2l$, $l \in \mathbb{N}$, then

$$\dim P_g^\pm = l \tag{21}$$

and if $\operatorname{ind} T(g) = 2l + 1$, $l \in \mathbb{Z}_+$, then

$$\dim P_g^\pm = l + \frac{1 \pm \sigma(g)}{2}. \tag{22}$$

Assuming that one of the indices κ_1 or κ_2 is greater than 1, we use (19), (21), (22) and the injectivity of the mapping φ to obtain that the kernel of $T(a)+H(b)$ contains a non-zero element. This contradicts the left-invertibility of the operator in question, so that the estimates (16) hold.

Assuming that $T(a) + H(b)$ is right-invertible, we note that its adjoint $T(\tilde{\overline{a}}) + H(\tilde{\overline{b}}) \in \mathfrak{M}(l^q)$ is left-invertible. Besides, the duo $(\overline{a}, \tilde{\overline{b}})$ is also a matching pair with the subordinated pair $(\overline{d}, \overline{c})$. Since $T(\overline{d})$ is right invertible, the assertion 1 yields

$$-\kappa_2 \leq 1, \quad -\kappa_1 \leq 1,$$

and the estimates (18) follows.

The remaining estimate (18) is a consequence of (19), (21), (22) and the assertions 1, 2. □

Consider now the case where $T(c)$ and $T(d)$ are respectively left- and right-invertible operators having non-zero indices. This situation is not covered by Theorem 3 and should be treated separately.

Theorem 4 *Let $T(a) + H(b) \in \mathfrak{M}^p_{TH}(\kappa_1, \kappa_2)$ and $\kappa_1 < 0$, $\kappa_2 > 0$.*

1. *If the operator $T(a) + H(b)$ is left-invertible, then:*
   - *(i) if κ_1 is an odd number and κ_2 an even one, then $\kappa_1 + \kappa_2 \leq \sigma(c)$;*
   - *(ii) if κ_1 and κ_2 are odd numbers, then $\kappa_1 + \kappa_2 \leq \sigma(c) + \sigma(d)$;*
   - *(iii) if κ_1 is an even number and κ_2 an odd one, then $\kappa_1 + \kappa_2 \leq \sigma(d)$;*
   - *(iv) if κ_1 and κ_2 are even numbers, then $\kappa_1 + \kappa_2 \leq 0$.*
2. *If the operator $T(a) + H(b)$ is right-invertible, then:*
   - *(i) if κ_1 is an odd number and κ_2 an even one, then $\kappa_1 + \kappa_2 \geq \sigma(c)$;*
   - *(ii) if κ_1 and κ_2 are odd numbers, then $\kappa_1 + \kappa_2 \geq \sigma(c) + \sigma(d)$;*
   - *(iii) if κ_1 is an even number and κ_2 an odd one, then $\kappa_1 + \kappa_2 \geq \sigma(d)$;*
   - *(iv) if κ_1 and κ_2 are even numbers, then $\kappa_1 + \kappa_2 \geq 0$.*

Proof In order to prove the necessary conditions for the one-sided invertibility of the operators from $\mathfrak{M}^p_{TH}(\kappa_1, \kappa_2)$, we need an additional characteristic for the indices of the subordinated operators $T(c)$ and $T(d)$. Let n and m be the integers such that

$$0 \leq \kappa_1 + 2n \leq 1, \quad 0 \leq 2m - \kappa_2 \leq 1.$$

Such n and m are uniquely defined and take values 0 or 1 depending on whether the corresponding index κ_j, $j = 1, 2$ is even or odd.

Now we can exploit the kernel and co-kernel descriptions for the operators from $\mathfrak{M}^p_{TH}(\kappa_1, \kappa_2)$ in the situation when the corresponding indices satisfy the condition $\kappa_1 < 0$, $\kappa_2 > 0$. In particular, to obtain necessary conditions for the left-invertibility of the operator $T(a)+H(b)$, one can use [9, Lemma 7]. We only prove the assertion 1(i). In this situation, we have

$$\kappa_1 + 2n = 1, \quad \kappa_2 = 2m.$$

Note that for the reader's convenience, here we write the related part of [9, Lemma 7] in the ready-to-use form. More exactly, let c_+ be the plus factor in the Wiener-Hopf factorization of the subordinated function c, and $\{(\widehat{c}_{+,j}^{-1})\}$ be the one-dimensional subspace of the space l^p generated by the sequence of the Fourier coefficients of the function c_+^{-1}. According to [9, Lemma 7] and Theorem 1, the kernel of the operator $T(a) + H(b)$ can be represented in the form

$$\ker(T(a) + H(b)) = T(t^{-n}) \left(\frac{1 - \sigma(c)}{2} \{(\widehat{c}_{+,j}^{-1})\} \dot{+} \varphi(\operatorname{im} P_d^+) \right). \tag{23}$$

Note that the operator φ in (20) is generated by the matching pair (a, b). The same representation (20) is used to determine the operator φ in (23), but the matching pair (a, b) should be replaced by the pair (at^{-n}, bt^n).

Consider now the subspace U of l^p defined as

$$U = \frac{1 - \sigma(c)}{2} \{(\widehat{c}_{+,j}^{-1})\} \dot{+} \varphi(\operatorname{im} P_d^+)$$

and assume that $\dim U > n$. Then, there is an element $u_0 \in U$, $u_0 \neq 0$, the first n coordinates of which are equal to zero—i.e.

$$u_0 = (0, \dots, 0, \xi_n, \xi_{n+1}, \dots).$$

Consequently, the kernel of the operator $T(a)+H(b)$ contains a non-zero element—viz. $T(t^{-n})u_0$, so that $T(a) + H(b)$ is not left-invertible. This contradicts the assumptions of Theorem 4. Thus,

$$n \geq \dim U.$$

Taking into account the definition of the numbers n, m and the relations (21), (22), we write the above inequality as

$$\frac{1 - \kappa_1}{2} \geq \frac{1 - \sigma(c)}{2} + \frac{\kappa_2}{2},$$

and the assertion 1(i) follows. The three other necessary left-invertibility conditions—viz. 1(ii)–1(iv), can be proven analogously.

In order to establish necessary conditions for the right-invertibility of an operator $T(a) + H(b)$ in the case $\kappa_1 < 0$, $\kappa_2 > 0$, we first have to describe the cokernel of the operator under consideration. This can be done by applying [9, Lemma 7] to the operator $T(\overline{a}) + H(\widetilde{\overline{b}})$. After that, the proof follows the scheme above. □

Combining the two parts of Theorem 4, we arrive at necessary conditions for the invertibility of operators $T(a) + H(b)$.

Corollary 2 *Let $T(a) + H(b) \in \mathfrak{M}^p_{TH}(\kappa_1, \kappa_2)$, $\kappa_1 < 0$, $\kappa_2 > 0$, and the operator $T(a) + H(b)$ be invertible. Then:*

1. *if κ_1 is an odd number and κ_2 an even one, then $\kappa_1 + \kappa_2 = \sigma(c)$;*
2. *if κ_1 and κ_2 are odd numbers, then $\kappa_1 + \kappa_2 = \sigma(c) + \sigma(d)$;*
3. *if κ_1 is an even number and κ_2 an odd one, then $\kappa_1 + \kappa_2 = \sigma(d)$;*
4. *if κ_1 and κ_2 are even numbers, then $\kappa_1 + \kappa_2 = 0$.*

4.2 Sufficient Conditions

As was already mentioned—cf. Sect. 2, the simultaneous left- or right-invertibility of the subordinated operators $T(c)$ and $T(d)$ yields the left- or right-invertibility of the corresponding Toeplitz plus Hankel operator $T(a) + H(b)$. However, the operator $T(a) + H(b)$ can be one-sided invertible or invertible even if the subordinated operators are not invertible from the same side. Let us note a few such cases.

Theorem 5 *Assume that $T(a) + H(b) \in \mathfrak{M}^p_{TH}(\kappa_1, \kappa_2)$. Then:*

1. *If $\kappa_2 < 0$, $\kappa_1 = 1$ and $\sigma(c) = 1$, then the operator $T(a) + H(b)$ is left invertible.*
2. *If $\kappa_1 > 0$, $\kappa_2 = -1$ and $\sigma(d) = 1$, then the operator $T(a) + H(b)$ is right invertible.*
3. *If $\kappa_1 = 1$, $\kappa_2 = -1$, and $\sigma(c) = \sigma(d) = 1$, then the operator $T(a) + H(b)$ is invertible.*

Proof Starting with the assertion 1, we note that since the subordinated operator $T(c)$ is right-invertible, the kernel of the operator $T(a) + H(b)$ has the form (19), i.e.

$$\ker(T(a) + H(b)) = \varphi_+(\operatorname{im} P_d^+) + \operatorname{im} P_c^-.$$

However, according to the assertion 3(ii) of Proposition 1, the set $\operatorname{im} P_c^-$ contains only the zero-element, and the left-invertibility of the operator $T(d)$ yields $\operatorname{im} P_d^+ = \{0\}$. Thus

$$\ker(T(a) + H(b)) = \{0\},$$

and the operator under consideration is left-invertible.

As far as the assertion 2. is concerned, we first note the relation

$$\operatorname{coker}(T(a) + H(b)) = \varphi_+(\operatorname{im} P_{\overline{c}}^+) + \operatorname{im} P_{\overline{d}}^-,$$

which can be obtained from [9, Lemma 5] by passing to the adjoint operator. Consequently, following the proof of the assertion 1, one shows that

$$\operatorname{coker}(T(a) + H(b)) = \{0\},$$

and the assertion 2. follows.

The assertion 3 is an obvious consequence of the statements 1. and 2. □

Let us also note another interesting situation where both subordinated operators are invertible from the same side, both have non-zero kernels or cokernels, but the corresponding Toeplitz plus Hankel operator is just invertible.

Theorem 6 *Assume that* $T(a) + H(b) \in \mathfrak{M}^p_{TH}(\kappa_1, \kappa_2)$. *Then:*

1. *If* $\kappa_1 = \kappa_2 = 1$ *and* $\sigma(c) = 1$, $\sigma(d) = -1$, *then the operator* $T(a) + H(b)$ *is invertible.*
2. *If* $\kappa_1 = \kappa_2 = -1$ *and* $\sigma(c) = -1$, $\sigma(d) = 1$, *then the operator* $T(a) + H(b)$ *is invertible.*

Proof Since the proof of both statements is similar, we only show the assertion 1. If $\kappa_1 = \kappa_2 = 1$, the both subordinated operators $T(c)$ and $T(d)$ are right-invertible, and the operator $T(a) + H(b)$ is also right invertible by assertion 2. of Theorem 1. The kernel of this operator has the form (19), and recalling the representations (15) for the basis of the image spaces of the projections $P^{\pm}$, we obtain that

$$\ker(T(a) + H(b)) = \{0\},$$

and the operator in question is invertible. □

Remark 3 Under the conditions of Theorem 5, the inverse operator $(T(a) + H(b))^{-1}$ can be constructed by the formulas (6)–(7). If the subordinated operators are as in Theorem 6, then the inverse operator can be written in the form (4) or (5), depending on whether $T(c)$ and $T(d)$ are left- or right-invertible.

5 Conclusion: A More General Class of Symbols

Analysing the relation (2), we note that since the symbol b is not uniquely defined by the operator $H(b)$, this identity can be replaced by another one, so that the methods used here can be extended to a considerably larger class of operators $T(a) + H(b)$. For example, the pair

$$a(t) = \frac{1}{2+t}, \quad b(t) = \frac{t^{n+1} + 2t + 1}{t(1+2t)}, \quad t \in \mathbb{T}, \quad n \in \mathbb{Z}_+,$$

does not satisfy the relation (2). However, representing b in the form

$$b(t) = \frac{t^n}{1+2t} + \frac{1}{t} = b_0(t) + \frac{1}{t},$$

we note that

$$T(a) + H(b) = T(a) + H(b_0),$$

and since the pair (a, b_0) already satisfies the relation (2), the methods of [6, 8, 10, 11] can be used to study the invertibility of the operator $T(a) + H(b)$ by replacing it via the operator $T(A) + H(b_0)$.

The above example suggests the following definition.

Definition 1 A duo (a, b) is called the generalized matching pair if there is a function $h \in \overline{H^\infty}$ such that $(a, b - h)$ is a matching pair.

It should be noted that determining of whether a given duo (a, b) is a generalized matching pair and finding a suitable function $h \in \overline{H^\infty}$ is a challenging problem and the authors are not aware of any possible solution. Nevertheless, if such a function h is known and $b-h$ belongs to a suitable algebra M^p, the invertibility of $T(a)+H(b)$ can be studied in more detail.

Acknowledgments The authors are very grateful to the referees and the editors for their valuable suggestions and comments which substantially improved the presentation of this work.

Competing Interests The work of the first author was partially supported by the Guangdong Provincial Key Laboratory of Interdisciplinary Research and Application for Data Science under UIC Grant 2022B1212010006 and by the NNSF of China (Grant Nos. 11731006 and K20911001.)

References

1. Baik, J., Rains, E.M.: Algebraic aspects of increasing subsequences. Duke Math. J. **109**, 1–65 (2001)
2. Basor, E.L., Ehrhardt, T.: Fredholm and invertibility theory for a special class of Toeplitz + Hankel operators. J. Spectral Theory **3**, 171–214 (2013)
3. Basor, E.L., Ehrhardt, T.: Assymptotic formulas for determinants of a special class of Toeplitz + Hankel matrices. In: Bini, D.A., Ehrhardt, T., Karlovich, A.Y., Spitkovsky, I. (eds.) Large Truncated Toeplitz Matrices, Toeplitz Operators, and Related Topics. The Albrecht Böttcher Anniversary Volume. Operator Theory. Advances and Applications, vol. 259, pp. 125–154. Birkhäuser, Basel (2017)
4. Basor, E.L, Chen, Y., Ehrhardt, T.: Painlevé V and time-dependent Jacobi polynomials. J. Phys. A **43**, 015204 (2010)
5. Böttcher, A., Silbermann, B.: Analysis of Toeplitz Operators. Springer Monographs in Mathematics. Springer, Berlin (2006)
6. Didenko, V.D., Silbermann, B.: Structure of kernels and cokernels of Toeplitz plus Hankel operators. Integr. Equ. Oper. Theory **80**, 1–31 (2014)
7. Didenko, V.D., Silbermann, B.: Generalized inverses and solution of equations with Toeplitz plus Hankel operators. Bol. Soc. Mat. Mex. **22**, 645–667 (2016)
8. Didenko, V.D., Silbermann, B.: Invertibility and inverses of Toeplitz plus Hankel operators. J. Oper. Theory **72**, 293–307 (2017)
9. Didenko, V.D., Silbermann, B.: Kernels of a class of Toeplitz plus Hankel operators with piecewise continuous generating functions. In: Kuo, F.Y., Dick, J., Woźniakowski, H. (eds.) Contemporary Computational Mathematics - A Celebration of the 80th Birthday of Ian Sloan, pp. 317–337, Springer, Cham (2018)

10. Didenko, V.D., Silbermann, B.: The invertibility of Toeplitz plus Hankel operators with subordinated operators of even index. Linear Algebra Appl. **578**, 425–445 (2019)
11. Didenko, V.D., Silbermann, B.: Invertibility issues for Toeplitz plus Hankel operators and their close relatives. In: Bastos, M.A., Castro, L., Karlovich, A.Y. (eds.) Operator Theory, Functional Analysis and Applications. Proceedings of the 30th International Workshop on Operator Theory and its Applications, IWOTA 2019, Lisbon, Portugal, July 22–26, 2019, pp. 113–156. Birkhäuser, Cham (2021)
12. Duduchava, R.: On discrete Wiener-Hopf equations in l^p spaces with weight. Soobsh. Akad. Nauk Gruz. SSR **67**, 17–20 (1972) (In Russian)
13. Duduchava, R.: The discrete Wiener-Hopf equations. Proc. A. Razmadze Math. Inst. **50**, 42–59 (1975) (In Russian)
14. Ehrhardt, T.: Factorization theory for Toeplitz+Hankel operators and singular integral operators with flip. Habilitation Thesis, Technische Universität Chemnitz, 2004
15. Ehrhardt, T.: Invertibility theory for Toeplitz plus Hankel operators and singular integral operators with flip. J. Funct. Anal. **208**, 64–106 (2004)
16. Forrester, P.J., Frankel, N.E.: Applications and generalizations of Fisher-Hartwig asymptotics. J. Math. Phys. **45**, 2003–2028 (2004)
17. Gohberg, I.C., Krupnik, N.Y.: On the algebra generated by Toeplitz matrices. Funkcional. Anal. i Priložen. **3**, 46–56 (1969) (In Russian)
18. Grudsky, S., Rybkin, A.: On positive type initial profiles for the KdV equation. Proc. Am. Math. Soc. **142**, 2079–2086 (2014)
19. Grudsky, S., Rybkin, A.: Soliton theory and Hankel operators. SIAM J. Math. Anal. **47**, 2283–2323 (2015)
20. Lebre, A.B., Meister, E. Teixeira, F.S.: Some results on the invertibility of Wiener-Hopf-Hankel operators. Z. Anal. Anwend. **11**, 57–76 (1992)
21. Meister, E., Penzel, F., Speck, F.O., Teixeira, F.S.: Two-media scattering problems in a half-space. In: Partial Differential Equations with Real Analysis. Dedicated to Robert Pertsch Gilbert on the Occasion of His 60th Birthday, pp. 122–146. Longman Scientific & Technical/John Wiley & Sons, Inc., Harlow/New York (1992)
22. Meister, E., Speck, F.O., Teixeira, F.S.: Wiener-Hopf-Hankel operators for some wedge diffraction problems with mixed boundary conditions. J. Integr. Equ. Appl. **4**, 229–255 (1992)
23. Teixeira, F.S.: Diffraction by a rectangular wedge: Wiener-Hopf-Hankel formulation. Integr. Equ. Oper. Theory **14**, 436–454 (1991)

On Pseudodifferential Operators with Slowly Oscillating Symbols on Variable Lebesgue Spaces with Khvedelidze Weights

Cláudio Fernandes and Oleksiy Karlovych

To Yuri Karlovich on the occasion of his 75th birthday

Abstract Let $p(\cdot)$ be a variable exponent in the class $LH^*(\mathbb{R})$ and ϱ be a Khvedelidze weight. We prove that if $a \in S^0_{1,0}(\mathbb{R} \times \mathbb{R})$ slowly oscillates at infinity in the first variable, then the condition

$$\lim_{R\to\infty} \inf_{|x|+|\xi|\geq R} |a(x,\xi)| > 0$$

is sufficient for the Fredholmness of the pseudodifferential operator $\mathrm{Op}(a)$ on the weighted variable Lebesgue space $L^{p(\cdot)}(\mathbb{R}, \varrho)$.

1 Introduction and Main Results

For a Banach space $\mathcal{X}$, let $\mathcal{B}(\mathcal{X})$ and $\mathcal{K}(\mathcal{X})$ denote the Banach algebra of all bounded linear operators on $\mathcal{X}$ and its closed two-sided ideal of all compact linear operators on $\mathcal{X}$, respectively. As usual, we denote by I the identity operator on $\mathcal{X}$. Recall that an operator $A \in \mathcal{B}(\mathcal{X})$ is said to be Fredholm if there is an operator $B \in \mathcal{B}(\mathcal{X})$ such that the operators $AB - I$ and $BA - I$ belong to $\mathcal{K}(\mathcal{X})$. In that case the operator B is called a regularizer for the operator A.

C. Fernandes (✉) · O. Karlovych
Centro de Matemática e Aplicações, Departamento de Matemática, Faculdade de Ciências e Tecnologia, Universidade Nova de Lisboa, Caparica, Portugal
e-mail: caf@fct.unl.pt; oyk@fct.unl.pt

A. Böttcher et al. (eds.), *Achievements and Challenges in the Field of Convolution Operators*, Operator Theory: Advances and Applications 306,
https://doi.org/10.1007/978-3-031-80486-1_8

Let $C_0^\infty(\mathbb{R})$ denote the set of all infinitely differentiable functions with compact support. Recall that, given $u \in C_0^\infty(\mathbb{R})$, a pseudodifferential operator $\mathrm{Op}(a)$ is formally defined by the formula

$$(\mathrm{Op}(a)u)(x) := \frac{1}{2\pi}\int_{\mathbb{R}} d\xi \int_{\mathbb{R}} a(x,\xi)u(y)e^{i(x-y)\xi}dy,$$

where the symbol a is assumed to be smooth in both the spatial variable x and the frequency variable ξ, and satisfies certain growth conditions (see, e.g., [27, Chap. VI]). An example of symbols one might consider is the class $S^m_{\rho,\delta}(\mathbb{R}\times\mathbb{R})$, introduced by Hörmander [12], consisting of $a \in C^\infty(\mathbb{R}\times\mathbb{R})$ satisfying

$$|\partial_\xi^\alpha \partial_x^\beta a(x,\xi)| \le C_{\alpha,\beta}\langle\xi\rangle^{m-\rho\alpha+\delta\beta}, \quad x,\xi \in \mathbb{R},$$

where $m \in \mathbb{R}$ and $0 \le \delta, \rho \le 1$ and the positive constants $C_{\alpha,\beta}$ depend only on $\alpha, \beta \in \mathbb{Z}_+ := \{0, 1, 2, \dots\}$. Here, as usual, $\partial_x := \partial/\partial_x$, $\partial_\xi := \partial/\partial_\xi$, and $\langle\xi\rangle := (1+\xi^2)^{1/2}$.

The aim of this paper is to initiate the study of Fredholmness of one-dimensional pseudodifferential operators on weighted variable Lebesgue spaces and to extend some results by Rabinovich and Samko [25] obtained by them in the nonweighted (and multidimensional) setting.

Let $p(\cdot) : \mathbb{R} \to [1,\infty]$ be a measurable a.e. finite function called a variable exponent. By $L^{p(\cdot)}(\mathbb{R})$ we denote the set of all complex-valued measurable functions f on $\mathbb{R}$ such that

$$I_{p(\cdot)}(f/\lambda) := \int_{\mathbb{R}} |f(x)/\lambda|^{p(x)}dx < \infty$$

for some $\lambda > 0$. This set becomes a Banach space when equipped with the norm

$$\|f\|_{L^{p(\cdot)}(\mathbb{R})} := \inf\left\{\lambda > 0 : I_{p(\cdot)}(f/\lambda) \le 1\right\}$$

(see, e.g., [6, Theorems 2.17 and 2.71]). It is easy to see that if p is constant, then $L^{p(\cdot)}(\mathbb{R})$ is nothing but the standard Lebesgue space $L^p(\mathbb{R})$. The space $L^{p(\cdot)}(\mathbb{R})$ is referred to as a variable Lebesgue space.

A measurable function $w : \mathbb{R} \to [0,\infty]$ is referred to as a weight whenever $0 < w(x) < \infty$ a.e. on $\mathbb{R}$. Given a variable a.e. finite exponent $p(\cdot) : \mathbb{R} \to [1,\infty]$ and a weight $w : \mathbb{R} \to [0,\infty]$, we define the weighted variable exponent space $L^{p(\cdot)}(\mathbb{R}, w)$ as the space of all measurable complex-valued functions f such that $fw \in L^{p(\cdot)}(\mathbb{R})$. The norm in this space is naturally defined by

$$\|f\|_{L^{p(\cdot)}(\mathbb{R},w)} := \|fw\|_{L^{p(\cdot)}(\mathbb{R})}.$$

We will consider our problem in the context of sufficiently regular variable exponents $p(\cdot)$ and so-called Khvedelidze weights. Let us give the corresponding definitions. Put

$$p_- := \operatorname*{ess\,inf}_{x\in\mathbb{R}} p(x), \quad p_+ := \operatorname*{ess\,sup}_{x\in\mathbb{R}} p(x).$$

We will assume that

$$1 < p_-, \quad p_+ < \infty. \tag{1}$$

Following [18, Section 1.1.1] (see also [6, Definition 2.2] and [8, Section 4.1]), a variable exponent $p(\cdot)$ is said to be locally log-Hölder continuous if there exists a constant $c_0 \in (0, \infty)$ such that

$$|p(x) - p(y)| \le \frac{c_0}{-\ln|x-y|} \tag{2}$$

for all $x, y \in \mathbb{R}$ satisfying $|x - y| \le 1/2$. A variable exponent $p(\cdot)$ is said to be log-Hölder continuous at infinity if there exist $c_1 \in (0, \infty)$ and $p_\infty \in (1, \infty)$ such that

$$|p(x) - p_\infty| \le \frac{c_1}{\ln(e + |x|)}$$

for all $x \in \mathbb{R}$. One says that $p(\cdot)$ is globally log-Hölder continuous on $\mathbb{R}$ if it is locally log-Hölder continuous and log-Hölder continuous at infinity. The class of all globally log-Hölder continuous variable exponents will be denoted by $LH(\mathbb{R})$.

Further, Kokilashvili, Paatashvili, and Samko introduced a slightly stronger condition than log-Hölder continuity at infinity (see [17, inequality (2.4)]). Following their work, we denote by $LH^*(\mathbb{R})$ the class of all locally log-Hölder continuous variable exponents $p(\cdot)$ such that there exist constants $c_2 \in (0, \infty)$ and $L \in (0, \infty)$ depending on $p(\cdot)$ and such that

$$|p(x) - p(y)| \le \frac{c_2}{-\ln|1/x - 1/y|} \tag{3}$$

for all $x, y \in \mathbb{R}$ satisfying $|x|, |y| > L$ and $|1/x - 1/y| \le 1/2$. It follows from [17, Remark 3.1] that

$$LH^*(\mathbb{R}) \subset LH(\mathbb{R}).$$

If $p(\cdot) \in LH^*(\mathbb{R})$, then the limit $\lim_{|x|\to\infty} p(x)$ exists. It will be denoted by $p(\infty)$.

The weights of the form

$$\varrho(x) := |x - i|^{\lambda_\infty} \prod_{j=1}^{m} |x - x_j|^{\lambda_j}, \quad x \in \mathbb{R}, \tag{4}$$

where $x_1 < \ldots < x_m$, $\lambda_1, \ldots, \lambda_m, \lambda_\infty \in \mathbb{R}$, will be called Khvedelidze weights. This class of weights was introduced by Khvedelidze [16], who studied the boundedness of the Cauchy singular integral operator on weighted Lebesgue spaces. For a variable exponent $p(\cdot) \in LH^*(\mathbb{R})$, the class of all Khvedelidze weights of the form (4) satisfying

$$0 < \frac{1}{p(x_j)} + \lambda_j < 1 \quad \text{for} \quad j = 1, \ldots, m, \quad 0 < \frac{1}{p(\infty)} + \lambda_\infty + \sum_{j=1}^{m} \lambda_j < 1, \tag{5}$$

will be denoted by $W_{p(\cdot)}(\mathbb{R})$. Note that Kokilashvili, Paatashvili, and Samko [17, Thoerem A] proved that if $p(\cdot) \in LH^*(\mathbb{R})$ and ϱ is a Khvedelidze weight of the form (4), then the Cauchy singular integral operator is bounded on $L^{p(\cdot)}(\mathbb{R}, \varrho)$ if and only if $\varrho \in W_{p(\cdot)}(\mathbb{R})$.

Our first result is the following theorem on the boundedness of pseudodifferential operators on weighted variable Lebesgue spaces.

Theorem 1 *Let* $0 < \rho \le 1$, $0 \le \delta < 1$, *and* $a \in S^{\rho-1}_{\rho,\delta}(\mathbb{R} \times \mathbb{R})$. *If* $p(\cdot) \in LH^*(\mathbb{R})$ *and* $\varrho \in W_{p(\cdot)}(\mathbb{R})$, *then* $\mathrm{Op}(a)$ *extends to a bounded operator on the weighted variable Lebesgue space* $L^{p(\cdot)}(\mathbb{R}, \varrho)$.

This result follows from a combination of results obtained in [9] and [14] (see Sect. 3 below).

Following [25, Definition 4.5], a symbol $a \in S^m_{1,0}(\mathbb{R} \times \mathbb{R})$ is said to be slowly oscillating at infinity in the first variable if

$$|\partial_\xi^\alpha \partial_x^\beta a(x, \xi)| \le C_{\alpha\beta}(x) \langle \xi \rangle^{m-\alpha},$$

where

$$\lim_{x \to \infty} C_{\alpha\beta}(x) = 0 \tag{6}$$

for each $\alpha \in \mathbb{Z}_+$ and each $\beta \in \mathbb{N}$. We denote by SO^m the class of all symbols slowly oscillating at infinity in the first variable. Finally, we denote by SO^m_0 the set of all symbols $a \in SO^m$, for which (6) holds for all indices $\alpha, \beta \in \mathbb{Z}_+$. Roughly speaking, allowing $\beta = 0$ in (6), one increases chances of the corresponding pseudodifferential operator $\mathrm{Op}(a)$ to be compact (cf. Proposition 2 below). The classes SO^m and SO^m_0 were introduced by Grushin [11].

Our main result is the following sufficient condition for the Fredholmness of pseudodifferential operators on weighted variable Lebesgue spaces.

Theorem 2 *Suppose* $p(\cdot) \in LH^*(\mathbb{R})$ *and* $\varrho \in W_{p(\cdot)}(\mathbb{R})$. *If* $a \in SO^0$ *and*

$$\lim_{R \to \infty} \inf_{|x|+|\xi| \ge R} |a(x, \xi)| > 0, \tag{7}$$

then the pesudodifferential operator $\mathrm{Op}(a)$ *is Fredholm on the weighted variable Lebesgue space* $L^{p(\cdot)}(\mathbb{R}, \varrho)$.

This result is a weighted and one-dimensional version of the theorem by Rabinovich and Samko [25, Theorem 6.1] (see also [18, Theorem 10.30] and [15, Theorem 1.3]).

The paper is organized as follows. In Sect. 2, we recall interpolation theorems for bounded and compact operators in the setting of Calderón products of Banach lattices. These theorems are far reaching generalizations of the Riesz-Thorin and Krasnosel'skii interpolation theorems. In Sect. 3, we recall the notion of a Banach function space and its associate space. Then we formulate the results on the boundedness of pseudodifferential operators on a Banach function space under the assumption that the Hardy-Littlewood maximal operator is bounded on the Banach function space and on its associate space obtained by the second author [14]. Finally, we recall our result with Medalha [9] saying that, under the assumptions $p(\cdot) \in LH^*(\mathbb{R})$ and $\varrho \in W_{p(\cdot)}(\mathbb{R})$, the Hardy-Littlewood maximal operator is bounded on the Banach function space $L^{p(\cdot)}(\mathbb{R}, \varrho)$ and on its associate space. These results lead to the proof of Theorem 1. Section 4.4 is devoted to the proof of the main result. First, we recall two important results from [9]. The first says that the Calderón product X_θ, $0 < \theta < 1$, of weighted variable Lebesgue spaces $L^{p_0(\cdot)}(\mathbb{R}, w_0)$ and $L^{p_1}(\mathbb{R}, w_1)$ is the weighted variable Lebesgue space $L^{p_\theta(\cdot)}(\mathbb{R}, w_\theta)$, where $1/p_\theta(\cdot) = (1-\theta)/p_0(\cdot) + \theta/p_1(\cdot)$ and $w_\theta = w_0^{1-\theta} w_1^{\theta}$. The second result says that for $p(\cdot) \in LH^*(\mathbb{R})$ and $\varrho \in W_{p(\cdot)}(\mathbb{R})$ there exist $p_0(\cdot) \in LH^*(\mathbb{R})$ and $\theta \in (0,1)$ such that $1/p(\cdot) = (1-\theta)/p_0(\cdot) + \theta/2$ and $\varrho_0 = \varrho^{1/(1-\theta)} \in W_{p_0(\cdot)}(\mathbb{R})$. So, by using the interpolation theorem from Sect. 2, one can conclude that if an operator is bounded on $L^{p(\cdot)}(\mathbb{R}, \varrho)$ under the assumptions $p(\cdot) \in LH^*(\mathbb{R})$ and $\varrho \in W_{p(\cdot)}(\mathbb{R})$ and is compact on $L^2(\mathbb{R})$, then it is compact on $L^{p(\cdot)}(\mathbb{R}, \varrho)$. Note also that a similar result for Lebesgue spaces with Muckenhoupt is due to Yuri Karlovich [13, Corollary 4.3]. Employing the above observation and following the scheme proposed by Rabinovich and Samko [25, Theorem 6.1], we prove Theorem 2.

2 Interpolation of Bounded and Compact Operators on Banach Lattices

2.1 Admissible Operators

Recall that a pair of complex Banach spaces $(\mathcal{X}_0, \mathcal{X}_1)$ is said to be a compatible couple if they are continuously embedded in a single complex topological vector space. The set

$$\mathcal{X}_0 + \mathcal{X}_1 = \{x = x_0 + x_1 : x_0 \in \mathcal{X}_0, x_1 \in \mathcal{X}_1\}$$

is a Banach space with respect to the norm

$$\|x\|_{\mathcal{X}_0+\mathcal{X}_1} := \inf\left\{\|x_0\|_{\mathcal{X}_0} + \|x_1\|_{\mathcal{X}_1} \ : \ x = x_0 + x_1,\ x_0 \in \mathcal{X}_0,\ x_1 \in \mathcal{X}_1\right\}.$$

Given a compatible couple $(\mathcal{X}_0, \mathcal{X}_1)$, a linear operator $T : \mathcal{X}_0 + \mathcal{X}_1 \to \mathcal{X}_0 + \mathcal{X}_1$ is said to be admissible if, for each $i = 0, 1$, the restriction of T to $\mathcal{X}_i$ takes values in $\mathcal{X}_i$ and is bounded from $\mathcal{X}_i$ to $\mathcal{X}_i$.

2.2 Banach Lattices and Their Calderón Products

The set of all Lebesgue measurable complex-valued functions on $\mathbb{R}$ is denoted by $\mathcal{M}(\mathbb{R})$. Let $\mathcal{M}_0(\mathbb{R})$ be the set of all a.e. finite functions in $\mathcal{M}(\mathbb{R})$. It is well known that $\mathcal{M}_0(\mathbb{R})$ is a complete separable metric vector space, where for every weight $w \in L^1(\mathbb{R})$ such that $\|w\|_{L^1(\mathbb{R})} = 1$, the metric is defined by

$$d_w(f, g) := \int_{\mathbb{R}} \frac{|f(x) - g(x)|}{1 + |f(x) - g(x)|} w(x)\, dx.$$

The topology in $\mathcal{M}_0(\mathbb{R})$ is independent of the choice of w (see, e.g., [3, Theorem 1.2.1]).

A subset $\mathcal{X}(\mathbb{R})$ of $\mathcal{M}_0(\mathbb{R})$ is said to be a Banach lattice if $\mathcal{X}(\mathbb{R})$ is a Banach space such that if $f, g \in \mathcal{M}_0(\mathbb{R})$ and $|f| \le |g|$ a.e., then $\|f\|_{\mathcal{X}(\mathbb{R})} \le \|g\|_{\mathcal{X}(\mathbb{R})}$. In view of [21, Chap. II, Theorem 1], a Banach lattice $\mathcal{X}(\mathbb{R})$ is continuously embedded into $\mathcal{M}_0(\mathbb{R})$. Hence, two Banach lattices $\mathcal{X}_0(\mathbb{R})$ and $\mathcal{X}_1(\mathbb{R})$ form a compatible couple.

Fix $\theta \in (0, 1)$. The Calderón product

$$\mathcal{X}_\theta(\mathbb{R}) := (\mathcal{X}_0(\mathbb{R}))^{1-\theta}(\mathcal{X}_1(\mathbb{R}))^\theta \tag{8}$$

of Banach lattices $\mathcal{X}_0(\mathbb{R})$ and $\mathcal{X}_1(\mathbb{R})$ is the set of all functions $f \in \mathcal{M}_0(\mathbb{R})$ such that

$$|f| \le \lambda |f_0|^{1-\theta} |f_1|^\theta \tag{9}$$

for some $\lambda > 0$ and $f_i \in \mathcal{X}_i(\mathbb{R})$ with $\|f_i\|_{\mathcal{X}_i(\mathbb{R})} \le 1$ and $i = 0, 1$. The norm in $\mathcal{X}_\theta(\mathbb{R})$ is the infimum of all $\lambda > 0$ for which inequality (9) is fulfilled. With this norm $\mathcal{X}_\theta(\mathbb{R})$ becomes a Banach lattice [4, p. 123].

2.3 Interpolation on Calderón Products

We recall that a Banach lattice $\mathcal{X}(\mathbb{R})$ is said to have the Fatou property if for any sequence $\{f_m\}$ of nonnegative functions in $\mathcal{X}(\mathbb{R})$ and any $f \in \mathcal{M}_0(\mathbb{R})$ such that $f_m \uparrow f$ as $m \to \infty$ and $\sup_{m\in\mathbb{N}} \|f\|_{\mathcal{X}(\mathbb{R})} < \infty$, one has $f \in \mathcal{X}(\mathbb{R})$ and $\|f_m\|_{\mathcal{X}(\mathbb{R})} \uparrow \|f\|_{\mathcal{X}(\mathbb{R})}$ as $m \to \infty$.

The following theorem is an extension of the classical interpolation theorems of Riesz-Thorin (see, e.g., [1, Chap. 4, Theorem 2.2]) and Krasnosel'skii [19] (see also [20, Theorem 3.19] and [13, Theorem 5.2]).

Theorem 3 *Let $\mathcal{X}_0(\mathbb{R})$ and $\mathcal{X}_1(\mathbb{R})$ be Banach lattices with the Fatou property, let*

$$T : \mathcal{X}_0(\mathbb{R}) + \mathcal{X}_1(\mathbb{R}) \to \mathcal{X}_0(\mathbb{R}) + \mathcal{X}_1(\mathbb{R})$$

be an admissible operator, and let $\theta \in (0, 1)$.

(a) *The restriction of T to the Calderón product $\mathcal{X}_\theta(\mathbb{R})$ defined by* (8) *takes values in $\mathcal{X}_\theta(\mathbb{R})$ and*

$$\|T\|_{\mathcal{B}(\mathcal{X}_\theta(\mathbb{R}))} \le \|T\|^{1-\theta}_{\mathcal{B}(\mathcal{X}_0(\mathbb{R}))} \|T\|^{\theta}_{\mathcal{B}(\mathcal{X}_1(\mathbb{R}))}.$$

(b) *If, in addition, $T \in \mathcal{K}(\mathcal{X}_1(\mathbb{R}))$, then $T \in \mathcal{K}(\mathcal{X}_\theta(\mathbb{R}))$.*

Part (a) under stronger assumptions was obtained by Calderón [4]. For the present form of part (a) we refer to [23, Theorem 3.11]. Part (b) follows from [5, Theorem 3.1].

3 Boundedness of Maximal and Pseudodifferential Operators on Weighted Variable Lebesgue Spaces

3.1 *Banach Function Spaces*

For the set of all Lebesgue measurable complex-valued function $\mathcal{M}(\mathbb{R})$, consider its subset $\mathcal{M}^+(\mathbb{R})$ of all functions whose values lie in $[0, \infty]$. The Lebesgue measure of a measurable set $E \subset \mathbb{R}$ is denoted by $|E|$ and we let χ_E stand for the characteristic function of E.

Following [22, p. 3] and [1, Chap. 1, Definition 1.1], a mapping $\rho : \mathcal{M}^+(\mathbb{R}) \to [0, \infty]$ is called a Banach function norm if, for all functions f, g, f_n $(n \in \mathbb{N})$ in $\mathcal{M}^+(\mathbb{R})$, for all constants $a \ge 0$, and for all measurable subsets E of $\mathbb{R}$, the following properties hold:

(A1) $\rho(f) = 0 \Leftrightarrow f = 0$ a.e., $\quad \rho(af) = a\rho(f), \quad \rho(f+g) \le \rho(f) + \rho(g)$,

(A2) $0 \le g \le f$ a.e. $\Rightarrow \rho(g) \le \rho(f)$ (the lattice property),

(A3) $0 \le f_n \uparrow f$ a.e. $\Rightarrow \rho(f_n) \uparrow \rho(f)$ (the Fatou property),

(A4) E is bounded $\Rightarrow \rho(\chi_E) < \infty$,

(A5) E is bounded $\Rightarrow \displaystyle\int_E f(x)\,dx \le C_E \rho(f)$

with $C_E \in (0, \infty)$ which may depend on E and ρ but is independent of f. When functions differing only on a set of measure zero are identified, the set $X(\mathbb{R})$ of all functions $f \in \mathcal{M}(\mathbb{R})$ for which $\rho(|f|) < \infty$ is called a Banach function space. For each $f \in X(\mathbb{R})$, the norm of f is defined by

$$\|f\|_{X(\mathbb{R})} := \rho(|f|).$$

Under the natural linear space operations and under this norm, the set $X(\mathbb{R})$ becomes a Banach space (see [22, Chap. 1, §1, Theorem 1] or [1, Chap. 1, Theorems 1.4 and 1.6]). If ρ is a Banach function norm, its associate norm ρ' is defined on $\mathcal{M}^+(\mathbb{R})$ by

$$\rho'(g) := \sup\left\{\int_{\mathbb{R}} f(x)g(x)\,dx \; : \; f \in \mathcal{M}^+(\mathbb{R}), \; \rho(f) \le 1\right\}, \quad g \in \mathcal{M}^+(\mathbb{R}).$$

It is a Banach function norm itself (see [22, Chap. 1, §1] or [1, Chap. 1, Theorem 2.2]). The Banach function space $X'(\mathbb{R})$ determined by the Banach function norm ρ' is called the associate space (Köthe dual) of $X(\mathbb{R})$. The associate space $X'(\mathbb{R})$ is naturally identified with a subspace of the (Banach) dual space $[X(\mathbb{R})]^*$.

Remark 1 We note that our definition of a Banach function space is slightly different from that found in [1, Chap. 1, Definition 1.1]. In particular, in Axioms (A4) and (A5) we assume that the set E is a bounded set, whereas it is sometimes assumed that E merely satisfies $|E| < \infty$. We do this so that the weighted variable Lebesgue spaces satisfy Axioms (A4) and (A5). Moreover, it is well known that all main elements of the general theory of Banach function spaces work with (A4) and (A5) stated for bounded sets [22] (see also the discussion at the beginning of Chapter 1 on page 2 of [1]).

3.2 *Boundedness of Pseudodifferential Operators on Banach Function Spaces*

Given $f \in L^1_{\mathrm{loc}}(\mathbb{R})$, the Hardy-Littlewood maximal operator is defined by

$$Mf(x) := \sup_{I \ni x} \frac{1}{|I|} \int_I |f(y)|dy$$

where the supremum is taken over all finite intervals $I \subset \mathbb{R}$ containing x.

Theorem 4 ([14, Theorem 1.1]) *Let $X(\mathbb{R})$ be a separable Banach function space such that the Hardy-Littlewood maximal operator M is bounded on $X(\mathbb{R})$ and on its associate space $X'(\mathbb{R})$. If $a \in S^{\rho-1}_{\rho,\delta}(\mathbb{R} \times \mathbb{R})$ with $0 < \rho \le 1$ and $0 \le \delta < 1$, then* $\mathrm{Op}(a)$ *extends to a bounded operator on $X(\mathbb{R})$.*

3.3 The Case of Weighted Variable Lebesgue Spaces

Let $p(\cdot) : \mathbb{R} \to [1, \infty]$ be an a.e. finite variable exponent. We will write $f \in L^{p(\cdot)}_{\mathrm{loc}}(\mathbb{R})$ if $f\chi_E \in L^{p(\cdot)}(\mathbb{R})$ for every bounded measurable set $E \subset \mathbb{R}$.

The following result is contained in [9, Theorems 2.2, 2.4, and 2.6].

Theorem 5 *Let $p(\cdot) \in LH^*(\mathbb{R})$, $\varrho \in W_{p(\cdot)}(\mathbb{R})$, and let $p'(\cdot)$ be the variable exponent defined by*

$$1/p(x) + 1/p'(x) = 1, \quad x \in \mathbb{R}.$$

Then

(a) $\varrho \in L^{p(\cdot)}_{\mathrm{loc}}(\mathbb{R})$ *and* $\varrho^{-1} \in L^{p'(\cdot)}_{\mathrm{loc}}(\mathbb{R})$;
(b) *the weighted variable Lebesgue spaces $L^{p(\cdot)}(\mathbb{R}, \varrho)$ and $L^{p'(\cdot)}(\mathbb{R}, \varrho^{-1})$ are separable and reflexive Banach function spaces;*
(c) *the associate space of $L^{p(\cdot)}(\mathbb{R}, \varrho)$ is isomorphic to $L^{p'(\cdot)}(\mathbb{R}, \varrho^{-1})$;*
(d) *the Hardy-Littlewood maximal operator M is bounded on the weighted variable Lebesgue spaces $L^{p(\cdot)}(\mathbb{R}, \varrho)$ and $L^{p'(\cdot)}(\mathbb{R}, \varrho^{-1})$.*

Now Theorem 1 on the boundedness of pseudodifferential operators on weighted variable Lebesgue spaces follows from Theorems 4 and 5.

4 Proof of the Main Result

4.1 Calderón Products of Weighted Variable Lebesgue Spaces

The abstract interpolation Theorem 3 can be applied in the setting of weighted variable exponent spaces because the Calderón product X_θ of weighted variable Lebesgue spaces $L^{p_0(\cdot)}(\mathbb{R}, w_0)$ and $L^{p_1(\cdot)}(\mathbb{R}, w_1)$ is a certain weighted variable Lebesgue space $L^{p_\theta(\cdot)}(\mathbb{R}, w_\theta)$ explicitly defined in the theorem below.

Theorem 6 ([9, Theorem 3.2]) *For $i = 0, 1$, let $p_i(\cdot)$ be variable exponents satisfying*

$$1 < (p_i)_-, (p_i)_+ < \infty$$

and w_i be weights satisfying $w_i \in L^{p_i(\cdot)}_{\mathrm{loc}}(\mathbb{R})$ and $w_i^{-1} \in L^{p_i'(\cdot)}_{\mathrm{loc}}(\mathbb{R})$. For $0 < \theta < 1$, let the variable exponent $p_\theta(\cdot)$ and the weight w_θ be defined by

$$\frac{1}{p_\theta(x)} = \frac{1-\theta}{p_0(x)} + \frac{\theta}{p_1(x)}, \quad w_\theta(x) = w_0(x)^{1-\theta} w_1(x)^\theta, \quad x \in \mathbb{R}.$$

Then the Calderón product

$$X_\theta := \left(L^{p_0(\cdot)}(\mathbb{R}, w_0)\right)^{1-\theta} \left(L^{p_1(\cdot)}(\mathbb{R}, w_1)\right)^{\theta}$$

of the weighted variable Lebesgue spaces $L^{p_0(\cdot)}(\mathbb{R}, w_0)$ *and* $L^{p_1(\cdot)}(\mathbb{R}, w_1)$ *coincides with the weighted variable Lebesgue space* $L^{p_\theta(\cdot)}(\mathbb{R}, w_\theta)$ *with the equivalence of the norms*

$$\|f\|_{X_\theta} \le \|f\|_{L^{p_\theta(\cdot)}(\mathbb{R}, w_\theta)} \le 2^{1/(p_\theta)_-} \|f\|_{X_\theta}, \quad f \in L^{p_\theta(\cdot)}(\mathbb{R}, w_\theta).$$

4.2 Perturbation of Variable Exponents and Weights

This subsection contains results on the perturbation of variable exponents $p(\cdot)$ in the class $LH^*(\mathbb{R})$ and weights $\varrho \in W_{p(\cdot)}(\mathbb{R})$. These results play a crucial role in the proof of Theorem 2.

Lemma 1 ([25, Corollary 2.3]) *For a variable exponent* $p(\cdot)$ *satisfying* (1), *let*

$$\theta_{p(\cdot)} := \min\{1, 2/p_+, 2 - 2/p_-\}. \tag{10}$$

Then for every $\theta \in (0, \theta_{p(\cdot)})$, *the variable exponent* $p_0(\cdot)$ *defined by*

$$\frac{1}{p(x)} = \frac{1-\theta}{p_0(x)} + \frac{\theta}{2}, \quad x \in \mathbb{R}, \tag{11}$$

satisfies $1 < (p_0)_-, (p_0)_+ < \infty$.

Lemma 2 ([9, Lemma 4.2]) *If* $p(\cdot)$ *belongs to the class* $LH^*(\mathbb{R})$ *and* $\theta_{p(\cdot)}$ *is defined by* (10), *then for every* $\theta \in (0, \theta_{p(\cdot)})$, *the variable exponent* $p_0(\cdot)$ *defined by* (11) *belongs to* $LH^*(\mathbb{R})$.

The following theorem can be viewed as a very modest attempt to extend a well-known result on the stability of Muckenhoupt weights (see, e.g., [2, Theorem 2.31] or [10, Theorem 7.25]). In its spirit, it is close to the proof of [13, Corollary 5.3] (see also [7, Corollary 3]).

Theorem 7 ([9, Theorem 4.3]) *Let* $p(\cdot) \in LH^*(\mathbb{R})$ *and* $\varrho \in W_{p(\cdot)}(\mathbb{R})$. *Suppose that* $\theta_{p(\cdot)}$ *is defined by* (10). *Then there exists* $\theta^*_{p(\cdot),\varrho} \in (0, \theta_{p(\cdot)}]$ *such that for every* $\theta \in (0, \theta^*_{p(\cdot),\varrho})$, *the weight* $\varrho_0 = \varrho^{1/(1-\theta)}$ *belongs to* $W_{p_0(\cdot)}(\mathbb{R})$, *where the variable exponent* $p_0(\cdot)$ *is defined by* (11).

Note that [13, Corollary 5.3] is also true for Muckenhoupt weights $w \in A_p$ over $\mathbb{R}^n$. On the other hand, the proof of Theorem 7 relies essentially on [17, Theorem A], a one-dimensional result. Since we do not have a multi-dimensional version of

Theorem 7 at our disposal, in this paper we restrict ourselves to the study of one-dimensional pseudodifferential operators.

4.3 Calculus of Pseudodifferential Operators

Let $m \in \mathbb{Z}$ and $OPSO^m$ be the class of all pseudodifferential operators $\mathrm{Op}(a)$ with $a \in SO^m$. By analogy with [11, Section 2] one can get the following composition formula (see also [24, Theorem 6.2.1] and [26, Chap. 4]).

Proposition 1 *If* $\mathrm{Op}(a_1) \in OPSO^{m_1}$ *and* $\mathrm{Op}(a_2) \in OPSO^{m_2}$*, then their product* $\mathrm{Op}(a_1)\,\mathrm{Op}(a_2) = \mathrm{Op}(\sigma)$ *belongs to* $OPSO^{m_1+m_2}$ *and its symbol* σ *is given by*

$$\sigma(x, \xi) = a_1(x, \xi)a_2(x, \xi) + c(x, \xi), \quad x, \xi \in \mathbb{R},$$

where $c \in SO_0^{m_1+m_2-1}$.

Proposition 2 ([11, Theorem 3.2]) *If* $c \in SO_0^{-1}$*, then* $\mathrm{Op}(c) \in \mathcal{K}(L^2(\mathbb{R}))$.

4.4 Proof of Theorem 2

The idea of the proof is borrowed from [11, Theorem 3.4] and [25, Theorem 6.1] (see also [15, Section 3.3]). Let $\varphi \in C_0^\infty(\mathbb{R} \times \mathbb{R})$ be such that $\varphi(x, \xi) = 1$ if $|x| + |\xi| \le 1$ and $\varphi(x, \xi) = 0$ if $|x| + |\xi| \ge 2$. For $R > 0$, put

$$\varphi_R(x, \xi) = \varphi(x/R, \xi/R), \quad x, \xi \in \mathbb{R}.$$

From (7) it follows that there exists an $R > 0$ such that

$$\inf_{|x|+|\xi|\ge R} |a(x, \xi)| > 0.$$

Then it is not difficult to check that

$$b_R(x, \xi) := \begin{cases} \dfrac{1 - \varphi_R(x, \xi)}{a(x, \xi)} & \text{if } |x| + |\xi| \ge R, \\ 0 & \text{if } |x| + |\xi| < R, \end{cases}$$

belongs to SO^0. It is also clear that $\varphi_R \in SO^0$.

From Proposition 1 it follows that there exists a function $c \in SO_0^{-1}$ such that

$$\mathrm{Op}(ab_R) - \mathrm{Op}(a)\,\mathrm{Op}(b_R) = \mathrm{Op}(c). \tag{12}$$

On the other hand, since

$$a(x,\xi)b_R(x,\xi) = 1 - \varphi_R(x,\xi), \quad x,\xi \in \mathbb{R},$$

we have

$$\mathrm{Op}(ab_R) = \mathrm{Op}(1-\varphi_R) = I - \mathrm{Op}(\varphi_R). \tag{13}$$

Combining (12)–(13), we get

$$I - \mathrm{Op}(a)\,\mathrm{Op}(b_R) = \mathrm{Op}(\varphi_R) + \mathrm{Op}(c) = \mathrm{Op}(\varphi_R + c). \tag{14}$$

It follows from Lemma 2 and Theorem 7 that there exists $\theta \in (0,1)$ such that the variable exponent $p_0(\cdot)$ defined by

$$\frac{1}{p(x)} = \frac{1-\theta}{p_0(x)} + \frac{\theta}{2}, \quad x \in \mathbb{R},$$

belongs to $LH^*(\mathbb{R})$ and the weight

$$\varrho_0(x) := \varrho(x)^{1/(1-\theta)}, \quad x \in \mathbb{R},$$

belongs to $W_{p_0(\cdot)}(\mathbb{R})$. We conclude from Theorem 1 that all pseudodifferential operators considered above are bounded on $L^{p_0(\cdot)}(\mathbb{R},\varrho_0)$ and $L^2(\mathbb{R})$. Since $\varphi_R + c \in SO_0^{-1}$, it follows from Proposition 2 that $\mathrm{Op}(\varphi_R + c) \in \mathcal{K}(L^2(\mathbb{R}))$. Then, by Theorem 3(b), $\mathrm{Op}(\varphi_R + c) \in \mathcal{K}(L^{p(\cdot)}(\mathbb{R},\varrho))$. Therefore, it follows from (14) that $\mathrm{Op}(b_R)$ is a right regularizer for $\mathrm{Op}(a)$. Analogously it can be shown that $\mathrm{Op}(b_R)$ is also a left regularizer for $\mathrm{Op}(a)$. Thus $\mathrm{Op}(a)$ is Fredholm on $L^{p(\cdot)}(\mathbb{R},\varrho)$. □

Acknowledgments This work is funded by national funds through the FCT - Fundação para a Ciência e a Tecnologia, I.P., under the scope of the projects UIDB/00297/2020 (https://doi.org/10.54499/UIDB/00297/2020) and UIDP/00297/2020 (https://doi.org/10.54499/UIDP/00297/2020) (Center for Mathematics and Applications).

References

1. Bennett, C., Sharpley, R: Interpolation of operators. Pure and Applied Mathematics, vol. 129. Academic Press, Inc., Boston (1988)
2. Böttcher A., Karlovich, Y.I.: Carleson curves, Muckenhoupt weights, and Toeplitz operators. Progress in Mathematics, vol. 154. Birkhäuser Verlag, Basel (1997)
3. Brudnyĭ, Y.A., Krugljak, N.Y.: Interpolation functors and interpolation spaces, vol. I. North-Holland Mathematical Library, vol. 47. North-Holland Publishing Co., Amsterdam (1991)
4. Calderón, A.P.: Intermediate spaces and interpolation, the complex method. Studia Math. **24**, 113–190 (1964)

5. Cobos, F., Kühn, T., Schonbek, T.: One-sided compactness results for Aronszajn-Gagliardo functors. J. Funct. Anal. **106**(2), 274–313 (1992)
6. Cruz-Uribe, D. V., Fiorenza, A.: Variable Lebesgue spaces. Foundations and harmonic analysis. Applied and Numerical Harmonic Analysis. Birkhäuser, Springer/Heidelberg (2013)
7. De la Cruz-Rodríguez, I., Karlovich, Y.I., Loreto-Hernández, I.: Commutators of convolution type operators with piecewise quasicontinuous data. Commun. Math. Anal. **17**(2), 131–150 (2014)
8. Diening, L., Harjulehto, P., Hästö, P., Růžička, M.: Lebesgue and Sobolev spaces with variable exponents. Lecture Notes in Mathematics, vol. 2017. Springer, Heidelberg (2011)
9. Fernandes, C., Karlovych, O., Medalha, S.: Invertibility of Fourier convolution operators with PC symbols on variable Lebesgue spaces with Khvedelidze weights. J. Math. Sci. (N.Y.) **266**(3), 419–434 (2022)
10. Grafakos, L.: Classical Fourier analysis. Graduate Texts in Mathematics, vol. 249, 3rd edn. Springer, New York (2014)
11. Grushin, V.V.: Pseudodifferential operators on $\mathbb{R}^n$ with bounded symbols. Funct. Anal. Appl. **4**, 202–212 (1970)
12. Hörmander, L.: Pseudo-differential operators and hypoelliptic equations. In: Singular Integrals (Proc. Sympos. Pure Math., Chicago, Ill., 1966), pp. 138–183. American Mathematical Society, Providence (1967)
13. Karlovich, Y.I.: Boundedness and compactness of pseudodifferential operators with non-regular symbols on weighted Lebesgue spaces. Integral Equ. Oper. Theory **73**(2), 217–254 (2012)
14. Karlovich, A.Y.: Boundedness of pseudodifferential operators on Banach function spaces. In: Operator Theory, Operator Algebras and Applications. Operator Theory: Advances and Applications, vol. 242, pp. 185–195. Birkhäuser/Springer, Basel (2014)
15. Karlovich, A.Y., Spitkovsky, I.M.: Pseudodifferential operators on variable Lebesgue spaces. In: Operator Theory, Pseudo-Differential Equations, and Mathematical Physics. Operator Theory: Advances and Applications, vol. 228, pp. 173–183. Birkhäuser/Springer Basel AG, Basel (2013)
16. Khvedelidze, B.V.: Linear discontinuous boundary problems in the theory of functions, singular integral equations and some of their applications. Akad. Nauk Gruzin. SSR. Trudy Tbiliss. Mat. Inst. Razmadze **23**, 3–158 (1956). In Russian
17. Kokilashvili, V., Paatashvili, V., Samko, S.: Boundedness in Lebesgue spaces with variable exponent of the Cauchy singular operator on Carleson curves. In: Modern Operator Theory and Applications. Operator Theory: Advances and Applications, vol. 170, pp. 167–186. Birkhäuser, Basel (2007)
18. Kokilashvili, V., Meskhi, A., Rafeiro, H., Samko, S.: Integral operators in non-standard function spaces, vol. 1. Variable Exponent Lebesgue and Amalgam Spaces. Operator Theory: Advances and Applications, vol. 248. Birkhäuser/Springer, Basel (2016)
19. Krasnosel'skiĭ, M.A.: On a theorem of M. Riesz. Dokl. Akad. Nauk SSSR **131**, 246–248 (1960). In Russian
20. Krasnosel'skiĭ, M.A., Zabreĭko, P.P., Pustyl'nik, E.I., Sobolevskiĭ, P.E.: Integral operators in spaces of summable functions. Monographs and Textbooks on Mechanics of Solids and Fluids, Mechanics: Analysis. Noordhoff International Publishing, Leiden (1976)
21. Kreĭn, S.G., Petunīn, Y.I., Semënov, E.M.: Interpolation of linear operators. Translations of Mathematical Monographs, vol. 54. American Mathematical Society, Providence (1982)
22. Luxemburg, W.A.J.: Banach function spaces. Thesis, Technische Hogeschool te Delft (1955)
23. Mastylo, M., Sawano, Y.: Complex interpolation and Calderón-Mityagin couples of Morrey spaces. Anal. PDE **12**(7), 1711–1740 (2019)
24. Rabinovich, V.S.: An introductory course on pseudodifferential operators. Textos de Matemática. Instituto Superior Técnico, Lisboa (1998)
25. Rabinovich, V., Samko, S.: Boundedness and Fredholmness of pseudodifferential operators in variable exponent spaces. Integral Equ. Oper. Theory **60**(4), 507–537 (2008)

26. Rabinovich, V., Roch, S., Silbermann, B.: Limit Operators and Their Applications in Operator Theory. Operator Theory: Advances and Applications, vol. 150. Birkhäuser Verlag, Basel (2004)
27. Stein, E.M.: Harmonic Analysis: Real-Variable Methods, Orthogonality, and Oscillatory Integrals. Princeton Mathematical Series, vol. 43. Princeton University Press, Princeton (1993)

Eigenvalues of the Laplacian Matrices of Cycles with One Overweighted Edge

Sergei M. Grudsky, Egor A. Maximenko, and Alejandro Soto-González

Dedicated to Yuri Karlovich on the occasion of his 75th birthday.

Abstract We study the individual behavior of the eigenvalues of the laplacian matrices of the cyclic graph of order n, where one edge has weight $\alpha \in \mathbb{C}$, with $\operatorname{Re}(\alpha) > 1$, and all the others have weights 1. This paper is a sequel to two previous ones where we considered $\operatorname{Re}(\alpha) \in [0, 1]$ and $\operatorname{Re}(\alpha) < 0$. Now, we prove that for $\operatorname{Re}(\alpha) > 1$ and $n > \operatorname{Re}(\alpha)/\operatorname{Re}(\alpha - 1)$, one eigenvalue is greater than 4 while the others belong to $[0, 4]$ and are distributed as the function $x \mapsto 4\sin^2(x/2)$. Additionally, we prove that as n tends to ∞, the outlier eigenvalue converges exponentially to $4\operatorname{Re}(\alpha)^2/(2\operatorname{Re}(\alpha)-1)$. We give exact formulas for half of the inner eigenvalues, while for the others we justify the convergence of Newton's method and the fixed-point iteration method. We find asymptotic expansions, as n tends to ∞, both for the eigenvalues belonging to $[0, 4]$ and the outliers. We also compute the eigenvectors and their norms.

S. M. Grudsky
CINVESTAV del IPN, Departamento de Matemáticas, Ciudad de México, Russia

Southern Federal University, Regional Mathematical Center, Rostov-on-Don, Russia
e-mail: grudsky@math.cinvestav.mx

E. A. Maximenko
Instituto Politécnico Nacional, Escuela Superior de Física y Matemáticas, Ciudad de México, Mexico
e-mail: emaximenko@ipn.mx

A. Soto-González (✉)
CINVESTAV del IPN, Departamento de Matemáticas, Ciudad de México, Mexico
e-mail: asoto@math.cinvestav.mx

A. Böttcher et al. (eds.), *Achievements and Challenges in the Field of Convolution Operators*, Operator Theory: Advances and Applications 306,
https://doi.org/10.1007/978-3-031-80486-1_9

1 Introduction

For every natural $n \geq 3$ and every α in $\mathbb{C}$, we consider the $n \times n$ complex laplacian matrix $L_{\alpha,n}$ with the following structure:

$$L_{\alpha,6} = \begin{bmatrix} 1+\overline{\alpha} & -1 & 0 & 0 & 0 & -\overline{\alpha} \\ -1 & 2 & -1 & 0 & 0 & 0 \\ 0 & -1 & 2 & -1 & 0 & 0 \\ 0 & 0 & -1 & 2 & -1 & 0 \\ 0 & 0 & 0 & -1 & 2 & -1 \\ -\alpha & 0 & 0 & 0 & -1 & 1+\alpha \end{bmatrix}.$$

If α is real, $L_{\alpha,n}$ is the laplacian matrix of $G_{\alpha,n}$, where $G_{\alpha,n}$ is the cyclic graph of order n, where the edge between the vertices 1 and n weighs α, and all other edges weigh 1. See [15] for the general theory on laplacian matrices. In Fig. 1, we show the case $n = 6$. The eigenvalues and eigenvectors of $L_{\alpha,n}$ are crucial to solve the heat and wave equations on $G_{\alpha,n}$. Moreover, matrices of the form $2I_n - L_{\alpha,n}$ are related to counting the paths in a cyclic graph with certain loops [5].

The matrices $L_{\alpha,n}$ can be considered as tridiagonal Toeplitz matrices with perturbations in the corners $(1, 1)$, $(1, n)$, $(n, 1)$ and (n, n). They can also be viewed as periodic Jacobi matrices. Some matrices of these classes and their applications were studied in [2–4, 6–8, 10, 11, 14, 16, 17, 19–21].

The present paper is a continuation of [12, 13]. In [12], we proved that for every α in $\mathbb{C}$ the characteristic polynomial of $L_{\alpha,n}$, defined by $D_{\alpha,n}(\lambda) := \det(\lambda I - L_{\alpha,n})$, equals the characteristic polynomial $D_{\mathrm{Re}(\alpha),n}$ of $L_{\mathrm{Re}(\alpha),n}$. This implies that the eigenvalues of $L_{\alpha,n}$ only depend on $\mathrm{Re}(\alpha)$. Therefore, to understand the behavior of the eigenvalues, it is sufficient to consider the case where $\alpha \in \mathbb{R}$ and the corresponding matrices $L_{\alpha,n}$ are real and symmetric. So, for every α in $\mathbb{C}$, the eigenvalues of $L_{\alpha,n}$ are real, and we enumerate them as follows:

$$\lambda_{\alpha,n,1} \leq \lambda_{\alpha,n,2} \leq \cdots \leq \lambda_{\alpha,n,n}.$$

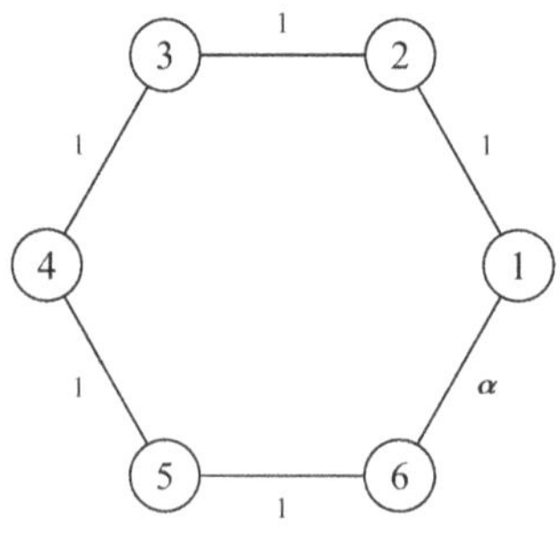

Fig. 1 Graph $G_{\alpha,6}$

It is a very well-known fact that the eigenvalues of the $n \times n$ tridiagonal Toeplitz matrix, with values $-1, 2, -1$ in the nonzero diagonals, are the numbers $g(j\pi/(n+1))$, $j = 1, \ldots, n$, where g is defined by

$$g(x) := 4\sin^2\frac{x}{2} \qquad (x \in [0, \pi]). \tag{1}$$

By the Cauchy interlacing theorem (see, e.g., [18, Theorem 4.2]), the eigenvalues of $L_{\alpha,n}$ are also asymptotically distributed by g on $[0, \pi]$, as n tends to infinity. This is also a simple consequence of the theory of generalized locally Toeplitz sequences [9].

In [12], we studied the individual behavior of the eigenvalues of the matrices $L_{\alpha,n}$ for α in $(0, 1)$. In that case, we showed that the eigenvalues of $L_{\alpha,n}$ belong to $[0, 4]$. We solved the characteristic equation by numerical methods and derived asymptotic formulas for all eigenvalues. In [13], we considered the case where $\alpha < 0$. In that scenery, we proved that if $n > (\alpha - 1)/\alpha$ then the minimal eigenvalue $\lambda_{\alpha,n,1}$ goes out of the interval $[0, 4]$; moreover, the sequence $(\lambda_{\alpha,n,1})_{n>(\alpha-1)/\alpha}$ strictly decreases and converges exponentially to $4\alpha^2/(2\alpha - 1)$.

In this paper, we consider the case where $\alpha > 1$ (or, more generally, $\operatorname{Re}(\alpha) > 1$). This means that the interaction between the vertices 1 and n is stronger than the interactions between the other neighbors in the cycle.

It turns out that, if n is even or if n is odd and satisfies $n > \alpha/(\alpha - 1)$, then the maximal eigenvalue $\lambda_{\alpha,n,n}$ is greater than 4, while the others belong to the interval $[0, 4]$ and behave similarly to the eigenvalues of $L_{\alpha,n}$ when $0 < \alpha < 1$, as discussed in [12].

We use the phrase "inner eigenvalues" for the eigenvalues belonging to the clustering set $[0, 4]$, and "outlier eigenvalue" for the one that does not belong to this set. See also our general definition of outlier eigenvalue in [13].

We show that if $\alpha > 1$, then the sequence of outlier eigenvalues $(\lambda_{\alpha,n,n})_{n\geq 3}$ converges exponentially to the number $\Omega_\alpha := 4\alpha^2/(2\alpha - 1)$. The major difference to the previous paper [13] is that the sequence of the outliers approaches the limit value from both directions:

$$\operatorname{sign}(\lambda_{\alpha,n,n} - \Omega_\alpha) = (-1)^n \qquad \left(n > \frac{\alpha}{\alpha - 1}\right). \tag{2}$$

The main results of this paper are stated in Sect. 2, while the majority of the content is dedicated to the corresponding proofs: we represent the characteristic polynomial in convenient forms and show the localization of the eigenvalues (Sect. 3), we study the asymptotic behavior of the inner eigenvalues and guarantee their computation with the Newton method (Sect. 4), then we focus our attention on the last eigenvalue $\lambda_{\alpha,n,n}$ (Sect. 5) and analyze its asymptotic behavior separately for both odd (Sect. 6) and even values of n (Sect. 7). Finally, we calculate the norms of the eigenvectors (Sect. 8) and show some numerical experiments (Sect. 9).

2 Main Results

As will be stated in Proposition 3.1, for every $\alpha \in \mathbb{C}$ we have that $D_{\alpha,n} = D_{\mathrm{Re}(\alpha),n}$. So, unless specified otherwise, we consider $\alpha > 1$.

We begin our analysis with the localization of the eigenvalues. For this purpose, define

$$\varkappa_\alpha := \frac{\alpha - 1}{\alpha}, \tag{3}$$

$$\Omega_\alpha := \frac{4\alpha^2}{2\alpha - 1}, \qquad \text{i.e.,} \qquad \Omega_\alpha = \frac{4}{1 - \varkappa_\alpha^2}. \tag{4}$$

Notice that $0 < \varkappa_\alpha < 1$ and $\Omega_\alpha > 4$. Also, for every j in $\{1, \dots, n\}$, we put

$$d_{n,j} := \frac{(j-1)\pi}{n}. \tag{5}$$

Theorem 2.1 (Localization of Eigenvalues) *Let $n \geq 3$. Then $\lambda_{\alpha,n,1} = 0$. For every j with $2 \leq j \leq n-1$,*

$$g\left(d_{n,j}\right) < \lambda_{\alpha,n,j} < g\left(d_{n,j+1}\right) \qquad (j \text{ odd}),$$
$$\lambda_{\alpha,n,j} = g\left(d_{n,j+1}\right) \qquad (j \text{ even}).$$

Furthermore, the localization of $\lambda_{\alpha,n,n}$ depends on n:

(1) if $n < \varkappa_\alpha^{-1}$ and n is odd, then $g(d_{n,n}) < \lambda_{\alpha,n,n} < g(\pi) = 4$;
(2) if $n = \varkappa_\alpha^{-1}$ and n is odd, then $\lambda_{\alpha,n,n} = 4$;
(3) if n is odd and $n > \varkappa_\alpha^{-1}$, then $4 < \lambda_{\alpha,n,n} < \Omega_\alpha$;
(4) if n is even, then $\Omega_\alpha < \lambda_{\alpha,n,n} \leq 4 + 2\alpha$.

According to Theorem 2.1, the eigenvalues $\lambda_{\alpha,n,j}$ with even indices j do not depend of α. This theorem also implies that the eigenvalues are asymptotically distributed as the function g on $[0, \pi]$:

$$\lim_{n\to\infty} \frac{\#\{j \in \{1, \dots, n\} \colon \lambda_{\alpha,n,j} \leq u\}}{n} = \frac{\mu(\{x \in [0, \pi] \colon g(x) \leq u\})}{\pi}. \tag{6}$$

Here, μ is the Lebesgue measure.

Statements (3) and (4) of Theorem 2.1 mean that for n large enough, we have two different localizations of the largest eigenvalue $\lambda_{\alpha,n,n}$ depending of the parity of n.

If n is odd, then the outlier eigenvalues of $L_{\alpha,n}$ and $L_{1-\alpha,n}$ are related by $\lambda_{\alpha,n,n} = 4 - \lambda_{1-\alpha,n,1}$ (Proposition 6.1). Therefore, in the analysis of $\lambda_{\alpha,n,n}$ for odd n, we can proceed very similarly to [12].

However, for even values of n, the equation for $\lambda_{\alpha,n,n}$ has a quite different form, see Theorem 2.2.

Motivated by Theorem 2.1, we use g defined by (1) as a change of variable in the characteristic equation when $\lambda_{\alpha,n,j} \in [0,4]$ and set

$$z_{\alpha,n,j} := \widetilde{g}^{-1}(\lambda_{\alpha,n,j}), \tag{7}$$

where $\widetilde{g}\colon [0,\pi] \to [0,4]$ is a restriction of g.

To state the main equation for inner eigenvalues, we define the function $\eta_\alpha\colon [0,\pi] \to \mathbb{R}$ by

$$\eta_\alpha(x) := 2\arctan\left(\varkappa_\alpha \tan\frac{x}{2}\right), \quad \text{i.e.,} \quad \eta_\alpha(x) = 2\arctan\left(\frac{\alpha-1}{\alpha}\tan\frac{x}{2}\right). \tag{8}$$

Since $\varkappa_\alpha$ is positive, η_α is positive, strictly increasing and takes values on $[0,\pi]$.

Theorem 2.2 (Main Equation for Inner Eigenvalues) *Let j be odd with $3 \le j \le n-1$. Then the number $z_{\alpha,n,j}$ is the unique solution in $[0,\pi]$ of the equation*

$$x = d_{n,j} + \frac{\eta_\alpha(x)}{n}. \tag{9}$$

The same Eq. (9) *also holds for $z_{\alpha,n,n}$, if n is odd and $n < \varkappa_\alpha^{-1}$.*

Now, we need a suitable change of variable associated to $\lambda_{\alpha,n,n}$. Thus, define $g_+\colon [0,\infty) \to [4,\infty)$ by

$$g_+(x) := 2 + 2\cosh(x) = 4\cosh^2\frac{x}{2} = 4 + 4\sinh^2\frac{x}{2}. \tag{10}$$

Let also

$$N_\alpha := \max\{3, \lfloor \varkappa_\alpha^{-1} \rfloor + 1\}. \tag{11}$$

So, if $n \ge 4$ is even or $n \ge N_\alpha$ is odd, then we use (10) as a change of variable and put

$$s_{\alpha,n} := g_+^{-1}(\lambda_{\alpha,n,n}). \tag{12}$$

In Fig. 2 we have glued together g and $x \mapsto g_+(x-\pi)$ into one spline.

Theorem 2.1 says that for every $n \ge N_\alpha$, $\lambda_{\alpha,n,n}$ is in a neighborhood of Ω_α, thus we define

$$\omega_\alpha := g_+^{-1}(\Omega_\alpha) = \log(2\alpha - 1). \tag{13}$$

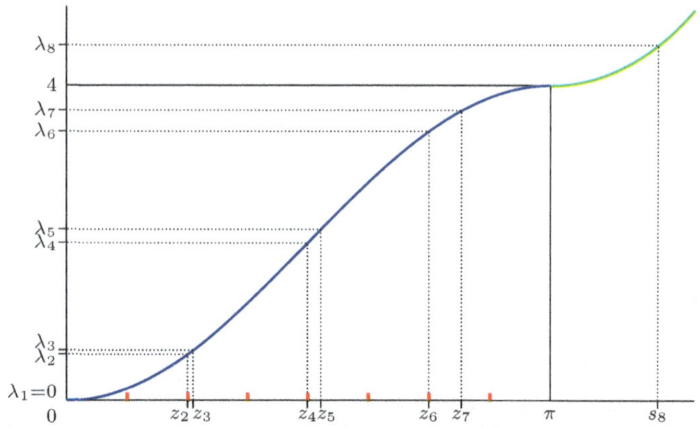

Fig. 2 Plot of g (blue), plot of $x \mapsto g_+(x-\pi)$ (green), points $z_{\alpha,n,j}$ and $s_{\alpha,n}$, and the corresponding values of $\lambda_{\alpha,n,j}$, for $\alpha = 3/2$ and $n = 8$. The red labels on the horizontal axis are $j\pi/n$

To get the main equation for the outlier eigenvalue, we define the real-valued functions $\psi_{\alpha,n}$ by

$$\psi_{\alpha,n}(x) := \begin{cases} 2\operatorname{arctanh}\left(\varkappa_\alpha \tanh \frac{nx}{2}\right), & \text{if } n \geq 3,\ n \text{ is odd},\ x \in [0, +\infty), \\ 2\operatorname{arctanh}\left(\varkappa_\alpha \coth \frac{nx}{2}\right), & \text{if } n \geq 4,\ n \text{ is even},\ x \in [\omega_\alpha, +\infty). \end{cases} \tag{14}$$

For every $n \geq 3$ and every $x \geq \omega_\alpha$,

$$\varkappa_\alpha \coth \frac{nx}{2} < \varkappa_\alpha \coth \frac{x}{2} \leq \varkappa_\alpha \coth \frac{\omega_\alpha}{2} = 1,$$

hence $\psi_{\alpha,n}$ is well defined. The two cases in (14) can be joined by elevating $\tanh(nx/2)$ to the power $(-1)^{n+1}$.

Theorem 2.3 (Main Equation for the Outlier Eigenvalue) *If n is odd and $n > \varkappa_\alpha^{-1}$, then $s_{\alpha,n}$ is the unique solution in $(0, \omega_\alpha)$ of the equation*

$$x = \psi_{\alpha,n}(x). \tag{15}$$

If n is even, then $s_{\alpha,n}$ is the unique solution in $(\omega_\alpha, +\infty)$ of the Eq. (15).

To get asymptotic expansions for the inner eigenvalues, we introduce the function $\Lambda_{\alpha,n} \colon [0, \pi] \to \mathbb{R}$ by

$$\Lambda_{\alpha,n}(x) := g(x) + \frac{g'(x)\eta_\alpha(x)}{n} + \frac{g'(x)\eta_\alpha(x)\eta'_\alpha(x) + \frac{1}{2}g''(x)\eta_\alpha(x)^2}{n^2}.$$

Then, for all $n \geq N_\alpha$ and all odd j with $3 \leq j \leq n-1$, we define $\lambda_{\alpha,n,j}^{\text{asympt}}$ by

$$\lambda_{\alpha,n,j}^{\text{asympt}} := \Lambda_{\alpha,n}(d_{n,j}). \tag{16}$$

Theorem 2.4 (Asymptotic Expansion of Inner Eigenvalues) *There exists $C_1(\alpha) > 0$ such that for every $n \geq N_\alpha$,*

$$\max_{\substack{3\leq j\leq n-1\\ j\ \text{odd}}} \left|\lambda_{\alpha,n,j} - \lambda_{\alpha,n,j}^{\text{asympt}}\right| \leq \frac{C_1(\alpha)}{n^3}. \tag{17}$$

To state the asymptotic expansion for $\lambda_{\alpha,n,n}$, we introduce the following numbers:

$$\begin{aligned} \beta_{\alpha,1} &:= \frac{16\alpha^2(\alpha-1)^2}{(2\alpha-1)^2}, \qquad \beta_{\alpha,2} := -\frac{64\alpha^3(\alpha-1)^3}{(2\alpha-1)^3}, \\ \beta_{\alpha,3} &:= \frac{32\alpha^2(1-\alpha)^2(2\alpha^2-2\alpha+1)}{(2\alpha-1)^3}. \end{aligned} \tag{18}$$

Equivalently,

$$\beta_{\alpha,1} = \frac{16\varkappa_\alpha^2}{(1-\varkappa_\alpha^2)^2}, \qquad \beta_{\alpha,2} = -\frac{64\varkappa_\alpha^3}{(1-\varkappa_\alpha^2)^3}, \qquad \beta_{\alpha,3} = \frac{32\varkappa_\alpha^2(\varkappa_\alpha^2+1)}{(1-\varkappa_\alpha^2)^3}. \tag{19}$$

Now, we define $\lambda_{\alpha,n,n}^{\text{asympt}}$ by

$$\lambda_{\alpha,n,n}^{\text{asympt}} := \Omega_\alpha + (-1)^n \beta_{\alpha,1} e^{-n\omega_\alpha} + \beta_{\alpha,2} n e^{-2n\omega_\alpha} + \beta_{\alpha,3} e^{-2n\omega_\alpha}. \tag{20}$$

Of course, $e^{-n\omega_\alpha}$ can also be written as $1/(2\alpha-1)^n$.

Theorem 2.5 (Asymptotic Expansion of the Last Eigenvalue) *As $n \to \infty$, the extreme eigenvalue $\lambda_{\alpha,n,n}$ of $L_{\alpha,n}$ converges exponentially to Ω_α. More precisely, there exists $C_2(\alpha) > 0$ such that for every $n \geq N_\alpha$,*

$$\left|\lambda_{\alpha,n,n} - \lambda_{\alpha,n,n}^{\text{asympt}}\right| \leq C_2(\alpha) n^2 e^{-3n\omega_\alpha}. \tag{21}$$

So, in the case when $\alpha > 1$ and n is large enough, the maximal eigenvalue goes out of $[0, 4]$ and converges rapidly to the number $\Omega_\alpha > 4$. While, the rest behaves asymptotically as the function g on $[0, \pi]$. The "right spectral gap" $\lambda_{\alpha,n,n} - \lambda_{\alpha,n,n-1}$ converges to $\Omega_\alpha - 4$.

Our last analysis focuses on the eigenvectors and their norms. Similarly to the situation with the eigenvalues, we have to separate the case $\lambda = 0$, the "trigonometric case" ($0 < \lambda \leq 4$), and the "hyperbolic case" ($\lambda > 4$).

Theorem 2.6 (Eigenvectors for $\operatorname{Re}(\alpha) > 1$**)** *Let $\alpha \in \mathbb{C}$ with $\operatorname{Re}(\alpha) > 1$ and $n \geq 3$. Then, $L_{\alpha,n}$ has the following eigenvectors.*

1. *$[1, \ldots, 1]^\top$ is an eigenvector associated to the eigenvalue $\lambda_{\alpha,n,1} = 0$.*
2. *For every j, $2 \leq j \leq n-1$, the vector $v_{\alpha,n,j} = [v_{\alpha,n,j,k}]_{k=1}^n$ with the following components is an eigenvector associated to $\lambda_{\alpha,n,j}$:*

$$v_{\alpha,n,j,k} := \sin(kz_{\alpha,n,j}) - (1-\overline{\alpha})\sin((k-1)z_{\alpha,n,j}) + \overline{\alpha}\sin((n-k)z_{\alpha,n,j}). \tag{22}$$

The same formula (22) also works for $j = n$, if n is odd and $n \leq \varkappa_{\operatorname{Re}(\alpha)}^{-1}$.

3. *If n is odd and $n > \varkappa_{\operatorname{Re}(\alpha)}^{-1}$, or n is even, then the vector $v_{\alpha,n,n} = [v_{\alpha,n,n,k}]_{k=1}^n$ with the following components is an eigenvector associated to $\lambda_{\alpha,n,n}$:*

$$\begin{aligned} v_{\alpha,n,n,k} := (-1)^k \big[&(-1)^n\overline{\alpha}\sinh((n-k)s_{\alpha,n}) + (1-\overline{\alpha})\sinh((k-1)s_{\alpha,n}) \\ &+ \sinh(ks_{\alpha,n})\big]. \end{aligned} \tag{23}$$

Finally, to present the asymptotic behavior of the norms of the eigenvectors given by (22), we need the following auxiliar function: for every x in $[0, \pi]$, we define

$$\nu_\alpha(x) := \frac{1-\operatorname{Re}(\alpha)}{2} g(x) + \frac{\operatorname{Re}(\alpha)}{2} g(\eta_{\operatorname{Re}(\alpha)}(x)) + \frac{|\alpha|^2 - \operatorname{Re}(\alpha)}{2} g(x - \eta_{\operatorname{Re}(\alpha)}(x)). \tag{24}$$

Theorem 2.7 (Norms of Eigenvectors for $\operatorname{Re}(\alpha) > 1$**)** *Let $\alpha \in \mathbb{C}$ with $\operatorname{Re}(\alpha) > 1$ and $n \geq N_{\operatorname{Re}(\alpha)}$.*

1. *If j is even and $2 \leq j \leq n-1$, then*

$$\|v_{\alpha,n,j}\|_2 = |\alpha - 1|\sqrt{2n}\,\sin\frac{j\pi}{2n}. \tag{25}$$

2. *If j is odd and $3 \leq j \leq n-1$, then as $n \to \infty$*

$$\|v_{\alpha,n,j}\|_2 = \sqrt{\nu_\alpha(d_{n,j})n} + O_\alpha\left(\frac{1}{\sqrt{n}}\right), \tag{26}$$

with $O_\alpha(1/\sqrt{n})$ uniformly on j.

3. *As $n \to \infty$,*

$$\|v_{\alpha,n,n}\|_2 = \frac{|\alpha|}{2\sqrt{2\operatorname{Re}(\alpha)(\operatorname{Re}(\alpha)-1)}} e^{n\omega_{\operatorname{Re}(\alpha)}} + O_\alpha(n). \tag{27}$$

In numerical computation of the eigenvectors, it is convenient to divide the expressions given in Theorem 2.6 by the norms' approximations from Theorem 2.7.

3 The Characteristic Polynomial and Eigenvalues' Localization

Recall that we denote the characteristic polynomial $\det(\lambda I - L_{\alpha,n})$ by $D_{\alpha,n}(\lambda)$. Aditionally, $\varkappa_\alpha$, ω_α are defined by (3), (13).

For every $m \geq 0$, let T_m and U_m the m-th degree Chebyshev polynomials of the first and second kind, respectively.

By cofactor expansion, it is easy to prove the following proposition.

Proposition 3.1 (Characteristic Polynomial of $L_{\alpha,n}$ for Complex α) *For $n \geq 3$ and $\alpha \in \mathbb{C}$,*

$$D_{\alpha,n}(\lambda) = (\lambda - 2\operatorname{Re}(\alpha))U_{n-1}\left(\frac{\lambda-2}{2}\right) - 2\operatorname{Re}(\alpha)U_{n-2}\left(\frac{\lambda-2}{2}\right) + 2(-1)^{n+1}\operatorname{Re}(\alpha). \tag{28}$$

Equivalently,

$$D_{\alpha,n}(\lambda) = U_n\left(\frac{\lambda-2}{2}\right) + 2(1-\operatorname{Re}(\alpha))U_{n-1}\left(\frac{\lambda-2}{2}\right) + (1-2\operatorname{Re}(\alpha))U_{n-2}\left(\frac{\lambda-2}{2}\right) + 2(-1)^{n+1}\operatorname{Re}(\alpha). \tag{29}$$

The next proposition is similar to [12, Proposition 14], but here we use the change of variable $\lambda = t^2$ instead of $\lambda = 4 - t^2$.

For $n \geq 3$ define

$$p_n(t) := \begin{cases} U_{n-1}(t/2), & \text{if } n \text{ is even}, \\ T_n(t/2), & \text{if } n \text{ is odd}, \end{cases} \tag{30}$$

$$q_{\alpha,n}(t) := \begin{cases} (1-\alpha)\frac{t}{2}T_n\left(\frac{t}{2}\right) + \alpha\frac{t^2-4}{4}U_{n-1}\left(\frac{t}{2}\right), & \text{if } n \text{ is even}, \\ (1-\alpha)\frac{t}{2}U_{n-1}\left(\frac{t}{2}\right) + \alpha T_n\left(\frac{t}{2}\right), & \text{if } n \text{ is odd}. \end{cases} \tag{31}$$

Proposition 3.2 *For every α in $\mathbb{R}$, every $n \geq 3$ and every t in $\mathbb{C}$,*

$$D_{\alpha,n}(t^2) = 4p_n(t)q_{\alpha,n}(t). \tag{32}$$

Proof Let $w = (t^2-2)/2$, i.e., $t^2 = 2w+2$. Then, (28) takes the following form:

$$D_{\alpha,n}(2w+2) = 2\Big((w+1-\alpha)U_{n-1}(w) - \alpha U_{n-2}(w) + (-1)^{n+1}\alpha\Big). \tag{33}$$

Let $n = 2m$. We apply $U_{2m-2}(w) = -U_{2m}(w) + 2wU_{2m-1}(w)$ on (33), obtaining

$$D_{\alpha,2m}(2w+2) = 2\Big(\alpha U_{2m}(w) + (w+1-\alpha-2\alpha w)U_{2m-1}(w) - \alpha\Big).$$

Now, we use the identities

$$\begin{aligned} U_{2m-1}(w) &= 2U_{m-1}(w)T_m(w), \\ U_{2m}(w) &= 2wU_{m-1}(w)T_m(w) + 2T_m^2(w) - 1, \\ U_{2m}(w) - U_{2m-1}(w) + 1 &= 2(w^2-1)U_{m-1}^2(w), \end{aligned}$$

deriving

$$D_{\alpha,2m}(2w+2) = 4(w+1)U_{m-1}(w)\Big((1-\alpha)T_m(w) + \alpha(w-1)U_{m-1}(w)\Big).$$

Considering the relations

$$T_{2m}\left(\frac{t}{2}\right) = T_m\left(\frac{t^2-2}{2}\right), \quad U_{2m+1}\left(\frac{t}{2}\right) = tU_m\left(\frac{t^2-2}{2}\right),$$

we obtain that the characteristic polynomial is the product of the polynomials (30) and (31).

If $n = 2m+1$, the analysis is similar. □

Remark 3.3 If $n \geq 3$ is odd, then the polynomial $q_{\alpha,n}$ coincides with the polynomial $q_{1-\alpha,n}$ written in [13].

We will apply the following elementary identities:

$$T_n\left(\sin\frac{x}{2}\right) = (-1)^{\frac{n}{2}}\cos\frac{nx}{2}, \quad U_n\left(\sin\frac{x}{2}\right) = (-1)^{\frac{n}{2}}\frac{\cos\frac{(n+1)x}{2}}{\cos\frac{x}{2}} \quad (n \text{ is even}), \tag{34}$$

$$T_n\left(\sin\frac{x}{2}\right) = (-1)^{\frac{n-1}{2}}\sin\frac{nx}{2}, \quad U_n\left(\sin\frac{x}{2}\right) = (-1)^{\frac{n-1}{2}}\frac{\sin\frac{(n+1)x}{2}}{\cos\frac{x}{2}} \quad (n \text{ is odd}). \tag{35}$$

Then, using the change of variable $t = 2\sin(x/2)$ in (30) and (31) yields

$$p_n\left(2\sin\frac{x}{2}\right) = \begin{cases} (-1)^{\frac{n}{2}+1}\dfrac{\sin\frac{nx}{2}}{\cos\frac{x}{2}}, & \text{if } n \text{ is even}, \\ (-1)^{\frac{n-1}{2}}\sin\frac{nx}{2}, & \text{if } n \text{ is odd}. \end{cases} \tag{36}$$

$$q_{\alpha,n}\left(2\sin\frac{x}{2}\right)=\begin{cases}(-1)^{\frac{n}{2}}\left((1-\alpha)\sin\frac{x}{2}\cos\frac{nx}{2}+\alpha\cos\frac{x}{2}\sin\frac{nx}{2}\right), & \text{if } n \text{ is even},\\ \frac{(-1)^{\frac{n-1}{2}}}{\cos\frac{x}{2}}\left((1-\alpha)\sin\frac{x}{2}\cos\frac{nx}{2}+\alpha\cos\frac{x}{2}\sin\frac{nx}{2}\right), & \text{if } n \text{ is odd}.\end{cases} \tag{37}$$

So, (32) becomes

$$D_{\alpha,n}(g(x))=(-1)^{n+1}\frac{4\sin\frac{x}{2}\sin\frac{nx}{2}}{\cos\frac{x}{2}}\left((1-\alpha)\cos\frac{nx}{2}+\alpha\frac{\sin\frac{nx}{2}}{\sin\frac{x}{2}}\cos\frac{x}{2}\right). \tag{38}$$

After the change of variable $t=2\cosh(x/2)$, formula (32) transforms to

$$\begin{aligned}D_{\alpha,n}(g_+(x))&=4\cosh\frac{x}{2}\frac{\sinh\frac{nx}{2}}{\sinh\frac{x}{2}}\left((1-\alpha)\cosh\frac{nx}{2}+\alpha\frac{\sinh\frac{x}{2}\sinh\frac{nx}{2}}{\cosh\frac{x}{2}}\right)\\ &\qquad (n \text{ is even}),\\ D_{\alpha,n}(g_+(x))&=4\cosh\frac{x}{2}\cosh\frac{nx}{2}\left((1-\alpha)\frac{\sinh\frac{nx}{2}}{\sinh\frac{x}{2}}+\alpha\frac{\cosh\frac{nx}{2}}{\cosh\frac{x}{2}}\right)\qquad (n \text{ is odd}).\end{aligned} \tag{39}$$

Proposition 3.4 (Trivial Eigenvalues of $L_{\alpha,n}$) *For every $n \geq 3$ and every even j with $0 \leq j \leq n-1$, the number $g(j\pi/n)$ is an eigenvalue of $L_{\alpha,n}$.*

Proof These eigenvalues come from the factor p_n in the decomposition (32). Indeed, the change of variable $\lambda=(2\sin(x/2))^2$ yields the factor $p_n\,(2\sin(x/2))$. According to (36), this expression vanishes for $x=2k\pi/n$, where k is an integer and $0\leq k\leq (n-1)/2$. □

Lemma 3.5 *If n is even, then* $\lim_{t\to+\infty} q_{\alpha,n}(t)=+\infty$.

Proof From the recurrent definition of Chebyshev polynomials, the leading term of $T_n(t/2)$ is $(1/2)t^n$, and the leading term of $U_{n-1}(t/2)$ is t^{n-1}. Therefore, by (31), the leading term of $q_{\alpha,n}(t)$ is $(1/4)t^{n+1}$. So, the leading coefficient is strictly positive, which implies the result. □

For every j with $1\leq j\leq n$, we define

$$I_{n,j}:=\left(\frac{(j-1)\pi}{n},\frac{j\pi}{n}\right)=(d_{n,j},d_{n,j+1}). \tag{40}$$

Proof of Theorem 2.1 For $1\leq j\leq n-1$, the proof is similar to the proof of [12, Theorem 1]. In particular, for odd j, we use Proposition 3.4.

1. If n is odd and satisfies $n < \varkappa_\alpha^{-1}$, then, using (37), it is easy to see that $q_{\alpha,n}(2\sin(x/2))$ changes of sign in $I_{n,n}$. Indeed, $q_{\alpha,n}(2\sin(d_{n,n}/2)) = -1$, and

$$\lim_{x\to\pi^-} q_{\alpha,n}\left(2\sin\frac{x}{2}\right) = -(n(1-\alpha)+\alpha) = (\alpha-1)(n-\varkappa_\alpha^{-1}).$$

2. If n is odd and satisfies $n = \varkappa_\alpha^{-1}$, then $q_{\alpha,n}(2) = (1-\alpha)n+\alpha$, hence $\lambda = 4$ is an eigenvalue of $L_{\alpha,n}$.
3. If $n \geq 3$ is odd and $n > \varkappa_\alpha^{-1}$, then $q_{\alpha,n}(t)$ takes values of opposite signs at the ends of the interval $[2, r_\alpha + r_\alpha^{-1}]$ where $r_\alpha := \sqrt{2\alpha-1}$:

$$q_{\alpha,n}(2) = (1-\alpha)n+\alpha < 0, \quad q_{\alpha,n}(r_\alpha + r_\alpha^{-1}) = \frac{r_\alpha^2+1}{2} r_\alpha^{-n} > 0.$$

   Then,

$$4 < \lambda_{\alpha,n,n} < \left(r_\alpha + \frac{1}{r_\alpha}\right)^2 = \Omega_\alpha.$$

4. For every even $n \geq 4$, $q_{\alpha,n}$ changes its sign in the interval $[r_\alpha + r_\alpha^{-1}, +\infty)$. Indeed, $\lim_{t\to+\infty} q_{\alpha,n}(t) = +\infty$ by Lemma 3.5, whereas

$$q_{\alpha,n}(r_\alpha + r_\alpha^{-1}) = -\frac{1}{4}(r_\alpha^2+1)\left(r_\alpha^{n+1} + r_\alpha^{-(n+1)}\right) < 0.$$

   Moreover, by the Gershgorin disks theorem (see, e.g., [18, Theorem 2.1]), the eigenvalues are bounded from above by $4+2\alpha$. Thus,

$$\Omega_\alpha = \left(r_\alpha + \frac{1}{r_\alpha}\right)^2 \leq \lambda_{\alpha,n,n} \leq 4+2\alpha.$$

Items 1, 2, and 3 could also be derived from [13, Lemmas 3.3, 3.4], taking into account Remark 3.3. □

4 Inner Eigenvalues

In this section we deal with the inner eigenvalues. The proofs of the upcoming propositions are very similar, if not identical, to the proofs given in [12, 13]. Recall that $\varkappa_\alpha$, η_α are defined by (3), (8).

Proof of Theorem 2.2 If $\lambda \in (0,4)$, we use the change of variable $\lambda = g(x)$, with $x \in (0,\pi)$. So, $D_{\alpha,n}(g(x))$ transforms into (38). Equivalently, we apply (32) with

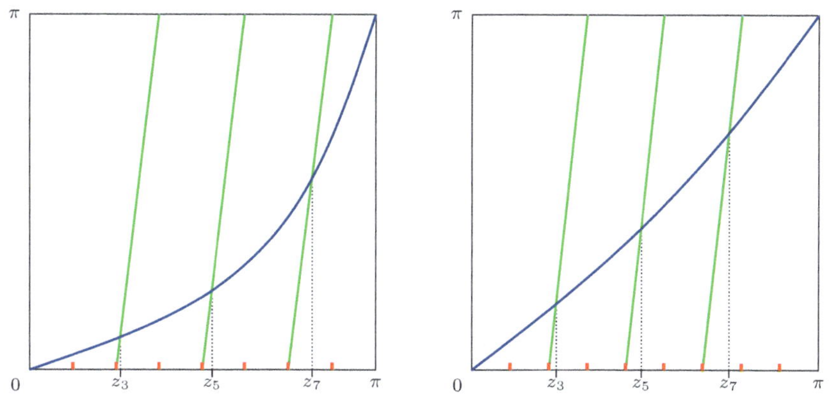

Fig. 3 Plot of η_α (blue) and the left-hand side of (42) (green) for $\alpha = 3/2$, $n = 8$ (left) and $\alpha = 4$, $n = 9$ (right)

$t = 2\sin(x/2)$. Then, $D_{\alpha,n}(g(x)) = 0$ reduces to $q_{\alpha,n}(2\sin(x/2)) = 0$, which is equivalent to

$$\tan\frac{nx}{2} = \varkappa_\alpha \tan\frac{x}{2}. \tag{41}$$

In particular, for odd j with $3 \le j \le n-1$, the solution $z_{\alpha,n,j}$ belonging to $I_{n,j}$ satisfies (41). □

Equation (9) from Theorem 2.2 can be rewritten in the form

$$nx - (j-1)\pi = \eta_\alpha(x). \tag{42}$$

Figure 3 shows η_α and the left-hand side of (42) for a couple of examples.

The first two derivatives of η_α are

$$\eta'_\alpha(x) = \frac{2\varkappa_\alpha}{1 + \varkappa_\alpha^2 + (1 - \varkappa_\alpha^2)\cos(x)}, \tag{43}$$

$$\eta''_\alpha(x) = \frac{2\varkappa_\alpha(1 - \varkappa_\alpha^2)\sin(x)}{(1 + \varkappa_\alpha^2 + (1 - \varkappa_\alpha^2)\cos(x))^2}. \tag{44}$$

Proposition 4.1 and Theorem 4.2 follow directly from the properties of η_α, similarly to [12, Propositions 21 and 22].

Proposition 4.1 *Each derivative of* η_α *is a bounded function on* $(0, \pi)$*. In particular,*

$$\sup_{0<x<\pi} |\eta'_\alpha(x)| = \varkappa_\alpha^{-1}, \qquad \sup_{0<x<\pi} |\eta''_\alpha(x)| \le \frac{\varkappa_\alpha^{-2} - 1}{2}.$$

Recall that N_α is defined by (11), and that for every j, the numbers $d_{n,j}$, $z_{n,j}$ are defined by (5) and (7), respectively.

Theorem 4.2 *Let* $n \ge N_\alpha$*,* j *be odd,* $3 \le j \le n-1$*. Then, the function* $x \mapsto d_j + \eta_\alpha(x)/n$ *is a contraction on* $\mathrm{cl}(I_{n,j})$*, and its fixed point is* $z_{\alpha,n,j}$*.*

In [12, Proposition 24], we proved some simple facts about the convergence of Newton's method for convex functions. Now we are going to state without proofs some similar facts for concave functions (see also [12, Remark 27]). Assume that $a, b \in \mathbb{R}$ with $a < b$; f is differentiable and $f' > 0$ on $[a, b]$; there exists c in $[a, b]$ such that $f(c) = 0$; $y^{(0)}$ is a point in $[a, b]$ and the sequence $(y^{(m)})_{m=0}^{\infty}$ is defined (when possible) by the recurrence relation

$$y^{(m+1)} = y^{(m)} - \frac{f\left(y^{(m)}\right)}{f'\left(y^{(m)}\right)}. \tag{45}$$

Proposition 4.3 (Linear Convergence of Newton's Method for Concave Functions) *If* f *is concave on* $[a, b]$*,* $a \le y^{(0)} \le c$*, then* $y^{(m)}$ *belongs to* $[a, c]$ *for every* $m \ge 0$*, the sequence* $(y^{(m)})_{m=0}^{\infty}$ *increases and converges to* c*, with*

$$c - y^{(m)} \le (b-a)\left(1 - \frac{f'(b)}{f'(a)}\right)^m. \tag{46}$$

For every $n \ge 4$ and every j odd with $3 \le j \le n$, we define $h_{\alpha,n,j}\colon \mathrm{cl}(I_{n,j}) \to \mathbb{R}$ by

$$h_{\alpha,n,j}(x) := nx - (j-1)\pi - \eta_\alpha(x).$$

Theorem 4.4 (Convergence of Newton's Method Applied to $h_{\alpha,n,j}$) *Let* $n \ge N_\alpha$*,* j *be odd,* $3 \le j \le n-1$ *and* $y_{\alpha,n,j}^{(0)} = d_{n,j}$*. Define the sequence* $(y_{\alpha,n,j}^{(m)})_{m=0}^{\infty}$ *by the recursive formula*

$$y_{\alpha,n,j}^{(m)} := y_{\alpha,n,j}^{(m-1)} - \frac{h_{\alpha,n,j}\left(y_{\alpha,n,j}^{(m-1)}\right)}{h'_{\alpha,n,j}\left(y_{\alpha,n,j}^{(m-1)}\right)} \quad (m \ge 1). \tag{47}$$

Then $(y_{\alpha,n,j}^{(m)})_{m=0}^{\infty}$ *is well defined and converges to* $z_{\alpha,n,j}$*, and the convergence is at least linear:*

$$z_{\alpha,n,j} - y_{\alpha,n,j}^{(m)} \leq \frac{\pi}{n}\left(\frac{\varkappa_\alpha^{-2}-1}{\varkappa_\alpha^{-1}n-1}\right)^m. \tag{48}$$

Moreover, if $n \geq 2N_\alpha$*, then the convergence is quadratic, and*

$$z_{\alpha,n,j} - y_{\alpha,n,j}^{(m)} \leq \frac{\pi}{n}\left(\frac{\pi\varkappa_\alpha^{-2}}{2n^2}\right)^{2^m-1}. \tag{49}$$

Proof Formulas (43) and (44) for η'_α and η''_α imply that $h'_{\alpha,n,j} > 0$ and $h''_{\alpha,n,j} < 0$ on $\operatorname{cl}(I_{n,j})$. Moreover, $y_{\alpha,n,j}^{(0)} = d_{n,j} < z_{\alpha,n,j} < d_{n,j+1}$. So, the assumptions of Proposition 4.3 are satisfied. Here are rough estimates of the derivatives of $h_{\alpha,n,j}$ at the extremes of $I_{n,j}$:

$$n - \varkappa_\alpha = h'_{\alpha,n,j}(0) \geq h'_{\alpha,n,j}(d_{n,j}) \geq h'_{\alpha,n,j}(d_{n,j+1}) \geq h'_{\alpha,n,j}(\pi) = n - \frac{1}{\varkappa_\alpha}.$$

Therefore,

$$1 - \frac{h'_{\alpha,n,j}(d_{n,j+1})}{h'_{\alpha,n,j}(d_{n,j})} \leq 1 - \frac{n-\varkappa_\alpha^{-1}}{n-\varkappa_\alpha} = \frac{\varkappa_\alpha^{-2}-1}{n\varkappa_\alpha^{-1}-1},$$

and we obtain (48).

Finally, if $n \geq 2N_\alpha$, then

$$\frac{\pi}{n}\cdot\frac{\max\limits_{0\leq x\leq\pi}|h''_{\alpha,n,j}(x)|}{2\min\limits_{0\leq x\leq\pi}|h'_{\alpha,n,j}(x)|} \leq \frac{\pi \max\limits_{0\leq x\leq\pi}|\eta''_\alpha(x)|}{2n\left(n - \max\limits_{0\leq x\leq\pi}|\eta'_\alpha(x)|\right)} \leq \frac{\pi(\varkappa_\alpha^{-2}-1)}{4n(n-\varkappa_\alpha^{-1})} \leq \frac{\pi\varkappa_\alpha^{-2}}{2n^2} < 1,$$

which implies the quadratic convergence with upper estimate (49); see, e.g., [1, Sect. 2.2] or [12, Proposition 26]. □

Proposition 4.5 *There exists* $C_1(\alpha) > 0$ *such that for every* $n \geq 3$ *and every* j *odd with* $3 \leq j \leq n-1$,

$$z_{\alpha,n,j} = d_{n,j} + \frac{\eta_\alpha(d_{n,j})}{n} + \frac{\eta_\alpha(d_{n,j})\,\eta'_\alpha(d_{n,j})}{n^2} + r_{\alpha,n,j},$$

where $|r_{\alpha,n,j}| \leq \frac{C_1(\alpha)}{n^3}$.

Proof of Theorem 2.4 Substituting (4.5) into g and using Taylor expansion of g around $d_{n,j}$, we obtain the asymptotic expansion (16) with error bound (17). □

5 Transformation of the Characteristic Equation for the Last Eigenvalue

Recall that $\varkappa_\alpha$, N_α, ω_α are defined by (3), (11), (13), respectively.

Proof of Theorem 2.3 If $\lambda \in (4, \infty)$, we make the change of variable $\lambda = g_+(x)$ with $x \in (0, \infty)$. In other words, we use (32) with $t = 2\cosh(x/2)$. Then, $p_n(2\cosh(x/2)) \neq 0$, and equation $D_{\alpha,n}(g_+(x)) = 0$ is equivalent to $q_{\alpha,n}(2\cosh(x/2)) = 0$, which takes the following form:

$$\tanh\frac{nx}{2} = \frac{1}{\varkappa_\alpha}\tanh\frac{x}{2} \quad (n \text{ is odd}), \tag{50}$$

$$\tanh\frac{nx}{2} = \varkappa_\alpha \coth\frac{x}{2} \quad (n \text{ is even}). \tag{51}$$

By Theorem 2.1, if n is odd and $n > \varkappa_\alpha$, then (50) has a unique solution on $(0, \omega_\alpha)$. If n is even and $n \geq 4$, then (51) has a unique solution on (ω_α, ∞). We apply arctanh to both sides of the Eqs. (50) and (51), and rewrite them as (15). □

6 Last Eigenvalue with Odd *n*

In this section, we suppose that n is odd and $n \geq N_\alpha$, and we study the behavior of $\lambda_{\alpha,n,n}$ and $s_{\alpha,n}$ which are related by (12), i.e., $\lambda_{\alpha,n,n} = g_+(s_{\alpha,n})$.

The main idea of this section is to exploit the symmetry between the last eigenvalue of $L_{\alpha,n}$ and the first eigenvalue of $L_{1-\alpha,n}$. Since $\alpha > 1$, the "dual" parameter $\alpha' := 1 - \alpha$ satisfies $\alpha' < 0$, and the matrices $L_{\alpha',n}$ with $\alpha' < 0$ were studied in [13].

As we showed in [13, proof of Theorem 2.2], $\lambda_{\alpha',n,1}$ can be computed as $g_-(s_{\alpha',n})$ where $g_-(x) := -4\sinh^2(x/2)$ and $s_{\alpha',n}$ is the unique solution of

$$\tanh\left(\frac{nx}{2}\right) = \varkappa_{\alpha'}\tanh\left(\frac{x}{2}\right). \tag{52}$$

Proposition 6.1 *Let n be odd such that $n \geq N_\alpha$. Then $\lambda_{\alpha,n,n} = 4 - \lambda_{1-\alpha,n,1}$.*

Proof Let $\alpha' := 1 - \alpha$. Notice that $\varkappa_{\alpha'} = \varkappa_\alpha^{-1}$. Therefore, Eqs. (50) and (52) coincide. They have the same solutions:

$$s_{\alpha,n} = s_{\alpha',n}. \tag{53}$$

Finally,

$$4 - \lambda_{\alpha',n,1} = 4 - g_-(s_{\alpha',n}) = g_+(s_{\alpha',n}) = g_+(s_{\alpha,n}) = \lambda_{\alpha,n,n}.$$

□

Let n be odd, and recall that $\psi_{\alpha,n}$ is defined by (14); another useful representation is

$$\psi_{\alpha,n}(x) := \operatorname{arctanh}\left(\varkappa_{1-\alpha}^{-1}\tanh\frac{nx}{2}\right).$$

It follows that $\psi_{\alpha,n}$ equals the function $\varphi_{1-\alpha,n}$ given in [13, (2.6)]. Therefore, the properties of (14) and (56) are the ones developed in [13, Propostions 5.1 and 5.3]. In particular, the first two derivatives of $\psi_{\alpha,n}$ are

$$\psi'_{\alpha,n}(x) = \frac{2n\varkappa_\alpha}{(1-\varkappa_\alpha^2)\cosh(nx)+1+\varkappa_\alpha^2}, \tag{54}$$

$$\psi''_{\alpha,n}(x) = -\frac{2n^2\varkappa_\alpha(1-\varkappa_\alpha^2)\sinh(nx)}{((1-\varkappa_\alpha^2)\cosh(nx)+1+\varkappa_\alpha^2)^2}. \tag{55}$$

Define

$$\ell_{\alpha,n} := \frac{2}{n}\operatorname{arccosh}\sqrt{\frac{n\varkappa_\alpha-\varkappa_\alpha^2}{1-\varkappa_\alpha^2}} = \frac{2}{n}\operatorname{arccosh}\sqrt{\frac{n\alpha(\alpha-1)-(\alpha-1)^2}{2\alpha-1}}.$$

Proposition 6.2 *Let n be odd such that $n \geq N_\alpha$. Then $\psi_{\alpha,n}$ has the following properties.*

1. $\psi'_{\alpha,n} > 0$ *and* $\psi''_{\alpha,n} < 0$ *on* $[0,+\infty)$.
2. $\psi'_{\alpha,n}(\ell_{\alpha,n}) = 1$; *moreover,* $\psi'_{\alpha,n} > 1$ *on* $[0,\ell_{\alpha,n})$ *and* $\psi'_{\alpha,n} < 1$ *on* $(\ell_{\alpha,n},+\infty)$.
3. $\lim_{x\to+\infty}\psi_{\alpha,n}(x) = \omega_\alpha$.
4. $s_{\alpha,n}$ *is the unique fixed point of* $\psi_{\alpha,n}$ *in* $(0,+\infty)$.
5. $\psi_{\alpha,n}(x) > x$ *for every* x *in* $(0,\ell_{\alpha,n}]$.
6. $\ell_{\alpha,n} < \psi_{\alpha,n}(\ell_{\alpha,n}) < s_{\alpha,n}$.

For every odd $n \geq N_\alpha$, we define $f_{\alpha,n}\colon [0,+\infty) \to \mathbb{R}$ by

$$f_{\alpha,n}(x) := x - \psi_{\alpha,n}(x) = x - 2\operatorname{arctanh}\left(\varkappa_\alpha\tanh\frac{nx}{2}\right). \tag{56}$$

Figure 4 shows $f_{\alpha,n}$.

The following theorem contains more detailed information than its analog [13, Theorem 5.4].

Theorem 6.3 (Convergence of Newton's Method Applied to $f_{\alpha,n}$ for Odd n) *Let $n \geq N_\alpha$ and n be odd. Then the sequence $(y_{\alpha,n}^{(m)})_{m=0}^\infty$ defined by*

$$y_{\alpha,n}^{(0)} := \omega_\alpha, \qquad y_{\alpha,n}^{(m)} := y_{\alpha,n}^{(m-1)} - \frac{f_{\alpha,n}\left(y_{\alpha,n}^{(m-1)}\right)}{f'_{\alpha,n}\left(y_{\alpha,n}^{(m-1)}\right)} \qquad (m \geq 1),$$

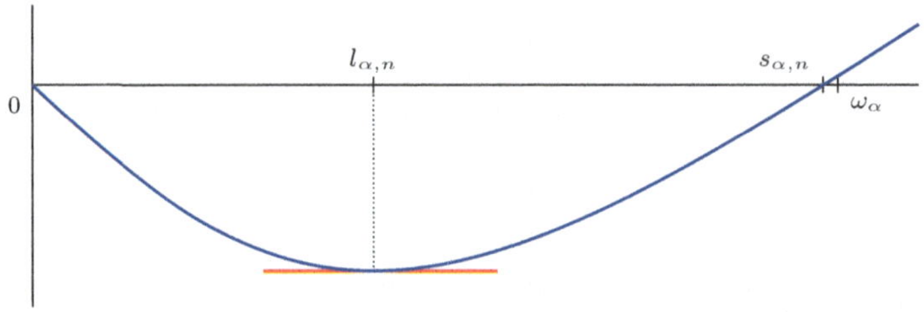

Fig. 4 Plot of $f_{\alpha,n}$ (blue) and tangent line to the graph of $f_{\alpha,n}$ at $\ell_{\alpha,n}$ (red), for $\alpha = 3/2$ and $n = 7$

takes values in $[s_{\alpha,n}, \omega_\alpha]$ *and converges to* $s_{\alpha,n}$. *The convergence is at least linear.*

Moreover, if n *is odd and large enough, then the convergence is quadratic, i.e., there exists* $Q_{\alpha,n}$ *in* $(0, 1/2)$ *such that for every* $m \geq 1$,

$$0 \leq y_{\alpha,n}^{(m)} - s_{\alpha,n} \leq \omega_\alpha Q_{\alpha,n}^{2^m-1}. \tag{57}$$

Proof By Proposition 6.2, $f'_{\alpha,n} > 0$ and $f''_{\alpha,n} > 0$ on $[\psi_{\alpha,n}(\ell_{\alpha,n}), \omega_\alpha]$. So, [13, Proposition 4.3] implies that the points $y_{\alpha,n}^{(m)}$ belong to the segment $[\psi_{\alpha,n}(\ell_{\alpha,n}), \omega_\alpha]$ (which is contained in $[s_{\alpha,n}, \omega_\alpha]$), and the convergence is at least linear.

It is easy to see that for n large enough, the dependence $n \mapsto \ell_{\alpha,n}$ is decreasing. Let n_0 be such a number that $\ell_{\alpha,n} \leq \ell_{\alpha,n_0}$ for every $n \geq n_0$.

Take $b_\alpha := \ell_{\alpha,n_0}$. Then for every $n > n_0$,

$$\ell_{\alpha,n} < b_\alpha < s_{\alpha,n} < \omega_\alpha.$$

Let $J_\alpha := [b_\alpha, \omega_\alpha]$. Since $\psi'_{\alpha,n}(b_\alpha) \to 0$ and $\sup_{J_\alpha} |\psi''| \to 0$ as $n \to \infty$, we choose n_1 such that for every $n > n_1$,

$$\psi'_{\alpha,n}(b_\alpha) < \frac{1}{2}, \qquad \sup_{J_\alpha} |\psi''_{\alpha,n}| < \frac{1}{2\omega_\alpha}.$$

Then, for $n > n_1$ and for every x in J_α,

$$\frac{1}{2} < f'_{\alpha,n}(b_\alpha) \leq f'_{\alpha,n}(x), \qquad |f''_{\alpha,n}(x)| < \frac{1}{2\omega_\alpha},$$

and

$$Q_{\alpha,n} := (\omega_\alpha - b_\alpha) \cdot \frac{\sup\limits_{J_\alpha} |f''_{\alpha,n}|}{2 \inf\limits_{J_\alpha} |f'_{\alpha,n}|} < \frac{1}{2}.$$

In fact, $Q_{\alpha,n}$ tends rapidly to 0 as n tends to ∞, but we have not found simple estimates. □

Define

$$\gamma_{1,\alpha} := \frac{4\varkappa_\alpha}{1-\varkappa_\alpha^2}, \qquad \gamma_{2,\alpha} := \frac{4\varkappa_\alpha(1+\varkappa_\alpha^2)}{(1-\varkappa_\alpha^2)^2}. \tag{58}$$

Theorem 6.4 (Asymptotic Expansion of $s_{\alpha,n}$ Where n Is Odd) *As n is odd and tends to infinity,*

$$s_{\alpha,n} = \omega_\alpha - \gamma_{1,\alpha} e^{-n\omega_\alpha} - \gamma_{1,\alpha}^2 n e^{-2n\omega_\alpha} + \gamma_{2,\alpha} e^{-2n\omega_\alpha} + O(n^2 e^{-3n\omega_\alpha}). \tag{59}$$

Proof Let $\alpha' := 1-\alpha$. In [13, Theorem 5.9], we proved that

$$s_{\alpha',n} = \omega_{\alpha'} - \gamma_{1,\alpha'} e^{-n\omega_{\alpha'}} - \gamma_{1,\alpha'}^2 n e^{-2n\omega_{\alpha'}} + \gamma_{2,\alpha'} e^{-2n\omega_{\alpha'}} + O(n^2 e^{-3n\omega_{\alpha'}}),$$

where

$$\omega_{\alpha'} = \log(1-2\alpha') = \log(2\alpha-1) = \omega_\alpha, \qquad \gamma_{1,\alpha'} = \gamma_{1,\alpha}, \qquad \gamma_{2,\alpha'} = \gamma_{2,\alpha}.$$

Now the result follows from (53). □

The asymptotic expansion of $\lambda_{\alpha,n,n}$ will be derived at the end of Sect. 7.

7 Last Eigenvalue for Even n

In this section, we study the behavior of the last eigenvalue $\lambda_{\alpha,n,n}$ supposing that n is even and $n \geq 4$. More precisely, we analyze the behavior of $s_{\alpha,n}$, defined by $\lambda_{\alpha,n,n} = g_+(s_{\alpha,n})$. Thus, in this section we suppose that n is even.

Define $r_\alpha := g_+{}^{-1}(4+2\alpha) = 2\,\mathrm{arcsinh}(\sqrt{\alpha/2})$. By Theorem 2.1, part 4, $s_{\alpha,n}$ is the unique solution of (51) in $(\omega_\alpha, r_\alpha)$.

Recall that $\psi_{\alpha,n}$ is defined by (14):

$$\psi_{\alpha,n}(x) = 2\,\mathrm{arctanh}\left(\varkappa_\alpha \coth\frac{nx}{2}\right) \qquad (x \geq \omega_\alpha).$$

Note that for $x \geq \omega_\alpha$,

$$\varkappa_\alpha \coth\frac{nx}{2} < \varkappa_\alpha \coth\frac{x}{2} \leq \varkappa_\alpha \coth\frac{\omega_\alpha}{2} = 1,$$

therefore $\psi_{\alpha,n}$ is well defined. A straightforward computation gives

$$\psi'_{\alpha,n}(x) = -\frac{2n\varkappa_\alpha}{(1-\varkappa_\alpha^2)\cosh(nx) - 1 - \varkappa_\alpha^2}, \tag{60}$$

$$\psi''_{\alpha,n}(x) = \frac{2n^2\varkappa_\alpha(1-\varkappa_\alpha^2)\sinh(nx)}{((1-\varkappa_\alpha^2)\cosh(nx) - 1 - \varkappa_\alpha^2)^2}. \tag{61}$$

Proposition 7.1 *Let n be even such that $n \geq 4$. Then $\psi_{\alpha,n}$ has the following properties.*

1. *$\psi'_{\alpha,n} < 0$ and $\psi''_{\alpha,n} > 0$ on $[\omega_\alpha, +\infty)$. So, $\psi_{\alpha,n}$ is a strictly decreasing convex function.*
2. $\lim_{x\to\infty} \psi_{\alpha,n}(x) = \omega_\alpha$.
3. *$s_{\alpha,n}$ is the unique fixed point of $\psi_{\alpha,n}$ and $\omega_\alpha < s_{\alpha,n} < r_\alpha$.*

Proof For every $x \geq \omega_\alpha$, due to the increasing property of cosh and the condition $n \geq 4$,

$$(1-\varkappa_\alpha^2)\cosh(nx) > (1-\varkappa_\alpha^2)\cosh(\omega_\alpha) = (1-\varkappa_\alpha^2)\frac{1+\varkappa_\alpha^2}{1-\varkappa_\alpha^2} = 1+\varkappa_\alpha^2.$$

Hence, the denominators of the fractions in the right-hand sides of (60) and (61) are strictly positive, and we get statement 1.

By definition of ω_α and $\varkappa_\alpha$,

$$\tanh\frac{\omega_\alpha}{2} = \frac{1-e^{-\omega_\alpha}}{1+e^{-\omega_\alpha}} = \frac{1-\frac{1}{2\alpha-1}}{1+\frac{1}{2\alpha-1}} = \frac{\alpha-1}{\alpha} = \varkappa_\alpha. \tag{62}$$

This equality implies statement 2. Finally, statement 3 is consequence of Theorems 2.1 and 2.3. □

We define $f_{\alpha,n}\colon [\omega_\alpha, \infty) \to \mathbb{R}$,

$$f_{\alpha,n}(x) := x - \psi_{\alpha,n}(x) = x - 2\operatorname{arctanh}\left(\varkappa_\alpha \coth\frac{nx}{2}\right). \tag{63}$$

We use the same notation $f_{\alpha,n}$ for two different functions, depending on the parity of n. Figure 5 shows $f_{\alpha,n}$.

Proposition 7.2 *For every even n with $n \geq 4$, $f'_{\alpha,n} > 1$ and $f''_{\alpha,n} < 0$ on $[\omega_\alpha, r_\alpha]$. Moreover, $s_{\alpha,n}$ is its only root in $(\omega_\alpha, r_\alpha)$.*

Proof Follows from Proposition 7.1. □

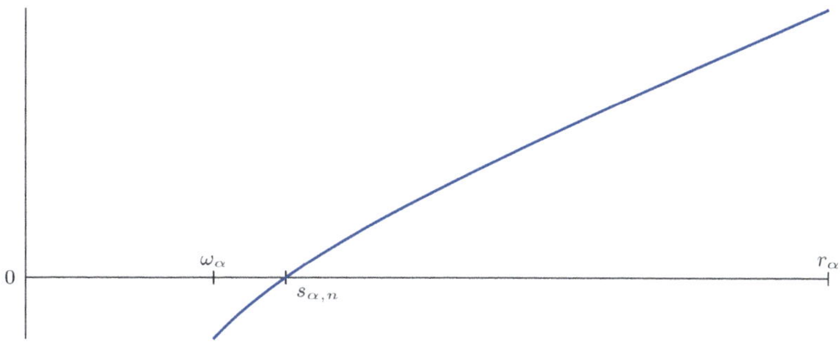

Fig. 5 Plot of $f_{\alpha,n}$ (blue), for $\alpha = 6/5$ and $n = 4$

Theorem 7.3 (Convergence of Newton's Method Applied to $f_{\alpha,n}$ for Even n) *Let $n \geq N_\alpha$ be even. Then the sequence $(y_{\alpha,n}^{(m)})_{m=0}^{\infty}$ defined by*

$$y_{\alpha,n}^{(0)} := \omega_\alpha, \qquad y_{\alpha,n}^{(m)} := y_{\alpha,n}^{(m-1)} - \frac{f_{\alpha,n}\left(y_{\alpha,n}^{(m-1)}\right)}{f'_{\alpha,n}\left(y_{\alpha,n}^{(m-1)}\right)} \qquad (m \geq 1),$$

takes values in $[\omega_\alpha, s_{\alpha,n}]$ and converges to $s_{\alpha,n}$.

Moreover, if n is even and large enough, then the convergence is quadratic, i.e., there exists $Q_{\alpha,n}$ in $(0, 1)$ such that for every $m \geq 1$,

$$0 \leq s_{\alpha,n} - \omega_\alpha \leq r_\alpha Q_{\alpha,n}^{2^m - 1}. \tag{64}$$

Proof By Propositions 7.2 and 4.3, the sequence $(y_{\alpha,n}^{(m)})_{m \geq 1}$ takes values in $[\omega_\alpha, s_{\alpha,n}]$ and converges at least linearly. Define

$$Q_{\alpha,n} := (r_\alpha - \omega_\alpha) \cdot \frac{\sup\limits_{[\omega_\alpha, r_\alpha]} |f''_{\alpha,n}|}{2 \inf\limits_{[\omega_\alpha, r_\alpha]} |f'_{\alpha,n}|}.$$

It follows from (61) that $\sup_{[\omega_\alpha, r_\alpha]} |f''_{\alpha,n}| \to 0$ as $n \to \infty$. Therefore, there exists n_0 such that $Q_{\alpha,n} < 1/2$, for every $n \geq n_0$. In fact, $Q_{\alpha,n}$ tends rapidly to 0 as n tends to ∞, but we have not found simple estimates. □

Lemma 7.4 *Let m, n be even such that $n > m \geq 4$. Then $s_{\alpha,m} > s_{\alpha,n}$.*

Proof Recall that $s_{\alpha,n}$ and $s_{\alpha,m}$ are the solutions of (15), respectively for n and m. In this lemma, we prefer to deal with the equivalent Eq. (51). Then

$$\tanh \frac{n s_{\alpha,m}}{2} \tanh \frac{s_{\alpha,m}}{2} > \tanh \frac{m s_{\alpha,m}}{2} \tanh \frac{s_{\alpha,m}}{2} = \frac{\alpha - 1}{\alpha} = \tanh \frac{n s_{\alpha,n}}{2} \tanh \frac{s_{\alpha,n}}{2}.$$

This implies $s_{\alpha,m} > s_{\alpha,n}$, since $x \mapsto \tanh \frac{nx}{2} \tanh \frac{x}{2}$ is a strictly increasing function on $[\omega_\alpha, \infty)$. □

Proposition 7.5 *Let n be even such that* $n \geq 4$. *Then*

$$0 \leq s_{\alpha,n} - \omega_\alpha \leq C_3(\alpha) e^{-n\omega_\alpha}, \tag{65}$$

where $C_3(\alpha) = \frac{(4+2\alpha)\alpha}{\alpha-1}$.

Proof By the mean value theorem applied to $x \mapsto \coth(x/2)$ on $[\omega_\alpha, s_{\alpha,n}]$, there exists $\xi \in (\omega_\alpha, s_{\alpha,n})$ such that

$$\coth \frac{s_{\alpha,n}}{2} - \coth \frac{\omega_\alpha}{2} = -\frac{1}{2\sinh^2 \frac{\xi}{2}} (s_{\alpha,n} - \omega_\alpha),$$

i.e.,

$$s_{\alpha,n} - \omega_\alpha = 2\sinh^2 \frac{\xi}{2} \left(\coth \frac{\omega_\alpha}{2} - \coth \frac{s_{\alpha,n}}{2} \right).$$

Now we apply the increasing property of sinh, identity (62), and the fact that $s_{\alpha,n}$ satisfies (51):

$$\begin{aligned} s_{\alpha,n} - \omega_\alpha &\leq 2\sinh^2 \frac{r_\alpha}{2} \left(\coth \frac{\omega_\alpha}{2} - \coth \frac{s_{\alpha,n}}{2} \right) = \frac{2\sinh^2 \frac{r_\alpha}{2}}{\varkappa_\alpha} \left(1 - \tanh \frac{n s_{\alpha,n}}{2} \right) \\ &\leq \frac{4\cosh^2 \frac{r_\alpha}{2}}{\varkappa_\alpha} e^{-n s_{\alpha,n}} \leq \frac{4\cosh^2 \frac{r_\alpha}{2}}{\varkappa_\alpha} e^{-n\omega_\alpha} = \frac{g_+(r_\alpha)}{\varkappa_\alpha} e^{-n\omega_\alpha}. \end{aligned}$$

The last expression simplifies to $C_3(\alpha) e^{-n\omega_\alpha}$. □

Recall that $\gamma_{1,\alpha}$ and $\gamma_{2,\alpha}$ are defined by (58).

Lemma 7.6 (Asymptotic Expansion of $\psi_{\alpha,1}$) *As t tends to infinity,*

$$\psi_{\alpha,1}(t) = \omega_\alpha + \gamma_{1,\alpha} e^{-t} + \gamma_{2,\alpha} e^{-2t} + O(e^{-3t}). \tag{66}$$

Proof The proof is analogous to the proof of [13, Lemma 5.8]. Since $\coth(t/2) = \frac{1+e^{-t}}{1-e^{-t}}$,

$$\psi_{\alpha,1}(t) = \sigma(e^{-t}), \quad \text{where} \quad \sigma(u) := 2\,\text{arctanh}\left(\varkappa_\alpha \frac{1+u}{1-u} \right).$$

We start with the Taylor–Maclaurin expansion of the rational function $u \mapsto (1+u)/(1-u)$ around 0:

$$\frac{1+u}{1-u} = 1 + \frac{2u}{1-u} = 1 + 2u + 2u^2 + O(u^3).$$

Then, we apply the Taylor expansion of arctanh around $\varkappa_\alpha$:

$$\operatorname{arctanh}(\varkappa_\alpha + y) = \operatorname{arctanh}(\varkappa_\alpha) + \frac{y}{1-\varkappa_\alpha^2} + \frac{\varkappa_\alpha\, y^2}{(1-\varkappa_\alpha^2)^2} + O(y^3).$$

In the last expansion, we substitute $y = 2\varkappa_\alpha(u + u^2 + O(u^3))$ and use the relation $O(y) = O(u)$:

$$\begin{aligned}\sigma(u) &= 2\operatorname{arctanh}\left(\varkappa_\alpha + 2\varkappa_\alpha(u+u^2+O(u^3))\right)\\ &= 2\operatorname{arctanh}(\varkappa_\alpha) + \frac{4\varkappa_\alpha}{1-\varkappa_\alpha^2}\left(u+u^2+O(u^3)\right)\\ &\quad + \frac{8\varkappa_\alpha^3}{(1-\varkappa_\alpha^2)^2}\left(u+u^2+O(u^3)\right)^2 + O(u^3).\end{aligned}$$

Simplifying and taking into account that $\tanh(\omega_\alpha/2) = \varkappa_\alpha$, we obtain the Taylor–Maclaurin expansion of σ around 0:

$$\sigma(u) = \omega_\alpha + \gamma_{1,\alpha}u + \gamma_{2,\alpha}u^2 + O(u^3).$$

Finally, we put $u = e^{-t}$ and obtain (66). □

Theorem 7.7 (Asymptotic Expansion of $s_{\alpha,n}$) *As n is even and tends to infinity,*

$$s_{\alpha,n} = \omega_\alpha + \gamma_{1,\alpha}e^{-n\omega_\alpha} - \gamma_{1,\alpha}^2 n e^{-2n\omega_\alpha} + \gamma_{2,\alpha}e^{-2n\omega_\alpha} + O(n^2e^{-3n\omega_\alpha}). \tag{67}$$

Proof The proof is analogous to the proof of [13, Theorem 5.9].

By formula (65) from Proposition 7.5, we have an asymptotic expansion of $s_{\alpha,n}$ with one exact term:

$$s_{\alpha,n} = \omega_\alpha + O(e^{-n\omega_\alpha}). \tag{68}$$

Therefore,

$$e^{-ns_{\alpha,n}} = e^{-n\omega_\alpha + O(ne^{-n\omega_\alpha})} = e^{-n\omega_\alpha}(1 + O(ne^{-n\omega_\alpha})) = e^{-n\omega_\alpha} + O(ne^{-2n\omega_\alpha}). \tag{69}$$

This also implies a rough upper bound for $e^{-ns_{\alpha,n}}$:

$$e^{-ns_{\alpha,n}} = O(e^{-n\omega_\alpha}). \tag{70}$$

The main idea of the following proof is to combine (68) with (15) and Lemma 7.6. We apply the asymptotic expansion (66) with two exact terms and with $ns_{\alpha,n}$ instead of t:

$$s_{\alpha,n} = \psi_{\alpha,n}(s_{\alpha,n}) = \psi_{\alpha,1}(ns_{\alpha,n}) = \omega_\alpha + \gamma_{1,\alpha}e^{-ns_{\alpha,n}} + O(e^{-2ns_{\alpha,n}}).$$

We simplify this expression using (69) and (70):

$$\begin{aligned} s_{\alpha,n} &= \omega_\alpha + \gamma_{1,\alpha}e^{-n\omega_\alpha} + O(ne^{-2n\omega_\alpha}) + O(e^{-2n\omega_\alpha}) \\ &= \omega_\alpha + \gamma_{1,\alpha}e^{-n\omega_\alpha} + O(ne^{-2n\omega_\alpha}). \end{aligned}$$

Now, we use this expansion to improve (69):

$$\begin{aligned} e^{-ns_{\alpha,n}} &= e^{-n\omega_\alpha}e^{-\gamma_{1,\alpha}ne^{-n\omega_\alpha}+O(n^2e^{-2n\omega_\alpha})} \\ &= e^{-n\omega_\alpha}\left(1-\gamma_{1,\alpha}ne^{-2n\omega_\alpha} + O(n^2e^{-2n\omega_\alpha})\right) \\ &= e^{-n\omega_\alpha} - \gamma_{1,\alpha}ne^{-2n\omega_\alpha} + O(n^2e^{-3n\omega_\alpha}). \end{aligned}$$

Next, we combine this expansion with (66):

$$\begin{aligned} s_{\alpha,n} &= \psi_{\alpha,n}(s_{\alpha,n}) = \psi_{\alpha,1}(ns_{\alpha,n}) = \omega_\alpha + \gamma_{1,\alpha}e^{-ns_{\alpha,n}} + \gamma_{2,\alpha}e^{-2ns_{\alpha,n}} + O(e^{-3ns_{\alpha,n}}) \\ &= \omega_\alpha + \gamma_{1,\alpha}\left(e^{-n\omega_\alpha} - \gamma_{1,\alpha}ne^{-2n\omega_\alpha} + O(n^2e^{-3n\omega_\alpha})\right) \\ &\quad + \gamma_{2,\alpha}\left(e^{-n\omega_\alpha} - \gamma_{1,\alpha}ne^{-2n\omega_\alpha} + O(n^2e^{-3n\omega_\alpha})\right)^2 + O(e^{-3n\omega_\alpha}). \end{aligned}$$

Simplifying this expression we get (67). □

In the next corollary, we join the asymptotic expansions (59) and (67).

Corollary 7.8 *As n tends to infinity,*

$$s_{\alpha,n} = \omega_\alpha + (-1)^n\gamma_{1,\alpha}e^{-n\omega_\alpha} - \gamma_{1,\alpha}^2ne^{-2n\omega_\alpha} + \gamma_{2,\alpha}e^{-2n\omega_\alpha} + O(n^2e^{-3n\omega_\alpha}). \tag{71}$$

Proof If n is odd, then (59) equals (71). If n is even, then (67) equals (71). □

Proof of Theorem 2.5 We expand g_+ by Taylor formula around ω_α:

$$g_+(\omega_\alpha + x) = g_+(\omega_\alpha) + g_+'(\omega_\alpha)x + \frac{g_+''(\omega_\alpha)}{2}x^2 + O(x^3).$$

Then we substitute the expansion (71) of $s_{\alpha,n}$ and simplify:

$$\begin{aligned}\lambda_{\alpha,n,n} &= g_+(s_{\alpha,n})\\ &= g_+\Big(\omega_\alpha + (-1)^n\gamma_{1,\alpha}e^{-n\omega_\alpha} - \gamma_{1,\alpha}^2 ne^{-2n\omega_\alpha} + \gamma_{2,\alpha}e^{-2n\omega_\alpha} + O(n^2e^{-3n\omega_\alpha})\Big)\\ &= g_+(\omega_\alpha) + (-1)^n\gamma_{1,\alpha}g_+'(\omega_\alpha)e^{-n\omega_\alpha} - \gamma_{1,\alpha}^2 g_+'(\omega_\alpha)ne^{-2n\omega_\alpha}\\ &\quad + \left(\gamma_{\alpha,2}g_+'(\omega_\alpha) + \frac{\gamma_{1,\alpha}^2 g_+''(\omega_\alpha)}{2}\right)e^{-2n\omega_\alpha} + O\left(n^2e^{-3n\omega_\alpha}\right).\end{aligned}$$

Recall that $g_+(\omega_\alpha) = \Omega_\alpha$. Hence, we obtain (20) and (21), with the following coefficients:

$$\beta_{\alpha,1} = g_+'(\omega_\alpha)\gamma_{1,\alpha},\quad \beta_{\alpha,2} = -g_+'(\omega_\alpha)\gamma_{1,\alpha}^2,\quad \beta_{\alpha,3} = g_+'(\omega_\alpha)\gamma_{\alpha,2} + \frac{1}{2}g_+''(\omega_\alpha)\gamma_{1,\alpha}^2.$$

Calculate the derivatives of g_+ at ω_α:

$$g_+'(\omega_\alpha) = 2\sinh(\omega_\alpha) = \frac{4\alpha(1-\alpha)}{1-2\alpha} = \frac{4\varkappa_\alpha}{1-\varkappa_\alpha^2},$$

$$g_+''(\omega_\alpha) = 2\cosh(\omega_\alpha) = \frac{2(2\alpha^2-2\alpha+1)}{2\alpha-1} = \frac{2(\varkappa_\alpha^2+1)}{1-\varkappa_\alpha^2}.$$

Combining with formulas (58), we write $\beta_{\alpha,1}$, $\beta_{\alpha,2}$, and $\beta_{\alpha,3}$ as (18) or (19). □

8 Norm of Eigenvectors

We recall that, due to Proposition 3.1, $\lambda_{\alpha,n,j} = \lambda_{\mathrm{Re}(\alpha),n,j}$ for every α in $\mathbb{C}$, every $n \geq 3$ and every $1 \leq j \leq n$. Nevertheless, it turns out that if $\mathrm{Im}(\alpha) \neq 0$, then the eigenvectors associated to $L_{\alpha,n}$ usually have complex components. So, in this section we suppose that α is a complex number such that $\mathrm{Re}(\alpha) > 1$. To simplify subindices, we put

$$\varkappa_\alpha := \varkappa_{\mathrm{Re}(\alpha)},\quad N_\alpha := N_{\mathrm{Re}(\alpha)},\quad \omega_\alpha := \omega_{\mathrm{Re}(\alpha)},\quad \Omega_\alpha := \Omega_{\mathrm{Re}(\alpha)},$$

$$\eta_\alpha := \eta_{\mathrm{Re}(\alpha)},\quad z_{\alpha,n,j} := z_{\mathrm{Re}(\alpha),n,j},\quad s_{\alpha,n} := s_{\mathrm{Re}(\alpha),n}.$$

Proof of Theorem 2.6 Formulas (22), (23) are consequences of [12, Proposition 8]. □

Recall that ν_α is defined by (24). For every x in $[0, \pi]$, we define

$$\xi_\alpha(x) := \frac{|\alpha-1|^2}{2} g(x)\cos(\eta_\alpha(x)) + \frac{|\alpha|^2}{2} g(\eta_\alpha(x))\cos(x) + \frac{|\alpha|^2 - \operatorname{Re}(\alpha)}{2} \left(g(x) + g(\eta_\alpha(x)) - g(x+\eta_\alpha(x))\right).$$

Proposition 8.1 (Exact Formulas for the Inner Eigenvectors) *Let $n \geq 3$ and $2 \leq j \leq n-1$. If j is even, then $\|v_{\alpha,n,j}\|_2$ is given by (25). If j is odd, then*

$$\|v_{\alpha,n,j}\|_2^2 = n\nu_\alpha(z_{\alpha,n,j}) + \frac{\sin(\eta_\alpha(z_{\alpha,n,j}))}{\sin(z_{\alpha,n,j})}\xi_\alpha(z_{\alpha,n,j}). \tag{72}$$

Proof These formulas are similar to [12, (66), (69)] and are proved in the same manner. □

We will use several identities for hyperbolic functions:

$$\sinh(x) \pm \sinh(y) = 2\sinh\left(\frac{x \pm y}{2}\right)\cosh\left(\frac{x \mp y}{2}\right), \tag{73}$$

$$2\sinh(x)\sinh(y) = \cosh(x+y) - \cosh(x-y), \tag{74}$$

$$2\sinh^2(x) = \cosh(2x) - 1, \tag{75}$$

$$\sum_{k=1}^{n} \cosh(2kx + y) = \frac{\sinh(nx)\cosh((n+1)x+y)}{\sinh(x)}. \tag{76}$$

For every $n \geq N_\alpha$, define

$$u_1(\alpha, n) := \frac{\lambda_{\alpha,n,n}}{2}\left(\frac{\sinh(2ns_{\alpha,n})}{2\sinh(s_{\alpha,n})} - n\right),$$

$$u_2(\alpha, n) := \begin{cases} 2|\alpha|^2 \cosh^2 \frac{(n-1)s_{\alpha,n}}{2}\, w(\alpha, n), & \text{if } n \text{ is even}, \\ 2|\alpha|^2 \sinh^2 \frac{(n-1)s_{\alpha,n}}{2}\, w(\alpha, n), & \text{if } n \text{ is odd}, \end{cases}$$

$$u_3(\alpha, n) := \begin{cases} -4\operatorname{Re}(\alpha) \cosh \frac{(n-1)s_{\alpha,n}}{2} \cosh \frac{ns_{\alpha,n}}{2} \cosh \frac{s_{\alpha,n}}{2}\, w(\alpha, n), & \text{if } n \text{ is even}, \\ -4\operatorname{Re}(\alpha) \sinh \frac{(n-1)s_{\alpha,n}}{2} \sinh \frac{ns_{\alpha,n}}{2} \cosh \frac{s_{\alpha,n}}{2}\, w(\alpha, n), & \text{if } n \text{ is odd}, \end{cases}$$

where $w(\alpha, n) := \frac{\sinh(ns_{\alpha,n})}{\sinh(s_{\alpha,n})} + (-1)^{n+1}n$.

Proposition 8.2 (Exact Formula for the Norm of the Last Eigenvector) *Let $n \geq N_\alpha$. Then*

$$\|v_{\alpha,n,n}\|_2^2 = u_1(\alpha, n) + u_2(\alpha, n) + u_3(\alpha, n). \tag{77}$$

Proof Let n be even. Then, from (23) and (73),

$$\begin{aligned}
v_{\alpha,n,n,k} &= (-1)^k \left(\overline{\alpha}\sinh((n-k)s_{\alpha,n}) + (1-\overline{\alpha})\sinh((k-1)s_{\alpha,n}) + \sinh(ks_{\alpha,n})\right) \\
&= (-1)^k \Big(\sinh((k-1)s_{\alpha,n}) + \sinh(ks_{\alpha,n}) \\
&\qquad + \overline{\alpha}\left(\sinh((n-k)s_{\alpha,n}) - \sinh((k-1)s_{\alpha,n})\right)\Big) \\
&= (-1)^k \Bigg(2\sinh\frac{(2k-1)s_{\alpha,n}}{2}\cosh\frac{s_{\alpha,n}}{2} \\
&\qquad + 2\overline{\alpha}\sinh\frac{(n+1-2k)s_{\alpha,n}}{2}\cosh\frac{(n-1)s_{\alpha,n}}{2}\Bigg).
\end{aligned}$$

Taking the squared absolute value and applying (74) and (75), yields

$$\begin{aligned}
|v_{\alpha,n,n,k}|^2 &= 4\cosh^2\frac{s_{\alpha,n}}{2}\sinh^2\frac{(2k-1)s_{\alpha,n}}{2} \\
&\quad + 4|\alpha|^2\cosh^2\frac{(n-1)s_{\alpha,n}}{2}\sinh^2\frac{(n+1-2k)s_{\alpha,n}}{2} \\
&\quad + 8\,\mathrm{Re}(\alpha)\cosh\frac{(n-1)s_{\alpha,n}}{2}\cosh\frac{s_{\alpha,n}}{2}\times \\
&\qquad \times\sinh\frac{(n+1-2k)s_{\alpha,n}}{2}\sinh\frac{(2k-1)s_{\alpha,n}}{2},
\end{aligned}$$

i.e., after a simplification,

$$\begin{aligned}
|v_{\alpha,n,n,k}|^2 &= \frac{\lambda_{\alpha,n,n}}{2}\left(\cosh(2ks_{\alpha,n} - s_{\alpha,n}) - 1\right) \\
&\quad + 2|\alpha|^2\cosh^2\frac{(n-1)s_{\alpha,n}}{2}\left(\cosh(2ks_{\alpha,n} - (n+1)s_{\alpha,n}) - 1\right) \\
&\quad + 4\,\mathrm{Re}(\alpha)\cosh\frac{(n-1)s_{\alpha,n}}{2}\cosh\frac{s_{\alpha,n}}{2}\times \\
&\qquad \times\left(\cosh\frac{ns_{\alpha,n}}{2} - \cosh\left(2ks_{\alpha,n} - \frac{n+2}{2}s_{\alpha,n}\right)\right).
\end{aligned}$$

Formula (77) is obtained summing the previous expression over k, considering the identity (76) in each term.

The proof is similar for odd n. □

Lemma 8.3 *As n tends to infinity,*

$$u_1(\alpha, n) = \frac{\operatorname{Re}(\alpha)}{4(\operatorname{Re}(\alpha) - 1)} e^{2n\omega_\alpha} + O(n e^{n\omega_\alpha}),$$

$$u_2(\alpha, n) = \frac{|\alpha|^2}{8\operatorname{Re}(\alpha)(\operatorname{Re}(\alpha) - 1)} e^{2n\omega_\alpha} + O(n e^{n\omega_\alpha}),$$

$$u_3(\alpha, n) = -\frac{\operatorname{Re}(\alpha)}{4(\operatorname{Re}(\alpha) - 1)} e^{2n\omega_\alpha} + O(n e^{n\omega_\alpha}).$$

Proof Proceed similarly to the proof of [13, Lemma 6.3]. □

Proof of Theorem 2.7 Formulas (26), (25) follow similarly to the proofs of [13, (2.20), (2.21)]. To prove (27), we apply Proposition 8.2 and Lemma 8.3. Finally, we take the square root and obtain (25). □

Remark 8.4 Using Theorems 2.6 and 2.7, it is possible to show that for n large enough, the inner components of the normalized eigenvector $v_{\alpha,n,n}/\|v_{\alpha,n,n}\|_2$ are very small:

$$\frac{1}{\|v_{\alpha,n,n}\|_2} v_{\alpha,n,n,k} = O_\alpha(e^{-k\omega_\alpha} + e^{-(n+1-k)\omega_\alpha}) = O_\alpha(e^{-\min\{k,n+1-k\}\omega_\alpha}).$$

Figure 6 shows the components of the normalized eigenvectors $v_{\alpha,n,n}/\|v_{\alpha,n,n}\|_2$ for some α and n, and Fig. 7 shows the logarithms of the absolute values of their components.

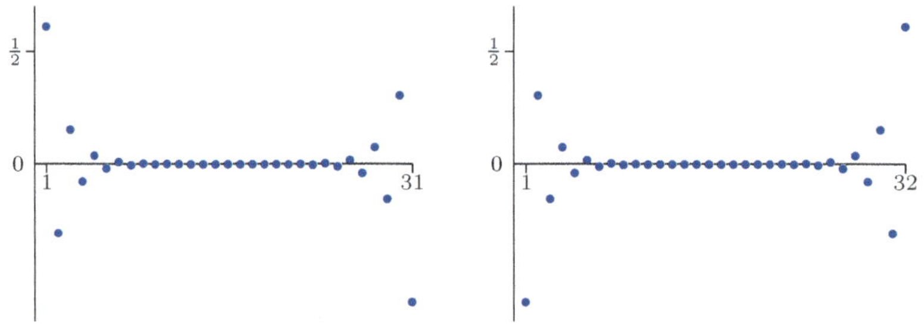

Fig. 6 Components of the eigenvectors $\frac{v_{\alpha,n,n}}{\|v_{\alpha,n,n}\|_2}$ for $\alpha = \frac{3}{2}$, $n = 31$ (left) and $n = 32$ (right)

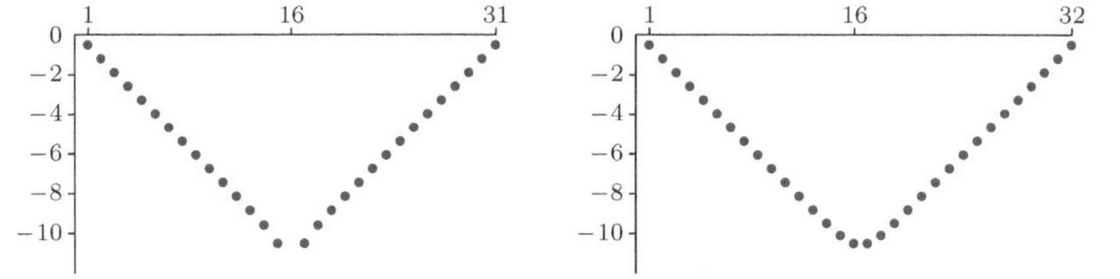

Fig. 7 Values of $\log|w_{\alpha,n,n,k}|$ where $w_{\alpha,n,n} := \frac{v_{\alpha,n,n}}{\|v_{\alpha,n,n}\|_2}$, for $\alpha = \frac{3}{2}$, $n = 31$ (left) and $n = 32$ (right). On the left picture, we skip the component with $k = 16$ because $w_{3/2,31,31,16}$ is very close to zero

9 Numerical Tests

With the help of SageMath, we have verified numerically (for many values of parameters) the representations (32), (39), (38) for the characteristic polynomial, and all the other exact formulas appearing in this paper.

We introduce the following notation for different approximations of the eigenvalues and eigenvectors.

- $\lambda^{\text{gen}}_{\alpha,n,j}$ are the eigenvalues computed with machine precision ($\approx$ 16 decimal digits), using a general eigenvalue algorithm from Sagemath.

All other computations are performed with 3322 binary digits ($\approx$ 1000 decimal digits).

- $z^{\text{N}}_{\alpha,n,j}$ is the numerical solution of the equation $h_{\alpha,n,j}(x) = 0$ computed by Newton's method, see Theorem 4.4.
- Similarly, $s^{\text{N}}_{\alpha,n}$ is the solution of $f_{\alpha,n}(x) = 0$ computed by Newton's method, see Theorems 6.3 and 7.3.
- $\lambda^{\text{N}}_{\alpha,n,j}$ is computed as $g(z^{\text{N}}_{\alpha,n,j})$ or $g(d_{n,j})$ or $g_+(s^{\text{N}}_{\alpha,n})$, depending on the case.
- $\lambda^{\text{bisec}}_{\alpha,n,j}$ is similar to $\lambda^{\text{N}}_{\alpha,n,j}$, but now we solve the corresponding equations by the bisection method.
- Using $z^{\text{bisec}}_{\alpha,n,j}$ we compute $v_{\alpha,n,j}$ by (22) and normalize it.
- Using $s^{\text{bisec}}_{\alpha,n}$ we compute $v_{\alpha,n,1}$ by (23) and normalize it.
- $\lambda^{\text{asympt}}_{\alpha,n,j}$ is the approximation given by (16) and (21).

We have constructed a large series of examples including all rational values α in $(1, 5]$ with denominators ≤ 4 and all n with $3 \leq n \leq 256$. In all these examples, we have obtained

$$\max_{1\leq j\leq n} \|L_{\alpha,n} v_{\alpha,n,j} - \lambda^{\text{bisec}}_{\alpha,n,j} v_{\alpha,n,j}\|_2 < 10^{-994}, \qquad \max_{1\leq j\leq n} |\lambda^{\text{gen}}_{\alpha,n,j} - \lambda^{\text{bisec}}_{\alpha,n,j}| < 10^{-13},$$

Table 1 Values of $\|R^{\text{asympt}}_{\alpha,n}\|_\infty$ and $n^3\|R^{\text{asympt}}_{\alpha,n}\|_\infty$ for some α and n

$\alpha = 4/3$			$\alpha = 13/4$		
n	$\|R^{\text{asympt}}_{\alpha,n}\|_\infty$	$n^3\|R^{\text{asympt}}_{\alpha,n}\|_\infty$	n	$\|R^{\text{asympt}}_{\alpha,n}\|_\infty$	$n^3\|R^{\text{asympt}}_{\alpha,n}\|_\infty$
256	4.37×10^{-6}	7.34×10^1	256	1.67×10^{-6}	2.80×10^1
512	5.68×10^{-7}	7.62×10^1	512	2.11×10^{-7}	2.83×10^1
1024	7.23×10^{-8}	7.76×10^1	1024	2.64×10^{-8}	2.84×10^1
2048	9.11×10^{-9}	7.83×10^1	2048	3.31×10^{-9}	2.84×10^1
4096	1.14×10^{-9}	7.86×10^1	4096	4.14×10^{-10}	2.85×10^1
8192	1.43×10^{-10}	7.88×10^1	8192	5.18×10^{-11}	2.85×10^1

Table 2 Values of $|R^{\text{asympt}}_{\alpha,n,n}|$ and $n^{-2}e^{3n\omega_\alpha}|R^{\text{asympt}}_{\alpha,n,n}|$ for some α and n

$\alpha = 4/3$			$\alpha = 13/4$		
n	$|R^{\text{asympt}}_{\alpha,n,n}|$	$n^{-2}e^{3n\omega_\alpha}|R^{\text{asympt}}_{\alpha,n,n}|$	n	$|R^{\text{asympt}}_{\alpha,n,n}|$	$n^{-2}e^{3n\omega_\alpha}|R^{\text{asympt}}_{\alpha,n,n}|$
64	1.91×10^{-39}	1.84	64	3.39×10^{-136}	1.17×10^3
128	2.00×10^{-81}	1.89	128	9.73×10^{-278}	1.18×10^3
192	1.15×10^{-123}	1.91	192	1.56×10^{-419}	1.19×10^3
256	5.23×10^{-166}	1.92	256	1.97×10^{-561}	1.19×10^3

where $v_{\alpha,n,j}$ was the normalized eigenvector. Moreover, in all examples with $n \geq N_\alpha$,

$$\max_{1\leq j\leq n} |\lambda^{\text{N}}_{\alpha,n,j} - \lambda^{\text{bisec}}_{\alpha,n,j}| < 10^{-997}.$$

For testing the asymptotic formulas, we have computed the errors

$$R^{\text{asympt}}_{\alpha,n,j} := \lambda^{\text{asympt}}_{\alpha,n,j} - \lambda^{\text{N}}_{\alpha,n,j}$$

and their maximums $\|R^{\text{asympt}}_{\alpha,n}\|_\infty = \max_{1\leq j\leq n} |R^{\text{asympt}}_{\alpha,n,j}|$. Table 1 shows that these errors indeed can be bounded by $O_\alpha(1/n^3)$.

We have done similar tests for many other values of α and n. Numerical experiments show that $n^3\|R^{\text{asympt}}_{\alpha,n}\|_\infty$ are bounded by some numbers depending on α.

Since $|R^{\text{asympt}}_{\alpha,n,j}|$ is smaller for the outiler eigenvalue, we show in Table 2 some numerical experiments for only this case.

Acknowledgments We are grateful to the referee for many corrections.

Funding The research of S.M. Grudsky was supported by CONAHCYT (Mexico) project "Ciencia de Frontera" FORDECYT-PRONACES/61517/2020 and by the Regional Mathematical Center of the Southern Federal University with the support of the Ministry of Science and Higher Education of Russia, agreement 075-02-2024-1427.

The research of E.A. Maximenko was supported by CONAHCYT (Mexico) project "Ciencia de Frontera" FORDECYT-PRONACES/61517/2020 and by IPN-SIP (Mexico) projects.

The reasearch of A. Soto-González was supported by CONAHCYT (Mexico) Ph.D. scholarship.

References

1. Atkinson, K.E.: An Introduction to Numerical Analysis, 2nd edn. Wiley, New York (1989)
2. Basak, A., Paquette, E., Zeitouni, O.: Spectrum of random perturbations of Toeplitz matrices with finite symbols. Trans. Am. Math. Soc. **373**, 4999–5023 (2020). https://doi.org/10.1090/tran/8040
3. Böttcher, A., Fukshansky, L., Garcia, S.R., Maharaj, H.: Toeplitz determinants with perturbations in the corners. J. Funct. Anal. **268**, 171–193 (2014). https://doi.org/10.1016/j.jfa.2014.10.023
4. Britanak, V., Yip, P.C., Rao, K.R.: Discrete Cosine and Sine Transforms: General Properties, Fast Algorithms and Integer Approximations. Academic Press, San Diego (2006)
5. Felsner, S., Heldt, D.: Lattice path enumeration and Toeplitz matrices. J. Integer Seq. **18**, Article 15. 1. 3 (2015). https://cs.uwaterloo.ca/journals/JIS/vol18.html
6. Fernandes, R., Da Fonseca, C.M.: The inverse eigenvalue problem for Hermitian matrices whose graphs are cycles. Linear Multilinear Alg. **57**, 673–682 (2009). https://doi.org/10.1080/03081080802187870
7. Da Fonseca, C.M., Kowalenko, V.: Eigenpairs of a family of tridiagonal matrices, three decades later. Acta Math. Hung. **160**, 376–389 (2020). https://doi.org/10.1007/s10474-019-00970-1
8. Da Fonseca, C.M., Veerman, J.J.P.: On the spectra of certain directed paths. Appl. Math. Lett. **22**, 1351–1355 (2009). https://doi.org/10.1016/j.aml.2009.03.006
9. Garoni, C., Serra-Capizzano, S.: Generalized Locally Toeplitz Sequences: Theory and Applications, vol. I. Springer, Cham (2017)
10. Grassmann, W.K., Tavakoli, J.: Spectrum of certain tridiagonal matrices when their dimension goes to infinity. Linear Algebra Appl. **431**, 1208–1217 (2009). https://doi.org/10.1016/j.laa.2009.04.013
11. Grudsky, S.M., Maximenko, E.A., Soto-González, A.: Eigenvalues of tridiagonal Hermitian Toeplitz matrices with perturbations in the off-diagonal corners. In: Karapetyants, A.N., Kravchenko, V.V., Liflyand, E., Malonek, H.R. (eds.) Operator Theory and Harmonic Analysis, OTHA 2020, Series Springer Proceedings in Mathematics & Statistics, vol. 357, pp. 179–202. Springer, Cham (2021). https://doi.org/10.1007/978-3-030-77493-6_11
12. Grudsky, S.M., Maximenko, E.A., Soto-González, A.: Eigenvalues of the laplacian matrices of the cycles with one weighted edge. Linear Algebra Appl. **653**, 86–115 (2022). https://doi.org/10.1016/j.laa.2022.07.011
13. Grudsky, S.M., Maximenko, E.A., Soto-González, A.: Eigenvalues of laplacian matrices of the cycles with one negative-weighted edge. Linear Algebra Appl. **697**, 249–276 (2024). https://doi.org/10.1016/j.laa.2023.12.003
14. Lin, Z., Wang, L., Han, Z., Fu, M.: Distributed formation control of multi-agent systems using complex Laplacian. IEEE Trans. Autom. Control **59**, 1765–1777 (2014). https://doi.org/10.1109/TAC.2014.2309031
15. Molitierno, J.J.: Applications of Combinatorial Matrix Theory to Laplacian Matrices of Graphs. CRC Press, Florida (2012)
16. Öteleş, A., Akbulak, M.: Positive integer powers of one type of complex tridiagonal matrix. Bull. Malays. Math. Sci. Soc. (2) **37**, 971–981 (2014). https://math.usm.my/bulletin/pdf/v37n3/v37n4p6.pdf
17. Reyes-Lega, A.F.: Some aspects of operator algebras in quantum physics. In: Cano, L., Cardona, A., Ocampo, H., Reyes-Lega, A.F. (eds.) Geometric, Algebraic and Topological Methods for Quantum Field Theory, pp. 1–74. World Scientific, Singapore (2016). https://doi.org/10.1142/9789814730884_0001

18. Stewart, G.W., Sun, J.: Matrix Perturbation Theory. Academic Press, Boston (1990)
19. Tavakolipour, H., Shakeri, F.: On tropical eigenvalues of tridiagonal Toeplitz matrices. Linear Algebra Appl. **539**, 198–218 (2017). https://doi.org/10.1016/j.laa.2017.11.009
20. Veerman, J.J.P., Hammond, D.K., Baldivieso, P.E.: Spectra of certain large tridiagonal matrices. Linear Algebra Appl. **548**, 123–147 (2018). https://doi.org/10.1016/j.laa.2018.03.005
21. Zhang, M., Jiang, X., Jiang, Z.: Explicit determinants, inverses and eigenvalues of four band Toeplitz matrices with perturbed rows. Special Matrices **7**, 52–66 (2019). https://doi.org/10.1515/spma-2019-0004

On the Algebra of Singular Integral Operators with Almost Periodic Coefficients

Oleksiy Karlovych and Márcio Valente

To Professor Yuri Karlovich on the occasion of his 75th birthday

Abstract Yuri Karlovich observed in the early 1990s that every Fredholm operator in the Banach algebra generated by the Cauchy singular integral operator S and the operators of multiplication by matrix-valued uniform almost periodic functions on the space $L_N^p(\mathbb{R})$ with $1 < p < \infty$ is invertible. We extend this result to the setting of reflexive rearrangement-invariant Banach function spaces with nontrivial Boyd indices.

1 Introduction and the Main Result

Given a Banach space $\mathcal{X}$, let $\mathcal{B}(\mathcal{X})$ denote the Banach algebra of all bounded linear operators on $\mathcal{X}$ and let $\mathcal{K}(\mathcal{X})$ denote the ideal of all compact operators on $\mathcal{X}$. As usual, A^* denotes the adjoint operator of $A \in \mathcal{B}(\mathcal{X})$. An operator $A \in \mathcal{B}(\mathcal{X})$ is said to be Fredholm on $\mathcal{X}$ if its image $\operatorname{Im} A$ is closed in $\mathcal{X}$ and

$$\dim \operatorname{Ker} A < \infty, \quad \dim(\mathcal{X}/\operatorname{Im} A) < \infty.$$

We denote by $\mathcal{X}_N$ the Banach space of all columns of height N with components in $\mathcal{X}$; the norm in $\mathcal{X}_N$ is defined by

$$\|(x_1, \dots, x_N)^\top\|_{\mathcal{X}_N} = \left(\sum_{\alpha=1}^{N} \|x_\alpha\|_{\mathcal{X}}^2\right)^{1/2}.$$

O. Karlovych · M. Valente (✉)
Centro de Matemática e Aplicações, Departamento de Matemática, Faculdade de Ciências e Tecnologia, Universidade Nova de Lisboa, Caparica, Portugal
e-mail: oyk@fct.unl.pt; mac.valente@campus.fct.unl.pt

A. Böttcher et al. (eds.), *Achievements and Challenges in the Field of Convolution Operators*, Operator Theory: Advances and Applications 306,
https://doi.org/10.1007/978-3-031-80486-1_10

Given a subalgebra B of $L^\infty(\mathbb{R})$, we denote by $B_{N\times N}$ the algebra of all $N \times N$ matrices with entries in B; we equip $B_{N\times N}$ with the norm

$$\|f\|_{B_{N\times N}} = \left\|(f_{\alpha\beta})_{\alpha,\beta=1}^N\right\|_{B_{N\times N}} = \left(\sum_{\alpha,\beta=1}^N \|f_{\alpha\beta}\|_B^2\right)^{1/2}.$$

An almost periodic polynomial is a function of the form

$$a(x) = \sum_{j=1}^m a_j e^{i\lambda_j x} \quad (x \in \mathbb{R}) \quad \text{with} \quad a_j \in \mathbb{C}, \quad \lambda_j \in \mathbb{R}.$$

The set of all almost periodic polynomials will be denoted by AP^0. The algebra AP of the uniform almost periodic functions is defined as the closure of AP^0 in $L^\infty(\mathbb{R})$. This definition is equivalent to Bohr's original definition of uniform almost periodic functions (see, e.g., [2, Ch. 1, §5]). Another equivalent definition of uniform almost periodic functions was given by Bochner (see, e.g., [2, Ch. 1, §2]).

Let $X(\mathbb{R})$ be a rearrangement-invariant Banach function space with the Boyd indices $0 \le \alpha_X \le \beta_X \le 1$ (see [1, Ch. 2] and Sect. 2 below for their definition). The archetypical example of such spaces is the Lebesgue space $L^p(\mathbb{R})$ with $1 \le p \le \infty$, whose Boyd indices are $\alpha_{L^p} = \beta_{L^p} = 1/p$ with the usual convention $1/\infty = 0$. Other interesting examples are Orlicz spaces $L^\Phi(\mathbb{R})$ and Lorentz spaces $L^{p,q}(\mathbb{R})$ (see, e.g., [1, Ch. 4]). In 1967, David Boyd proved that the Cauchy singular integral operator

$$(Sf)(x) := \lim_{\varepsilon\to 0} \frac{1}{\pi i} \int_{\mathbb{R}\setminus(x-\varepsilon,x+\varepsilon)} \frac{f(t)}{t-x} dt$$

is bounded on a rearrangement-invariant Banach function space $X(\mathbb{R})$ if and only if

$$0 < \alpha_X \le \beta_X < 1$$

(see [5] and also [1, Theorem 5.18]). In the latter case, one says that the Boyd indices are nontrivial.

If $a \in AP_{N\times N}$, then the operator aI of multiplication by a is bounded on $X_N(\mathbb{R}) := [X(\mathbb{R})]_N$. The operator S is defined on $X_N(\mathbb{R})$ elementwise. Let $\mathcal{A}_N(AP, S; X(\mathbb{R}))$ denote the smallest Banach subalgebra of the Banach algebra $\mathcal{B}(X_N(\mathbb{R}))$ that contains the operator S and the operators of multiplication aI by matrix-valued functions $a \in AP_{N\times N}$.

In the early 1990s, Yuri Karlovich observed the following (cf. [8, Corollary 2]).

Theorem 1 *Let $N \in \mathbb{N}$, $1 < p < \infty$, and $A \in \mathcal{A}_N(AP, S; L^p(\mathbb{R}))$. Then the operator A is Fredholm on $L^p_N(\mathbb{R})$ if and only if it is invertible on $L^p_N(\mathbb{R})$.*

We also refer to a more general result contained in [4, Corollary 18.11].

The aim of this short paper is to extend the above result to the setting of reflexive rearrangement-invariant Banach function spaces with nontrivial Boyd indices. Our main result is the following.

Theorem 2 (Main Result) *Let $N \in \mathbb{N}$ and let $X(\mathbb{R})$ be a reflexive rearrangement-invariant Banach function space with nontrivial Boyd indices. If an operator A belongs to $\mathcal{A}_N(AP, S; X(\mathbb{R}))$, then A is Fredholm on $X_N(\mathbb{R})$ if and only if it is invertible on $X_N(\mathbb{R})$.*

The paper is organized as follows. In Sect. 2, we recall the definitions of a Banach function space and its associate space, of a rearrangement-invariant Banach function space and its Boyd indices. In Sect. 3, we compute the limit operators (see [14, 18]) of the Cauchy singular integral operator, of the multiplication operator by an almost periodic polynomial and of compact operators on rearrangement-invariant Banach function spaces. Armed with these results, we prove Theorem 2 by extending the proof of Theorem 1 to this larger class of function spaces. We conclude our paper with the open question formulated in Sect. 4 concerning the possibility of extension of Theorem 2 to the case of the Lorentz spaces $L^{p,1}(\mathbb{R})$ and Marcinkiewicz spaces $L^{p,\infty}(\mathbb{R})$ with $1 < p < \infty$ (as it is well known, these spaces are nonreflexive and their Boyd indices are equal to $1/p$).

2 Rearrangement-Invariant Banach Function Spaces and Their Boyd Indices

2.1 Banach Function Spaces

Denote by m and $\overline{m}$ the Lebesgue measure on $\mathbb{R}$ and $\mathbb{R}_+ := [0, \infty)$, respectively. Let $(\mathbb{S}, \mu)$ be one of the measure spaces $(\mathbb{R}, m)$ or $(\mathbb{R}_+, \overline{m})$. The set of all μ-measurable complex-valued functions on $\mathbb{S}$ is denoted by $\mathcal{M}(\mathbb{S}, \mu)$. Let $\mathcal{M}^+(\mathbb{S}, \mu)$ be the subset of all functions in $\mathcal{M}(\mathbb{S}, \mu)$ whose values lie in $[0, \infty]$. The measure and the characteristic (indicator) function of a measurable set $E \subset \mathbb{S}$ are denoted by $\mu(E)$ and χ_E, respectively. Following [1, Ch. 1, Definition 1.1], a mapping $\rho : \mathcal{M}^+(\mathbb{S}, \mu) \to [0, \infty]$ is called a *Banach function norm* if, for all functions f, g, f_n $(n \in \mathbb{N})$ in $\mathcal{M}^+(\mathbb{S}, \mu)$, for all constants $a \geq 0$, and for all measurable subsets E of $\mathbb{S}$, the following axioms hold:

(A1) $\rho(f) = 0 \Leftrightarrow f = 0$ a.e., $\quad \rho(af) = a\rho(f), \quad \rho(f + g) \leq \rho(f) + \rho(g)$,

(A2) $0 \leq g \leq f$ a.e. $\Rightarrow \rho(g) \leq \rho(f)$ (the lattice property),

(A3) $0 \leq f_n \uparrow f$ a.e. $\Rightarrow \rho(f_n) \uparrow \rho(f)$ (the Fatou property),

(A4) $\mu(E) < \infty \Rightarrow \rho(\chi_E) < \infty$,

(A5) $\mu(E) < \infty \Rightarrow \int_E f(x)\,dx \leq C_E \rho(f)$

with $C_E \in (0, \infty)$, which may depend on E and ρ but is independent of f. When functions differing only on a set of measure zero are identified, the set $X(\mathbb{S})$ of all functions $f \in \mathcal{M}(\mathbb{S}, \mu)$ for which $\rho(|f|) < \infty$ is called a *Banach function space*. For each $f \in X(\mathbb{S})$, the norm of f is defined by $\|f\|_{X(\mathbb{S})} := \rho(|f|)$. Under the natural linear space operations and under this norm, the set $X(\mathbb{S})$ becomes a Banach space (see [1, Ch. 1, Theorems 1.4 and 1.6]). If ρ is a Banach function norm, its associate norm ρ' is defined on $\mathcal{M}^+(\mathbb{S}, \mu)$ by

$$\rho'(g) := \sup\left\{\int_{\mathbb{S}} f(x)g(x)\,dx : f \in \mathcal{M}^+(\mathbb{S}, \mu),\ \rho(f) \le 1\right\}, \quad g \in \mathcal{M}^+(\mathbb{S}, \mu).$$

It is a Banach function norm itself [1, Ch. 1, Theorem 2.2]. The Banach function space $X'(\mathbb{S})$ determined by the Banach function norm ρ' is called the associate space (Köthe dual) of $X(\mathbb{S})$. The associate space $X'(\mathbb{S})$ is naturally identified with a subspace of the (Banach) dual space $X^*(\mathbb{S})$.

The following duality result is a consequence of [1, Ch. 1, Corollaries 4.3 and 5.6].

Theorem 3 *Let $N \in \mathbb{N}$ and $X(\mathbb{R})$ be a separable Banach function space. For every continuous linear functional G on $X_N(\mathbb{R})$ there exists a unique function $g = (g_1, \dots, g_N) \in X_N(\mathbb{R})$ such that*

$$G(f) = \sum_{\alpha=1}^{N} \int_{\mathbb{R}} f_\alpha(x)\overline{g_\alpha(x)}dx =: \langle f, g\rangle \tag{1}$$

for all $f = (f_1, \dots, f_N) \in X_N(\mathbb{R})$. Moreover, the norms $\|\cdot\|_{(X_N(\mathbb{R}))^*}$ and $\|\cdot\|_{X'_N(\mathbb{R})}$ are equivalent.

We also note that if $X(\mathbb{R})$ is reflexive, then both $X(\mathbb{R})$ and $X'(\mathbb{R})$ are separable (see [1, Ch. 1, Corollaries 4.4 and 5.6]).

2.2 Rearrangement-Invariant Banach Function Spaces

Let $\mathcal{M}_0(\mathbb{S}, \mu)$ and $\mathcal{M}_0^+(\mathbb{S}, \mu)$ be the classes of a.e. finite functions in $\mathcal{M}(\mathbb{S}, \mu)$ and $\mathcal{M}^+(\mathbb{S}, \mu)$, respectively. The distribution function μ_f of $f \in \mathcal{M}_0(\mathbb{S}, \mu)$ is given by

$$\mu_f(\lambda) := \mu\{x \in \mathbb{S} : |f(x)| > \lambda\}, \quad \lambda \ge 0.$$

The non-increasing rearrangement of $f \in \mathcal{M}_0(\mathbb{S}, \mu)$ is the function defined by

$$f^*(t) := \inf\{\lambda : \mu_f(\lambda) \le t\}, \quad t \ge 0.$$

We here use the standard convention that $\inf \emptyset = +\infty$. Now let $(\mathbb{S}, \mu), (\mathbb{T}, \nu) \in \{(\mathbb{R}, m), (\mathbb{R}_+, \overline{m})\}$. Two functions $f \in \mathcal{M}_0(\mathbb{S}, \mu)$ and $g \in \mathcal{M}_0(\mathbb{T}, \nu)$ are said to be equimeasurable if $\mu_f(\lambda) = \nu_g(\lambda)$ for all $\lambda \geq 0$.

A Banach function norm $\rho : \mathcal{M}^+(\mathbb{S}, \mu) \to [0, \infty]$ is called *rearrangement-invariant* if for every pair of equimeasurable functions $f, g \in \mathcal{M}_0^+(\mathbb{S}, \mu)$, the equality $\rho(f) = \rho(g)$ holds. In that case, the Banach function space $X(\mathbb{S})$ generated by ρ is said to be a rearrangement-invariant Banach function space (or simply a rearrangement-invariant space). Lebesgue spaces $L^p(\mathbb{S})$, $1 \leq p \leq \infty$, Orlicz spaces $L^\Phi(\mathbb{S})$, and Lorentz spaces $L^{p,q}(\mathbb{S})$ are classical examples of rearrangement-invariant Banach function spaces (see, e.g., [1] and the references therein). By [1, Ch. 2, Proposition 4.2], if a Banach function space $X(\mathbb{S})$ is rearrangement-invariant, then its associate space $X'(\mathbb{S})$ is also rearrangement-invariant.

2.3 *Submultiplicative Functions and Their Indices*

A measurable function $\varrho : (0, \infty) \to (0, \infty)$ is said to be submultiplicative if $\varrho(x_1 x_2) \leq \varrho(x_1)\varrho(x_2)$ for all $x_1, x_2 \in (0, \infty)$. The behavior of a measurable submultiplicative function ϱ in neighborhoods of zero and infinity is described by the quantities

$$\alpha(\varrho) := \sup_{x \in (0,1)} \frac{\log \varrho(x)}{\log x} = \lim_{x \to 0} \frac{\log \varrho(x)}{\log x}, \tag{2}$$

$$\beta(\varrho) := \inf_{x \in (1,\infty)} \frac{\log \varrho(x)}{\log x} = \lim_{x \to \infty} \frac{\log \varrho(x)}{\log x}, \tag{3}$$

where $-\infty < \alpha(\varrho) \leq \beta(\varrho) < \infty$ (see [12, Ch. II, Theorem 1.3]). The numbers $\alpha(\varrho)$ and $\beta(\varrho)$ are called the lower and upper indices of the measurable submultiplicative function ϱ.

2.4 *Dilation Operators on the Luxemburg Representation and Boyd Indices*

Let $X(\mathbb{R})$ be a rearrangement-invariant Banach function space generated by a rearrangement-invariant Banach function norm ρ over $(\mathbb{R}, m)$. By the Luxemburg representation theorem (see [1, Ch. 2, Theorem 4.10] or [16, Theorem 7.8.3]), there exists a unique rearrangement-invariant Banach function norm $\overline{\rho}$ over $(\mathbb{R}_+, \overline{m})$ such that

$$\rho(f) = \overline{\rho}(f^*), \quad f \in \mathcal{M}_0^+(\mathbb{R}, m).$$

The rearrangement-invariant Banach function space over $(\mathbb{R}_+, \overline{m})$ generated by $\overline{\rho}$ is denoted by $\overline{X}(\mathbb{R}_+)$ and is called the Luxemburg representation of $X(\mathbb{R})$. For $t > 0$, let E_t be the dilation operator defined on the set $\mathcal{M}_0(\mathbb{R}_+, \overline{m})$ by

$$(E_t\varphi)(s) = \varphi(ts), \quad 0 < s < \infty. \tag{4}$$

It follows from [1, Ch. 3, Proposition 5.11] that the operators E_t are bounded on $\overline{X}(\mathbb{R}_+)$ for all $t > 0$. The operator norm of the operator $E_{1/t}$ on the Luxemburg representation $\overline{X}(\mathbb{R}_+)$ will be denoted by

$$h(t, X) := \|E_{1/t}\|_{\mathcal{B}(\overline{X}(\mathbb{R}_+))}, \quad t > 0.$$

By [1, Ch. 3, Proposition 5.11], the function $h(\cdot, X)$ is nondecreasing (and hence, measurable), submultiplicative on $(0, \infty)$, and

$$h_X(t) \le \max\{1, t\}, \quad 0 < t < \infty. \tag{5}$$

The indices of $h(\cdot, X)$ are called the *Boyd indices* [6] of the rearrangement-invariant Banach function space $X(\mathbb{R})$ and are denoted by

$$\alpha_X := \alpha(h(\cdot, X)), \quad \beta_X := \beta(h(\cdot, X)).$$

So, $\alpha_X \le \beta_X$. Equalities (2)–(3) and inequality (5) imply that the Boyd indices of $X(\mathbb{R})$ satisfy $0 \le \alpha_X, \beta_X \le 1$. We refer to the survey paper [15] and the monographs [1, 12] for the properties of the Boyd indices of rearrangement-invariant Banach function spaces.

3 Proof of the Main Result

3.1 Injection and Surjection Moduli

Let $\mathcal{X}$ be a Banach space and $A \in \mathcal{B}(\mathcal{X})$. Following [17, Sections B.3.1 and B.3.4], consider its injection modulus

$$\mathcal{J}(A; \mathcal{X}) := \sup\left\{c \ge 0 : \ \|Af\|_{\mathcal{X}} \ge c\|f\|_{\mathcal{X}} \text{ for all } f \in \mathcal{X}\right\}$$

and its surjection modulus

$$\mathcal{Q}(A; \mathcal{X}) := \sup\left\{c \ge 0 : \ cB_{\mathcal{X}} \subset AB_{\mathcal{X}}\right\}$$

where B_X is the closed unit ball of $\mathcal{X}$. Sometimes these characteristics are also called lower norms of A (see, e.g., [13, Section 1.3]). Fundamental properties of the injection and surjection moduli are collected in the following statements.

Lemma 1 (See, e.g., [17, Section B.3.8]) *If $A \in \mathcal{B}(\mathcal{X})$, then*

$$\mathcal{J}(A;\mathcal{X}) = \mathcal{Q}(A^*;\mathcal{X}^*), \quad \mathcal{Q}(A;\mathcal{X}) = \mathcal{J}(A^*;\mathcal{X}^*).$$

Theorem 4 (See, e.g., [13, Theorem 1.3.2]) *An operator $A \in \mathcal{B}(\mathcal{X})$ is invertible if and only if*

$$\mathcal{J}(A;\mathcal{X}) > 0, \quad \mathcal{Q}(A;\mathcal{X}) > 0.$$

If A is invertible, then

$$\mathcal{J}(A;\mathcal{X}) = \mathcal{Q}(A;\mathcal{X}) = \frac{1}{\|A^{-1}\|_{\mathcal{B}(\mathcal{X})}}.$$

3.2 *Translations, Singular Integral Operators, and Their Adjoints*

Let $X(\mathbb{R})$ be a rearrangement-invariant Banach function space. Given $f \in X(\mathbb{R})$, consider its translation by $h \in \mathbb{R}$ defined by

$$(T_h f) := f(x+h), \quad x \in \mathbb{R}.$$

Since the functions f and $T_h f$ are equimeasurable, $\|f\|_{X(\mathbb{R})} = \|T_h f\|_{X(\mathbb{R})}$. So, the translation operator $T_h : f \mapsto T_h f$ is an isometry on $X(\mathbb{R})$. Moreover, it is invertible and its inverse is $T_h^{-1} = T_{-h}$. For $N \in \mathbb{N}$ the translation operator T_h on $X_N(\mathbb{R})$ is defined elementwise. Since Theorem 3 is at our disposal, the proof of the following statement is straightforward.

Lemma 2 *Let $N \in \mathbb{N}$ and $X(\mathbb{R})$ be a reflexive Banach function space. If $h \in \mathbb{R}$, then*

$$T_h^* = T_{-h} \in \mathcal{B}(X'_N\mathbb{R})).$$

For $a \in L^\infty_{N\times N}(\mathbb{R})$, let a^* denote the complex conjugate of the transpose matrix function $a^\top$.

Since Theorem 3 is available, the following lemma can be proved by literal repetition of the proof of [9, Lemma 3.9].

Lemma 3 *Let $N \in \mathbb{N}$ and $X(\mathbb{R})$ be a reflexive Banach function space. If a belongs to $L^\infty_{N\times N}(\mathbb{R})$, then*

$$(aI)^* = a^* I \in \mathcal{B}(X'_N(\mathbb{R})).$$

Lemma 4 *Let $N \in \mathbb{N}$. If $X(\mathbb{R})$ is a reflexive rearrangement-invariant Banach function space with nontrivial Boyd indices, then $S^* = S \in \mathcal{B}(X'_N(\mathbb{R}))$.*

Proof Since the operator S is defined elementwise, it is enough to prove the statement for $N = 1$. In the latter case it follows from the duality relations for the Boyd indices: $\alpha_{X'} = 1 - \beta_X$ and $\beta_{X'} = 1 - \alpha_X$ (see [1, Ch. 3, Proposition 5.13]), the Lorentz-Shimogaki theorem (see [1, Ch. 3, Theorem 5.16]), and [10, Theorem 3.8(b)] (see also [9, Lemma 3.11]). □

3.3 Limit Operators

The proof of our main result relies on the method of limit operators (see [3, 14, 18]). It is based on the observation that for a given bounded linear operator A on a Banach space $\mathcal{X}$ and a cleverly chosen sequence of isometries $\{V_n\}_{n=1}^\infty$ on $\mathcal{X}$, the strong limit of the sequence $V_n^{-1} A V_n$ (if it exists) preserves some important information about A and can be much simpler than the original operator A. This strong limit is called the limit operator of the operator A with respect to the sequence of isometries $\{V_n\}$.

Since the translation operator T_h is for every $h \in \mathbb{R}$ an isometry on a rearrangement-invariant Banach function space, we will consider limit operators with respect to sequences of isometries $\{T_{h_n}\}_{n=1}^\infty$ for a given sequence of real numbers $\{h_n\}_{n=1}^\infty$.

We start by considering limit operators of compact operators. The following result is a consequence of [11, Corollary 2] and [19, Lemma 1.4.6].

Theorem 5 *Let $N \in \mathbb{N}$ and $X(\mathbb{R})$ be a reflexive rearrangement-invariant Banach function space. If $K \in \mathcal{K}(X_N(\mathbb{R}))$ and $\{h_n\}_{n=1}^\infty \subset \mathbb{R}$ is a sequence such that $h_n \to +\infty$ as $n \to \infty$, then*

$$\lim_{n\to\infty} \left\| T_{h_n}^{-1} K T_{h_n} f \right\|_{X_N(\mathbb{R})} = 0,$$

for all $f \in X_N(\mathbb{R})$.

The following corollary of the Kronecker theorem on almost periodic functions (see, e.g., [4, Theorem 1.12]) will allow us to calculate limit operators of operators of multiplication by almost periodic polynomials.

Lemma 5 (See [4, Lemma 10.2]) *Let $N \in \mathbb{N}$. If $a_1, \ldots, a_M \in AP^0_{N\times N}$ is a finite collection of almost periodic polynomials, then there exists a sequence $\{h_n\}_{n=1}^{\infty} \subset \mathbb{R}$ such that $h_n \to +\infty$ as $n \to \infty$ and*

$$\lim_{n\to\infty} \|a_j(\cdot + h_n) - a_j(\cdot)\|_{L^\infty_{N\times N}(\mathbb{R})} = 0$$

for all $j \in \{1, \ldots, M\}$.

For $N \in \mathbb{N}$ and a rearrangement-invariant Banach function space $X(\mathbb{R})$ with nontrivial Boyd indices, let $\mathcal{A}^0_N(AP^0, S; X(\mathbb{R}))$ denote the nonclosed algebra consisting of the operators of the form

$$\sum_{i\in\mathcal{I}} c_i \prod_{j\in\mathcal{J}} A_{ij},$$

where $\mathcal{I}, \mathcal{J} \subset \mathbb{N}$ are finite ordered sets, $c_i \in \mathbb{C}$ for $i \in \mathcal{I}$, and

$$A_{ij} \in \left\{aI : a \in AP^0_{N\times N}\right\} \cup \{S\}, \quad (i, j) \in \mathcal{I} \times \mathcal{J}.$$

The following result is the main ingredient of the proof of Theorem 2.

Theorem 6 *Let $N \in \mathbb{N}$ and $X(\mathbb{R})$ be a reflexive rearrangement-invariant Banach function space with nontrivial Boyd indices. For every $B \in \mathcal{A}^0_N(AP^0, S; X(\mathbb{R}))$, there exists a sequence $\{h_n\}_{n=1}^{\infty} \subset \mathbb{R}$ such that $h_n \to +\infty$ as $n \to \infty$,*

$$\lim_{n\to\infty} \left\| T_{h_n}^{-1} B T_{h_n} f - Bf \right\|_{X_N(\mathbb{R})} = 0$$

for all $f \in X_N(\mathbb{R})$ and

$$\lim_{n\to\infty} \left\| \left(T_{h_n}^{-1} B T_{h_n}\right)^* g - B^* g \right\|_{X'_N(\mathbb{R})} = 0$$

for all $g \in X'_N(\mathbb{R})$.

Proof Fix $B \in \mathcal{A}^0_N(AP^0, S; X(\mathbb{R}))$. Taking into account that the condition

$$\lim_{n\to\infty} \left\| \left(T_{h_n}^{-1} B T_{h_n}\right) f - Bf \right\|_{X_N(\mathbb{R})} = 0, \quad f \in X_N(\mathbb{R}),$$

is equivalent to

$$\lim_{n\to\infty} \left\| \left(T_{h_n}^{-1} B_{\alpha\beta} T_{h_n}\right) f - B_{\alpha\beta} f \right\|_{X(\mathbb{R})} = 0, \quad f \in X(\mathbb{R}),$$

for all $\alpha, \beta \in \{1, \dots, N\}$, it suffices to consider only the case when $N = 1$. By definition of $\mathcal{A}_1^0(AP^0, S; X(\mathbb{R}))$, we can write

$$B = \sum_{i \in \mathcal{I}} c_i \prod_{j \in \mathcal{J}} A_{i,j}.$$

where $A_{i,j} = S$ or $A_{i,j} = aI$ for some $a \in AP^0$.

Claim 1 For every sequence $\{h_n\}_{n=1}^{\infty} \subset \mathbb{R}$,

$$\lim_{n\to\infty} \left\| \left(T_{h_n}^{-1} S T_{h_n}\right) f - Sf \right\|_{X(\mathbb{R})} = 0, \qquad f \in X(\mathbb{R}),$$

$$\lim_{n\to\infty} \left\| \left(T_{h_n}^{-1} S T_{h_n}\right)^* g - S^* g \right\|_{X'(\mathbb{R})} = 0, \qquad g \in X'(\mathbb{R}).$$

To see this it is enough to point out that for all $h \in \mathbb{R}$ and $f \in X(\mathbb{R})$,

$$\begin{aligned}
[(T_h^{-1} S T_h) f](x) &= (S T_h f)(x - h) \\
&= \lim_{\varepsilon \to 0} \frac{1}{\pi i} \int_{\mathbb{R} \setminus (x-h-\varepsilon, x-h+\varepsilon)} \frac{f(t+h)}{t - x + h}\, dt \\
&= \lim_{\varepsilon \to 0} \frac{1}{\pi i} \int_{\mathbb{R} \setminus (x-\varepsilon, x+\varepsilon)} \frac{f(y)}{y - x}\, dy \\
&= (Sf)(x).
\end{aligned}$$

Analogously, by making use of Lemmas 2 and 4, one can simply repeat the argument to prove the statement in the dual space. This proves Claim 1.

Claim 2 For every finite sequence $a_1, \dots, a_M \in AP^0$, there is a sequence $\{h_n\}_{n=1}^{\infty} \subset \mathbb{R}$ such that $h_n \to +\infty$ as $n \to \infty$ and, for each $j \in \{1, \dots M\}$,

$$\lim_{n\to\infty} \left\| \left(T_{h_n}^{-1} a_j I T_{h_n}\right) f - (a_j I) f \right\|_{X(\mathbb{R})} = 0, \qquad f \in X(\mathbb{R}), \tag{6}$$

$$\lim_{n\to\infty} \left\| \left(T_{h_n}^{-1} a_j I T_{h_n}\right)^* g - (a_j I)^* g \right\|_{X'(\mathbb{R})} = 0, \qquad g \in X'(\mathbb{R}). \tag{7}$$

Consider any finite sequence $a_1, \dots, a_M \in AP^0$. By Lemma 5, there is a sequence $\{h_n\}_{n=1}^{\infty} \subset \mathbb{R}$ such that $h_n \to +\infty$ as $n \to \infty$ and, for each $j \in \{1, \dots M\}$,

$$\lim_{n\to\infty} \left\| T_{-h_n} a_j - a_j \right\|_{L^\infty(\mathbb{R})} = 0.$$

Fix $f \in X(\mathbb{R})$. It follows from Axiom (A2) in the definition of a Banach function space that

$$\begin{aligned}\left\| \left(T_{h_n}^{-1} a_j I T_{h_n} - a_j I \right) f \right\|_{X(\mathbb{R})} &= \left\| (T_{-h_n} a_j - a_j) f \right\|_{X(\mathbb{R})} \\ &\le \left\| T_{-h_n} a_j - a_j \right\|_{L^\infty(\mathbb{R})} \|f\|_{X(\mathbb{R})}.\end{aligned}$$

Passing to the limit as $n \to \infty$, the squeeze theorem yields (6). Equality (7) is proved analogously with the aid of Lemmas 2 and 3. This finishes the proof of Claim 2.

Now that the auxiliary results have been established, we are ready to proceed with the proof. Consider the set

$$\mathcal{U} := \left\{ (i, j) \in \mathcal{I} \times \mathcal{J} : A_{i,j} = aI \text{ for some } a \in AP^0 \right\}.$$

If $\mathcal{U} = \emptyset$, then the result follows immediately from Claim 1. Suppose now that $\mathcal{U} \neq \emptyset$. For each $(i, j) \in \mathcal{U}$, consider an element $a_{i,j} \in AP^0$ such that $A_{i,j} = a_{i,j} I$. Repeating this process, we obtain a finite sequence $(a_{i,j})_{(i,j)\in\mathcal{U}}$ in AP^0. Applying Claim 2 to this sequence, we find that there exists a sequence $\{h_n\}_{n=1}^\infty \subset \mathbb{R}$ such that $h_n \to +\infty$ as $n \to \infty$ and, for each $(i, j) \in \mathcal{U}$,

$$\lim_{n\to\infty} \left\| \left(T_{h_n}^{-1} a_{i,j} I T_{h_n} \right) f - (a_{i,j} I) f \right\|_{X(\mathbb{R})} = 0, \qquad f \in X(\mathbb{R}),$$

$$\lim_{n\to\infty} \left\| \left(T_{h_n}^{-1} a_{i,j} I T_{h_n} \right)^* g - (a_{i,j} I)^* g \right\|_{X'(\mathbb{R})} = 0, \qquad g \in X'(\mathbb{R}).$$

Combining the above remarks with Claims 1 and 2, we arrive at the conclusion that for all $(i, j) \in \mathcal{I} \times \mathcal{J}$, $f \in X(\mathbb{R})$ and $g \in X'(\mathbb{R})$,

$$\lim_{n\to\infty} \left(T_{h_n}^{-1} A_{i,j} T_{h_n} \right) f = A_{i,j} f \ \text{ in } X(\mathbb{R}),$$

$$\lim_{n\to\infty} \left(T_{h_n}^{-1} A_{i,j} T_{h_n} \right)^* g = A_{i,j}^* g \ \text{ in } X'(\mathbb{R}).$$

Faced with this, we are left with appealing to the definition of the operator B and the basic operations among limit operators (see, e.g., [14, Proposition 3.4] or [18, Proposition 1.1.2]) to get that for all $f \in X(\mathbb{R})$ and $g \in X'(\mathbb{R})$,

$$\lim_{n\to\infty} \left(T_{h_n}^{-1} B T_{h_n} \right) f = Bf \ \text{ in } X(\mathbb{R}),$$

$$\lim_{n\to\infty} \left(T_{h_n}^{-1} B T_{h_n} \right)^* g = B^* g \ \text{ in } X'(\mathbb{R}),$$

which completes the proof. □

3.4 Proof of Theorem 2

Suppose $A \in \mathcal{A}_N(AP, S; X(\mathbb{R}))$ is Fredholm on $X_N(\mathbb{R})$. By [7, Ch. XI, Theorem 5.1], there exists an operator $R \in \mathcal{B}(X_N(\mathbb{R}))$, called a regularizer of A, such that the operators $K_1 = I - RA$ and $K_2 = I - AR$ are compact on $X_N(\mathbb{R})$. It follows from the definition of the algebras $\mathcal{A}_N(AP, S; X(\mathbb{R}))$ and $AP_{N\times N}$ that there exists $B \in \mathcal{A}_N^0(AP^0, S; X(\mathbb{R}))$ such that

$$\|A - B\|_{\mathcal{B}(X_N(\mathbb{R}))} < \frac{1}{4\|R\|_{\mathcal{B}(X_N(\mathbb{R}))}}, \tag{8}$$

$$\|A^* - B^*\|_{\mathcal{B}(X'_N(\mathbb{R}))} < \frac{1}{4\|R^*\|_{\mathcal{B}(X'_N(\mathbb{R}))}}. \tag{9}$$

By Theorem 6, there exists a sequence $\{h_n\}_{n=1}^{\infty} \subset \mathbb{R}$ such that $h_n \to +\infty$ as $n \to \infty$,

$$\lim_{n\to\infty} \left\| T_{h_n}^{-1} B T_{h_n} f \right\|_{X_N(\mathbb{R})} = \|Bf\|_{X_N(\mathbb{R})} \tag{10}$$

for all $f \in X_N(\mathbb{R})$ and

$$\lim_{n\to\infty} \left\| \left(T_{h_n}^{-1} B T_{h_n}\right)^* g \right\|_{X'_N(\mathbb{R})} = \|B^* g\|_{X'_N(\mathbb{R})} \tag{11}$$

for all $g \in X'_N(\mathbb{R})$. It follows from Theorem 5 that

$$\lim_{n\to\infty} \left\| T_{h_n}^{-1} K_1 T_{h_n} f \right\|_{X_N(\mathbb{R})} = 0 \tag{12}$$

for all $f \in X_N(\mathbb{R})$ and

$$\lim_{n\to\infty} \left\| \left(T_{h_n}^{-1} K_2 T_{h_n}\right)^* g \right\|_{X'_N(\mathbb{R})} = 0 \tag{13}$$

for all $g \in X'_N(\mathbb{R})$.

Let $f \in X_N(\mathbb{R})$. Then for every $n \in \mathbb{N}$,

$$\begin{aligned} f &= T_{h_n}^{-1} T_{h_n} f = T_{h_n}^{-1}(RA + K_1) T_{h_n} f \\ &= T_{h_n}^{-1}(R(A - B) + RB + K_1) T_{h_n} f \\ &= \left(T_{h_n}^{-1} R T_{h_n}\right)\left(T_{h_n}^{-1}(A - B) T_{h_n}\right) f \\ &\quad + \left(T_{h_n}^{-1} R T_{h_n}\right)\left(T_{h_n}^{-1} B T_{h_n}\right) f + \left(T_{h_n}^{-1} K_1 T_{h_n}\right) f. \end{aligned}$$

Taking into account that $\left\|T_{h_n}^{\pm 1}\right\|_{\mathcal{B}(X_N(\mathbb{R}))} = 1$, we obtain

$$\begin{aligned}\|f\|_{X_N(\mathbb{R})} \leq & \|R\|_{\mathcal{B}(X_N(\mathbb{R}))}\|A - B\|_{\mathcal{B}(X_N(\mathbb{R}))}\|f\|_{X_N(\mathbb{R})} \\ & + \|R\|_{\mathcal{B}(X_N(\mathbb{R}))}\left\|T_{h_n}^{-1}BT_{h_n}f\right\|_{X_N(\mathbb{R})} + \left\|T_{h_n}^{-1}K_1T_{h_n}f\right\|_{X_N(\mathbb{R})}. \end{aligned} \tag{14}$$

It follows from (8), (10), (12), and (14) that

$$\begin{aligned}\|f\|_{X_N(\mathbb{R})} &\leq \|R\|_{\mathcal{B}(X_N(\mathbb{R}))}\|A - B\|_{\mathcal{B}(X_N(\mathbb{R}))}\|f\|_{X_N(\mathbb{R})} + \|R\|_{\mathcal{B}(X_N(\mathbb{R}))}\|Bf\|_{X_N(\mathbb{R})} \\ &\leq 2\|R\|_{\mathcal{B}(X_N(\mathbb{R}))}\|A - B\|_{\mathcal{B}(X_N(\mathbb{R}))}\|f\|_{X_N(\mathbb{R})} + \|R\|_{\mathcal{B}(X_N(\mathbb{R}))}\|Af\|_{X_N(\mathbb{R})} \\ &< \frac{1}{2}\|f\|_{X_N(\mathbb{R})} + \|R\|_{\mathcal{B}(X_N(\mathbb{R}))}\|Af\|_{X_N(\mathbb{R})}.\end{aligned}$$

Hence

$$\|f\|_{X_N(\mathbb{R})} \leq 2\|R\|_{\mathcal{B}(X_N(\mathbb{R}))}\|Af\|_{X_N(\mathbb{R})}.$$

This inequality implies that

$$\mathcal{J}(A; X_N(\mathbb{R})) \geq \frac{1}{2\|R\|_{\mathcal{B}(X_N(\mathbb{R}))}} > 0. \tag{15}$$

Analogously, it follows from

$$I = (AR + K_2)^* = R^*A^* + K_2^*$$

and (9), (11), and (13) that

$$\mathcal{J}(A^*; X'_N(\mathbb{R})) \geq \frac{1}{2\|R^*\|_{\mathcal{B}(X'_N(\mathbb{R}))}} > 0. \tag{16}$$

Combining (15) and (16) with Lemma 1 and Theorem 4, we conclude that the operator A is invertible on the space $X_N(\mathbb{R})$. □

4 Open Question for the Lorentz Space $L^{p,1}(\mathbb{R})$ and the Marcinkiewicz Space $L^{p,\infty}(\mathbb{R})$

The non-increasing rearrangement of a function $f \in \mathcal{M}_0(\mathbb{R}, m)$ is defined by

$$f^*(x) := \inf\{\lambda : m_f(\lambda) \leq x\}, \quad x \in [0, \infty).$$

For $x \in (0, \infty)$, put

$$f^{**}(x) = \frac{1}{x} \int_0^x f^*(y)\,dy.$$

Suppose $1 < p < \infty$ and $1 \le q \le \infty$. The Lorentz space $L^{p,q}(\mathbb{R})$ consists of all functions $f \in \mathcal{M}_0(\mathbb{R}, m)$ for which the quantity

$$\|f\|_{L^{p,q}} = \begin{cases} \left(\int_0^\infty \left(x^{1/p} f^{**}(x) \right)^q \frac{dx}{x} \right)^{1/q}, & \text{if } 1 \le q < \infty, \\ \sup\limits_{0<x<\infty} \left(x^{1/p} f^{**}(x) \right), & \text{if } q = \infty, \end{cases}$$

is finite. The space $L^{p,\infty}(\mathbb{R})$ is frequently called the weak-L^p space or the Marcinkiewicz space.

According to [1, Ch. 4, Theorem 4.6], if $1 < p < \infty$ and $1 \le q \le \infty$, then $L^{p,q}(\mathbb{R})$ is a rearrangement-invariant Banach function space with respect to $\|\cdot\|_{L^{p,q}}$ with the Boyd indices

$$\alpha_{L^{p,q}} = \beta_{L^{p,q}} = 1/p.$$

In this case, the definition of the algebra $\mathcal{A}_N(AP, S; L^{p,q}(\mathbb{R}))$ makes sense.

By [1, Ch. 4, Corollary 4.8] (see also [16, Corollary 8.5.4]), the space $L^{p,q}(\mathbb{R})$ is separable provided $1 < p < \infty$ and $1 \le q < \infty$. As a result of this, it follows that $L^{p,q}(\mathbb{R})$ is reflexive if $1 < p, q < \infty$ (see [16, Corollary 8.5.5]). Note that Corollaries 8.5.4 and 8.5.5 in [16] both contain the incorrect condition $1 \le q \le \infty$, which should be replaced by $1 \le q < \infty$ and $1 < q < \infty$, respectively.

So, Theorem 2 immediately implies the following.

Corollary 1 *Let $N \in \mathbb{N}$, $1 < p, q < \infty$, and $A \in \mathcal{A}_N(AP, S; L^{p,q}(\mathbb{R}))$. Then the operator A is Fredholm on $L_N^{p,q}(\mathbb{R})$ if and only if it is invertible on $L_N^{p,q}(\mathbb{R})$.*

It would be interesting to answer the following question.

Question 1 Let $N \in \mathbb{N}$, $1 < p < \infty$ and $q \in \{1, \infty\}$. Is it true that every operator $A \in \mathcal{A}_N(AP, S; L^{p,q}(\mathbb{R}))$ is Fredholm on $L_N^{p,q}(\mathbb{R})$ if and only if it is invertible on $L_N^{p,q}(\mathbb{R})$?

We would like to conclude this paper with the observation that if the answer to the above question is positive, then the corresponding proof should be different from the proof of Theorem 2.

Indeed, the proof of Theorem 2 relies on Theorem 5 based on the fact that if $X(\mathbb{R})$ is a reflexive rearrangement-invariant space, then the sequence of translation operators $\{T_{h_n}\}_{n=1}^\infty$ weakly converges to the zero operator on $X(\mathbb{R})$ for every sequence $\{h_n\}_{n=1}^\infty$ of real numbers such that $|h_n| \to +\infty$ as $n \to \infty$ (see [11, Corollary 2]). The same fact is still true for the nonreflexive separable space $L^{p,1}(\mathbb{R})$ with $1 < p < \infty$ (see [11, Corollary 3]). However, it is not true for the nonreflexive and nonseparable space $L^{p,\infty}(\mathbb{R})$ with $1 < p < \infty$: for any sequence $\{h_n\}_{n=1}^{\infty}$ of real numbers such that $|h_n| \to +\infty$ as $n \to \infty$, the sequence of translation operators $\{T_{h_n}\}_{n=1}^{\infty}$ does not converge weakly to the zero operator on the Marcinkiewicz space $L^{p,\infty}(\mathbb{R})$ (see [11, Theorem 2(b)] with $\varphi(t) = t^{1/p}$).

Acknowledgments This work is funded by national funds through the FCT—Fundação para a Ciência e a Tecnologia, I.P., under the scope of the projects UIDB/00297/2020 (https://doi.org/10.54499/UIDB/00297/2020) and UIDP/00297/2020 (https://doi.org/10.54499/UIDP/00297/2020) (Center for Mathematics and Applications). The second author is funded by national funds through the FCT—Fundação para a Ciência e a Tecnologia, I.P., under the scope of the PhD scholarship 2022.12247.BD.

References

1. Bennett, C., Sharpley, R.: Interpolation of Operators, volume 129 of Pure and Applied Mathematics. Academic Press, Boston, MA (1988)
2. Besicovitch, A.S.: Almost Periodic Functions. Dover Publications, New York (1955)
3. Böttcher, A., Karlovich, Y.I., Rabinovich, V.S.: The method of limit operators for one-dimensional singular integrals with slowly oscillating data. J. Operator Theory **43**(1), 171–198 (2000)
4. Böttcher, A., Karlovich, Y.I., Spitkovsky, I.M.: Convolution Operators and Factorization of Almost Periodic Matrix Functions, volume 131 of Operator Theory: Advances and Applications. Birkhäuser Verlag, Basel (2002)
5. Boyd, D.W.: The Hilbert transform on rearrangement-invariant spaces. Canadian J. Math. **19**, 599–616 (1967)
6. Boyd, D.W.: Indices of function spaces and their relationship to interpolation. Canadian J. Math. **21**, 1245–1254 (1969)
7. Gohberg, I., Goldberg, S., Kaashoek, M.A.: Classes of Linear Operators. Vol. I, volume 49 of Operator Theory: Advances and Applications. Birkhäuser Verlag, Basel (1990)
8. Karlovich, Y.I.: On the Haseman problem. Demonstratio Math. **26**(3–4), 581–595 (1993)
9. Karlovich, A.Y., Spitkovsky, I.M.: On singular integral operators with semi-almost periodic coefficients on variable Lebesgue spaces. J. Math. Anal. Appl. **384**(2), 706–725 (2011)
10. Karlovich, A.Y., Spitkovsky, I.M.: The Cauchy singular integral operator on weighted variable Lebesgue spaces. In: Concrete Operators, Spectral Theory, Operators in Harmonic Analysis and Approximation, volume 236 of Operator Theory: Advances and Applications, pp. 275–291. Birkhäuser/Springer, Basel (2014)
11. Karlovych, O., Shargorodsky, E.: On the weak convergence of shift operators to zero on rearrangement-invariant spaces. Rev. Mat. Comput. **36**(1), 91–124 (2023)
12. Kreĭn, S.G., Petunīn, Y.I., Semënov, E.M.: Interpolation of Linear Operators, volume 54 of Translations of Mathematical Monographs. American Mathematical Society, Providence, RI (1982)
13. Kurbatov, V.G.: Functional-differential Operators and Equations, volume 473 of Mathematics and Its Applications. Kluwer Academic Publishers, Dordrecht (1999)
14. Lindner, M.: An introduction to the limit operator method. In: Infinite Matrices and Their Finite Sections. Frontiers in Mathematics. Birkhäuser Verlag, Basel (2006)
15. Maligranda, L.: Indices and interpolation. Dissertationes Math. (Rozprawy Mat.) **234**, 49 (1985)
16. Pick, L., Kufner, A., John, O., Fučík, S.: Function Spaces. Vol. 1, volume 14 of De Gruyter Series in Nonlinear Analysis and Applications. Walter de Gruyter, Berlin, extended edition (2013)

17. Pietsch, A.: Operator Ideals, volume 20 of North-Holland Mathematical Library. North-Holland Publishing, Amsterdam (1980). Translated from German by the author
18. Rabinovich, V., Roch, S., Silbermann, B.: Limit Operators and Their Applications in Operator Theory, volume 150 of Operator Theory: Advances and Applications. Birkhäuser Verlag, Basel (2004)
19. Roch, S., Santos, P.A., Silbermann, B.: Non-commutative Gelfand Theories: A Tool-kit for Operator Theorists and Numerical Analysts. Universitext. Springer-Verlag London, London (2011)

Approximate Reconstruction of a One-Dimensional Parabolic Equation from Boundary Data

Vladislav V. Kravchenko

To Yuri Karlovich on the occasion of his 75th birthday

Abstract A method for recovering the spatially-dependent coefficient $q(x)$ in the parabolic equation $w_t - w_{xx} + q(x)w = 0$, $x \in (0, L)$, $t > 0$, from a knowledge of the boundary data $w(0,t)$, $w_x(0,t)$, $w(L,t)$ and under the condition $w(x, 0) = 0$, is developed. It is based on Neumann series of Bessel functions (NSBF) representations for solutions of the related Sturm-Liouville equation. With the aid of the Laplace transform and NSBF representations, the inverse problem is reduced to a system of linear algebraic equations for the NSBF coefficients. The coefficient $q(x)$ is recovered from an arithmetic combination of the first two unknowns of this system. The approach leads to an efficient numerical algorithm. Numerical efficiency is illustrated by test examples.

1 Introduction

The problem of recovering the unknown coefficient $q(x)$ in the parabolic equation

$$w_t - w_{xx} + q(x)w = 0 \tag{1}$$

has attracted considerable interest [2, 3, 7–11, 13–15, 17, 19, 26, 28–32] due to numerous applications of corresponding inverse coefficient problems in industry, in particular, in detecting mechanical imperfections by non-destructive testing, monitoring of the heat distribution in chemical reactors among many others (see, e.g., [13, 14, 19, 29] and references therein). Various types of initial-boundary conditions have been considered. In particular, inverse problems involving some

V. V. Kravchenko (✉)
Departamento de Matemáticas, Cinvestav, Unidad Querétaro, Querétaro, Qro., Mexico
e-mail: vkravchenko@math.cinvestav.edu.mx

A. Böttcher et al. (eds.), *Achievements and Challenges in the Field of Convolution Operators*, Operator Theory: Advances and Applications 306,
https://doi.org/10.1007/978-3-031-80486-1_11

homogeneous boundary conditions at the endpoints of the spatial variable interval $x \in (0, L)$ and prescribed $w(x, 0)$ as well as $w(x, T)$ or $w(x_0, t)$, where $T > 0$ and x_0 is a fixed point in the interval $[0, L]$, were considered, e.g., in [3, 9, 32]. Typically, one prescribes primary initial-boundary conditions, which would allow one to uniquely determine $w(x, t)$ if $q(x)$ were known, for example, the initial-boundary data $w(x, 0)$, $w(0, t)$ and $w(L, t)$. One then sets additional boundary conditions in order to attempt to reconstruct $q(x)$, and this has often been the flux on one of the lines $x = 0$ or $x = L$, for example, $w_x(0, t)$. In [26], in the case when $q \in C^1[0, L]$ is real valued, it was shown that if $w(x, 0) = 0$, $w(L, t) = 0$ and $w(0, t)$, $w_x(0, t)$ are prescribed, then there is at most one solution pair $(q(x), w(x, t))$. Additionally, we refer to [19, Sect. 4.6] and [29, Sect. 3.10] for the proof of the uniqueness under less restrictive conditions on a real valued coefficient $q(x)$.

In the present work, we develop a method for solving the problem (IP1) consisting of recovering a complex-valued coefficient $q(x) \in \mathcal{L}_1(0, L)$ under the condition $w(x, 0) = 0$ and from the prescribed $w(0, t)$, $w_x(0, t)$, $w(L, t)$. The method is based on reducing the original inverse problem to an inverse coefficient problem (IP2) for the Sturm-Liouville equation

$$- y''(x) + q(x)y(x) = \rho^2 y(x), \quad x \in (0, L). \tag{2}$$

Here, the potential $q(x)$ needs to be recovered from the prescribed functions $y(\rho, 0) = a(\rho)$, $y'(\rho, 0) = b(\rho)$ and $y(\rho, L) = l(\rho)$, where $y(\rho, x)$ denotes a solution of (2) for a corresponding value of the complex parameter ρ. The problem IP2 was studied in [25] in the case of a real valued potential (see also [12], where the prescribed data $y(\rho, L) = l(\rho)$ was substituted by $y'(\rho, L) + Hy(\rho, L) = l(\rho)$, $H \in \mathbb{R}$), where a method for its numerical solution, based on the Neumann series of Bessel functions (NSBF) representations and on the reduction to a two-spectrum inverse problem was developed. More recently, another approach for solving IP2 with complex-valued potentials was developed in [22], also based on the NSBF, but not requiring computation of the spectra.

The problem IP2 is obtained from IP1 by applying the Laplace transform, which naturally leads to the equation

$$- y''(x) + q(x)y(x) = -\lambda y(x), \tag{3}$$

where $\operatorname{Re} \lambda > 0$, while the NSBF representations enjoy most remarkable convergence properties for $\rho \in \mathbb{R}$ in (2) or, more generally, when $|\operatorname{Im} \rho| < C$, where the positive constant C should not be too large. Thus, the minus sign on the right-hand side of (3) complicates the direct application of the approach based on the NSBF representations. However, we overcome this difficulty by shifting the potential in (1) and using corresponding properties of the Laplace transform.

The approach developed in the present work is applicable to a large variety of inverse coefficient problems for partial differential equations. Its numerical efficiency is illustrated by numerical examples.

2 Problem Setting

Consider the one-dimensional parabolic equation

$$w_t(x,t) - w_{xx}(x,t) + q(x)w(x,t) = 0, \quad 0 < x < L, \quad t > 0, \tag{4}$$

where the coefficient $q \in \mathcal{L}_1(0, L)$ is complex valued. Assume

$$w(x,0) = 0, \quad 0 \leq x \leq L. \tag{5}$$

The inverse coefficient problem (IP1) consists of recovering the coefficient $q(x)$ from the prescribed functions

$$w(0,t) = \alpha(t), \quad w_x(0,t) = \beta(t), \quad w(L,t) = \gamma(t), \quad t \geq 0, \tag{6}$$

where $\alpha, \beta, \gamma \in \mathcal{L}_1(0, \infty)$ are complex valued.

Let us take the Laplace transform of (4)–(6). Denote

$$v(x,\lambda) = \mathcal{L}[w(x,t)] := \int_0^\infty w(x,t)e^{-\lambda t}dt,$$

$A(\lambda) := v(0,\lambda)$, $B(\lambda) := v_x(0,\lambda)$, $G(\lambda) := v(L,\lambda)$. Then (4)–(6) can be written as

$$-v'' + q(x)v = -\lambda v, \quad 0 < x < L, \tag{7}$$

$$v(0,\lambda) = A(\lambda), \quad v'(0,\lambda) = B(\lambda), \quad v(L,\lambda) = G(\lambda). \tag{8}$$

Note that $A(\lambda)$, $B(\lambda)$, $G(\lambda)$ are analytic in the half-plane $\mathrm{Re}\,\lambda > 0$. Moreover, without loss of generality, we assume that they are analytic in a larger half-plane $\Pi_\mu := \{\lambda : \mathrm{Re}\,\lambda > -\mu\}$, where $\mu > 0$ can be arbitrarily large. Indeed, fix $\mu > 0$ and instead of Eq. (4), consider the equation

$$u_t(x,t) - u_{xx}(x,t) + \widetilde{q}(x)u(x,t) = 0, \quad 0 < x < L,\ t > 0, \tag{9}$$

where $\widetilde{q}(x) := q(x) + \mu$. Solutions of (4) and (9) are related by

$$u(x,t) = e^{-\mu t}w(x,t).$$

$w(x,t)$ satisfies (4)–(6) iff $u(x,t)$ satisfies (9), (5), and

$$u(0,t) = \widetilde{\alpha}(t), \quad u_x(0,t) = \widetilde{\beta}(t), \quad u(L,t) = \widetilde{\gamma}(t), \quad t \geq 0,$$

where $\widetilde{\alpha}, \widetilde{\beta}, \widetilde{\gamma}$ are obtained from α, β, γ, respectively, by multiplying them by $e^{-\mu t}$. Thus, $\widetilde{\alpha}, \widetilde{\beta}, \widetilde{\gamma} \in \mathcal{L}_1(0, \infty; e^{\mu t})$, and their images under the Laplace transform are analytic functions in Π_μ.

Thus, we consider the inverse problem of recovering a potential $q(x)$ in (7) from the prescribed $A(\lambda)$, $B(\lambda)$, $G(\lambda)$ in (8), which are analytic functions in the half-plane Π_μ.

In particular, we have that the functions $A(\lambda)$, $B(\lambda)$, $G(\lambda)$ are known for $\lambda \in (-\mu, \infty)$. Thus, taking $\rho = \sqrt{-\lambda}$, $\lambda \in (-\mu, \infty)$, we may consider the problem (IP2) of recovering $q(x)$ in the equation

$$-y'' + q(x)y = \rho^2 y, \quad x \in (0, L) \tag{10}$$

from the knowledge of the boundary data

$$y(\rho, 0) = a(\rho), \quad y'(\rho, 0) = b(\rho), \quad y(\rho, L) = \ell(\rho), \tag{11}$$

for $\rho = i\tau$, $\tau > 0$ and for $\rho \in [0, \sqrt{\mu})$. Here $a(\rho) := A(-\rho^2)$, $b(\rho) := B(-\rho^2)$, $\ell(\rho) := G(-\rho^2)$.

3 Solution of Inverse Problem

It was shown in the previous section that the inverse problem IP1 reduces to Problem IP2. For solving the latter, we apply the approach proposed in [22]. Here, we briefly describe it. We denote by $\varphi(\rho, x)$ and $S(\rho, x)$ the solutions of (10) satisfying the initial conditions

$$\varphi(\rho, 0) = 1, \quad \varphi'(\rho, 0) = 0,$$

$$S(\rho, 0) = 0, \quad S'(\rho, 0) = 1.$$

Additionally, we denote by $T(\rho, x)$ the solution of (10) which satisfies the initial conditions at $x = L$:

$$T(\rho, L) = 0 \quad \text{and} \quad T'(\rho, L) = 1.$$

Note that

$$T(\rho, x) = \varphi(\rho, L)S(\rho, x) - S(\rho, L)\varphi(\rho, x). \tag{12}$$

Notice that equality (12) is nothing more than a generalization of the trigonometric identity

$$\sin(\rho(x - L)) = \cos(\rho L)\sin(\rho x) - \sin(\rho L)\cos(\rho x),$$

which arises from (12) when $q \equiv 0$.

In terms of the solutions $\varphi(\rho, x)$ and $S(\rho, x)$, the solution $y(\rho, x)$, satisfying (10) and the initial conditions from (11) at the origin, has the form

$$y(\rho, x) = a(\rho)\varphi(\rho, x) + b(\rho)S(\rho, x).$$

Thus, the third equality in (11) leads to the following equality for the functions $\varphi(\rho, L)$ and $S(\rho, L)$:

$$a(\rho)\varphi(\rho, L) + b(\rho)S(\rho, L) = \ell(\rho). \tag{13}$$

Below, we show that both $\varphi(\rho, L)$ and $S(\rho, L)$ can be recovered from this equality, considered at a sufficiently large number of points ρ_k. This is the first step of the algorithm for solving IP2. We compute the NSBF coefficients of $\varphi(\rho, L)$ and $S(\rho, L)$, that gives us the possibility to compute $\varphi(\rho, L)$ and $S(\rho, L)$ (approximately) for any $\rho \in \mathbb{R}$.

In the second step, we use $\varphi(\rho, L)$ and $S(\rho, L)$ computed at a sufficiently large number of the points $\rho \in \mathbb{R}$ and substitute them into the identity (12). For every $x \in [0, L]$, this leads to a system of linear algebraic equations for three sets of the NSBF coefficients corresponding to three solutions $\varphi(\rho, x)$, $S(\rho, x)$ and $T(\rho, x)$. Moreover, in order to obtain $q(x)$ we do not need the whole sets of the NSBF coefficients. It is worth mentioning here that the final step, the recovery of $q(x)$, depends on the NSBF that are used. The first NSBF for the solutions of (10) were obtained in [24] (see also [20] and [18]). For example, for the solution $\varphi(\rho, x)$, that NSBF has the form

$$\varphi(\rho, x) = \cos(\rho x) + \sum_{n=0}^{\infty} g_n(x)\mathbf{j}_{2n}(\rho x),$$

where $\mathbf{j}_k(z)$ stands for the spherical Bessel function of order k (for their definition, see, e.g., [1]). Here, the first NSBF coefficient has the form $g_0(x) = \varphi(0, x) - 1$. Hence, for recovering $q(x)$, one can take into account that $\varphi''(0, x) - q(x)\varphi(0, x) = 0$ and thus

$$q(x) = \frac{\varphi''(0, x)}{\varphi(0, x)} = \frac{g_0''(x)}{g_0(x) + 1}.$$

A drawback of this procedure is of course the necessity to take the second derivative (numerically) in the last step of the algorithm. This was one of the motivations for developing another NSBF representation for the solutions [22], such that allows us to recover $q(x)$ from an arithmetic combination of the first two NSBF coefficients, without any numerical differentiation involved. The limitation of this alternative NSBF representation is that it is valid for the potentials from $C^1[0, L]$, and in general, it is not applicable to potentials from $\mathcal{L}_1(0, L)$. This becomes clear from the procedure of its deduction [23]. Thus, if no additional information on the regularity of $q(x)$ is available, one can apply the NSBF representations from [24] (for their use in inverse coefficient problems, we refer to [21] and [4–6, 12, 25]). Here, we proceed with the development of the method under the condition $q \in C^1[0, L]$.

We use the following series representations for the solutions $\varphi(\rho, x)$, $S(\rho, x)$ and $T(\rho, x)$.

Theorem 1 ([22]) *Let $q \in C^1[0, L]$. Then the solutions $\varphi(\rho, x)$, $S(\rho, x)$ and $T(\rho, x)$ of* (10) *admit the series representations*

$$\varphi(\rho, x) = \cos(\rho x) + \frac{\sin(\rho x)}{\rho}\omega(x) - \frac{x\mathbf{j}_1(\rho x)}{\rho}q^-(x) - \frac{1}{\rho^2}\sum_{n=1}^{\infty}\varphi_n(x)\mathbf{j}_{2n}(\rho x), \tag{14}$$

$$S(\rho, x) = \frac{\sin(\rho x)}{\rho} + \frac{\omega(x)}{\rho^2}\left(\frac{3\mathbf{j}_1(\rho x)}{\rho x} - \cos(\rho x)\right) + \frac{q^+(x)}{\rho^3}(\sin(\rho x) - 3\mathbf{j}_1(\rho x)) - \frac{1}{\rho^3}\sum_{n=1}^{\infty}\sigma_n(x)\mathbf{j}_{2n+1}(\rho x), \tag{15}$$

$$T(\rho, x) = \frac{\sin(\rho(x-L))}{\rho} - \frac{\omega_L(x)}{\rho^2}\left(\frac{3\mathbf{j}_1(\rho(x-L))}{\rho(x-L)} - \cos(\rho(x-L))\right) + \frac{q_L^+(x)}{\rho^3}(\sin(\rho(x-L)) - 3\mathbf{j}_1(\rho(x-L))) - \frac{1}{\rho^3}\sum_{n=1}^{\infty}\theta_n(x)\mathbf{j}_{2n+1}(\rho(x-L)), \tag{16}$$

where $\mathbf{j}_k(z)$ stands for the spherical Bessel function of order k (see, e.g., [1]),

$$\omega(x) := \frac{1}{2}\int_0^x q(s)ds, \quad \omega_L(x) := \frac{1}{2}\int_x^L q(s)ds,$$

$$q^{\pm}(x) := \frac{q(x) \pm q(0)}{4} - \frac{\omega^2(x)}{2}, \quad q_L^+(x) := \frac{q(x) + q(L)}{4} - \frac{\omega_L^2(x)}{2}.$$

For every $x \in [0, L]$, the series converge uniformly on any compact set of the complex plane of the variable ρ, and for any $\rho \in \mathbb{C}\setminus\{0\}$ the remainders of their partial sums admit the estimates

$$|\varphi(\rho, x) - \varphi_N(\rho, x)| \le \frac{\varepsilon_N(x)}{|\rho|^2}\sqrt{\frac{\sinh(2\operatorname{Im}\rho x)}{\operatorname{Im}\rho}},$$

$$|S(\rho, x) - S_N(\rho, x)| \le \frac{\varepsilon_N(x)}{|\rho|^3}\sqrt{\frac{\sinh(2\operatorname{Im}\rho x)}{\operatorname{Im}\rho}},$$

$$|T(\rho, x) - T_N(\rho, x)| \le \frac{\varepsilon_N(x)}{|\rho|^3}\sqrt{\frac{\sinh(2\operatorname{Im}\rho x)}{\operatorname{Im}\rho}},$$

where the subindex N indicates that in (14)–(16) *the sum is taken up to N, and $\varepsilon_N(x)$ is a positive function tending to zero when $N \to \infty$.*

Remark 1 For $\rho = 0$, the following equalities hold

$$\varphi(0, x) = \varphi_1(0, x), \quad S(0, x) = S_1(0, x), \quad T(0, x) = T_1(0, x). \tag{17}$$

More explicit relations can be obtained by using the asymptotics of the spherical Bessel functions: $\mathbf{j}_k(z) \sim \frac{z^k}{(2k+1)!!}$, when $z \to 0$. Equalities (17) turn into the formulas

$$\varphi(0, x) = 1 + x\omega(x) - \frac{x^2}{3}\left(q^-(x) + \frac{\varphi_1(x)}{5}\right),$$

$$S(0, x) = x + \frac{2}{5}x^2\omega(x) - \frac{x^3}{15}\left(q^+(x) + \frac{\sigma_1(x)}{7}\right),$$

$$T(0, x) = x - L - \frac{2}{5}(x - L)^2\omega_L(x) - \frac{(x - L)^3}{15}\left(q_L^+(x) + \frac{\theta_1(x)}{7}\right).$$

Now, the method for recovering $q(x)$ from the known functions (11) consists of two steps. First, we substitute the series representations (14) and (15) evaluated at $x = L$ into (13), which gives us the following system of linear algebraic equations

$$\left(\frac{a_k \sin(\rho_k L)}{\rho_k} + \frac{b_k}{\rho_k^2}\left(\frac{3\mathbf{j}_1(\rho_k L)}{\rho_k L} - \cos(\rho_k L)\right)\right)\omega(L)$$

$$-\frac{a_k L\mathbf{j}_1(\rho_k L)}{\rho_k}q^-(L) - \frac{a_k}{\rho_k^2}\sum_{n=1}^{\infty}\varphi_n(L)\mathbf{j}_{2n}(\rho_k L)$$

$$+\frac{b_k}{\rho_k^3}\left(\sin\left(\rho_k L\right)-3\mathbf{j}_1(\rho_k L)\right)q^+(L)-\frac{b_k}{\rho_k^3}\sum_{n=1}^{\infty}\sigma_n(L)\mathbf{j}_{2n+1}(\rho_k L)$$
$$=\ell_k-a_k\cos\left(\rho_k L\right)-\frac{b_k}{\rho_k}\sin\left(\rho_k L\right)$$

for $\rho_k \in [0, \sqrt{\mu})$, $k = 1, 2, \ldots$. Here

$$a_k := a(\rho_k), \quad b_k := b(\rho_k), \quad \ell_k := \ell(\rho_k), \quad k = 1, 2, \ldots.$$

This leads to a finite system of linear algebraic equations for computing the NSBF coefficients $\omega(L)$, $q^-(L)$, $q^+(L)$, $\varphi_n(L)$, $\sigma_n(L)$, $n = 1, \ldots, N$,

$$\left(\frac{a_k\sin\left(\rho_k L\right)}{\rho_k}+\frac{b_k}{\rho_k^2}\left(\frac{3\mathbf{j}_1(\rho_k L)}{\rho_k L}-\cos\left(\rho_k L\right)\right)\right)\omega(L)$$
$$-\frac{a_k L\mathbf{j}_1(\rho_k L)}{\rho_k}q^-(L)-\frac{a_k}{\rho_k^2}\sum_{n=1}^{N}\varphi_n(L)\mathbf{j}_{2n}(\rho_k L)$$
$$+\frac{b_k}{\rho_k^3}\left(\sin\left(\rho_k L\right)-3\mathbf{j}_1(\rho_k L)\right)q^+(L)-\frac{b_k}{\rho_k^3}\sum_{n=1}^{N}\sigma_n(L)\mathbf{j}_{2n+1}(\rho_k L)$$
$$=\ell_k-a_k\cos\left(\rho_k L\right)-\frac{b_k}{\rho_k}\sin\left(\rho_k L\right) \tag{18}$$

for $k = 1, \ldots, K$.

Remark 2 Note that in general we consider overdetermined systems of linear algebraic equations with $K \geq 2N + 3$. In practice, a least-squares solution of an overdetermined system gives better results and allows us to make use of all available data, while keeping the number of the coefficients relatively small (in practice, $N = 4$ or 5 may prove sufficient).

Remark 3 The parameter $\omega(L)$ arises as a factor in the second term of the asymptotics of Sturm-Liouville eigenvalues of (10). Often, numerical techniques for solving inverse Sturm-Liouville problems require its prior knowledge as, for example, in [33] or [16]. Here, $\omega(L)$ is obtained immediately, directly from the input data of the problem, together with the parameters $q^-(L)$ and $q^+(L)$. Their combination gives the values of $q(x)$ at the end points:

$$q(0) = 2\left(q^+(L) - q^-(L)\right) \quad \text{and} \quad q(L) = 2\left(q^+(L) + q^-(L) + \omega^2(L)\right). \tag{19}$$

The knowledge of the coefficients $\omega(L)$, $q^-(L)$, $q^+(L)$, $\varphi_n(L)$, $\sigma_n(L)$, $n = 1, \ldots, N$ allows us to compute the functions $\varphi_N(\rho, L)$ and $S_N(\rho, L)$ for any value of ρ. Estimates from Theorem 1 show that the accuracy of the approximation of the exact solutions $\varphi(\rho, L)$ and $S(\rho, L)$ by the approximate ones $\varphi_N(\rho, L)$ and $S_N(\rho, L)$ does not deteriorate for large values of $\rho \in \mathbb{R}$ and even improves.

Second, we convert the knowledge at $x = L$ of the solutions $\varphi(\rho, x)$ and $S(\rho, x)$ into the knowledge of $q(x)$ on the whole interval. This is done by considering identity (12). We use it for constructing the main system of linear algebraic equations. Assume $\varphi(\rho, L)$ and $S(\rho, L)$ to be computed on a set of points $\{\gamma_k\}_{k=1}^{K_1}$, which in general may be different from $\{\rho_k\}_{k=1}^{K}$, which were used in the first step. Denote

$$S_k := S(\gamma_k, L) \quad \text{and} \quad F_k := \varphi(\gamma_k, L). \tag{20}$$

Now, for all $x \in (0, L)$, substitution of the series representations (14), (15) and (16) into (12) leads to a system of linear algebraic equations for the functions

$$\omega(x), \qquad Q(x) := \frac{q(x)}{4} - \frac{\omega^2(x)}{2}, \qquad q_0 := \frac{q(0)}{4},$$

$\omega_L(x)$, $q_L^+(x)$ and $\{\varphi_n(x), \sigma_n(x), \theta_n(x)\}_{n=1}^{N}$, $3N + 5 \leq K_1$:

$$\begin{aligned} &A_{k1}(x)\omega(x) + A_{k2}(x)Q(x) + A_{k3}(x)q_0 + A_{k4}(x)\omega_L(x) + A_{k5}(x)q_L^+(x) \\ &\quad + \sum_{n=1}^{N} B_{kn}(x)\varphi_n(x) - \sum_{n=1}^{N} C_{kn}(x)\sigma_n(x) + \sum_{n=1}^{N} D_{kn}(x)\theta_n(x) \\ &= \frac{\sin(\gamma_k(x-L))}{\gamma_k} + S_k \cos(\gamma_k x) - \frac{F_k \sin(\gamma_k x)}{\gamma_k}, \qquad k = 1, \ldots, K_1, \end{aligned} \tag{21}$$

where

$$A_{k1}(x) := -\frac{S_k \sin(\gamma_k x)}{\gamma_k} + \frac{F_k}{\gamma_k^2}\left(\frac{3\mathbf{j}_1(\gamma_k x)}{\gamma_k x} - \cos(\gamma_k x)\right),$$

$$A_{k2}(x) := \frac{F_k}{\gamma_k^3}(\sin(\gamma_k x) - 3\mathbf{j}_1(\gamma_k x)) + \frac{S_k x \mathbf{j}_1(\gamma_k x)}{\gamma_k},$$

$$A_{k3}(x) := -\frac{S_k x \mathbf{j}_1(\gamma_k x)}{\gamma_k} + \frac{F_k}{\gamma_k^3}(\sin(\gamma_k x) - 3\mathbf{j}_1(\gamma_k x)),$$

$$A_{k4}(x) := \frac{1}{\gamma_k^2}\left(\frac{3\mathbf{j}_1(\gamma_k(x-L))}{\gamma_k(x-L)} - \cos(\gamma_k(x-L))\right),$$

$$A_{k5}(x) := -\frac{1}{\gamma_k^3}(\sin(\gamma_k(x-L)) - 3\mathbf{j}_1(\gamma_k(x-L))),$$

$$B_{kn}(x) := \frac{S_k}{\gamma_k^2}\mathbf{j}_{2n}(\gamma_k x), \quad C_{kn}(x) := \frac{F_k}{\gamma_k^3}\mathbf{j}_{2n+1}(\gamma_k x),$$

$$D_{kn}(x) := \frac{1}{\gamma_k^3}\mathbf{j}_{2n+1}(\gamma_k(x-L)).$$

The potential $q(x)$, $x \in (0, L)$ is obtained from the equality

$$q(x) = 4Q(x) + 2\omega^2(x).$$

The values $q(0)$ and $q(L)$ are obtained in the first step, see Remark 3.

Thus, in order to obtain the potential $q(x)$ at a point $x \in (0, L)$, one should solve consecutively two systems of linear algebraic equations: system (18) and (21).

4 Numerical Examples

Example 1 Let c be a positive constant. Consider the equation

$$w_t(x,t) - w_{xx}(x,t) + cw(x,t) = 0, \quad 0 < x < 1, \quad t > 0,$$

subject to the conditions (5) and

$$w(0,t) = te^{-ct}, \quad w_x(0,t) = -2\sqrt{\frac{t}{\pi}}e^{-ct},$$

$$w(1,t) = te^{-ct}\left(\left(1+\frac{1}{2t}\right)\operatorname{erfc}\left(\frac{1}{2\sqrt{t}}\right) - \frac{1}{\sqrt{\pi t}}e^{-\frac{1}{4t}}\right), \quad t \ge 0$$

(see [27, 1.1.1-6, Example 3]). The Laplace transform of these functions can be calculated explicitly, so that (8) has the form

$$A(\lambda) = \frac{1}{(\lambda+c)^2}, \quad B(\lambda) = -\frac{1}{(\lambda+c)^{\frac{3}{2}}}, \quad G(\lambda) = \frac{e^{-\sqrt{\lambda+c}}}{(\lambda+c)^2}. \tag{22}$$

The numerical solution of the inverse problem was performed for $c = \pi^2$. The values of the functions $A(\lambda)$, $B(\lambda)$ and $G(\lambda)$ were calculated at 101 values of λ with a negative real part: $\lambda_k = -\rho_k^2 = -(\tau_k + 0.1i)^2$, where the numbers τ_k were distributed uniformly from 0 to 10, and at 11 positive values, distributed uniformly from $\lambda = 0.0001$ to $\lambda = 1$. The value of N in (18) was chosen as $N = 4$. Thus, $\omega(L)$, $q^-(L)$, $q^+(L)$ and the sets $\{\varphi_n(L)\}_{n=1}^4$, $\{\sigma_n(L)\}_{n=1}^4$ were computed. It is worth mentioning that at this stage, the absolute error of the computed $\omega(L)$ was $6.5 \cdot 10^{-7}$, while the absolute error of $q(0)$ and $q(L)$, obtained by (19), resulted in $9 \cdot 10^{-6}$ and $2 \cdot 10^{-4}$, respectively. With the coefficients $\omega(L)$, $q^-(L)$, $q^+(L)$

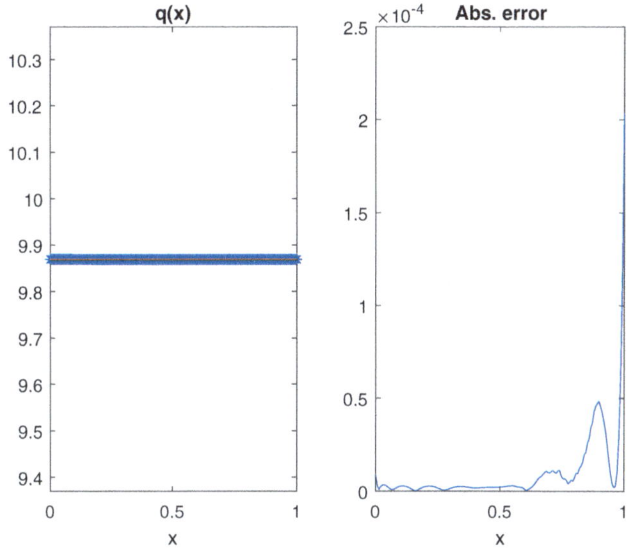

Fig. 1 Potential from Example 1, with $c = \pi^2$, recovered from the values of the functions $A(\lambda)$, $B(\lambda)$ and $G(\lambda)$ (22), calculated at 101 values of λ with a negative real part: $\lambda_k = -\rho_k^2 = -(\tau_k + 0.1i)^2$, where the numbers τ_k were distributed uniformly from 0 to 10, and at 11 positive values, distributed uniformly from $\lambda = 0.0001$ to $\lambda = 1$. The maximum absolute error resulted in $2 \cdot 10^{-4}$ (at $x = 1$)

and $\{\varphi_n(L), \sigma_n(L)\}_{n=1}^4$ computed in the first step, the values (20) were computed at 1001 points γ_k logarithmically equally spaced on the segment [0.1, 1300]. These values were used to write the main system (21) with $N = 4$. System (21) was solved at 151 points x_j, uniformly distributed on the segment [0, 1], with obvious simplifications at the endpoints of the segment. Figure 1 presents the recovered potential and the distribution of the absolute error. The maximum absolute error resulted in $2 \cdot 10^{-4}$.

Example 2 Consider the equation

$$w_t(x,t) - w_{xx}(x,t) + \left(x^2 + c\right) w(x,t) = 0, \quad 0 < x < 1, \quad t > 0. \tag{23}$$

Here c is a complex number. In order to construct an exact solution, the transformation from [27, subsect. 1.3.1-1] can be used, which relates solutions of this equation to solutions of the heat equation $u_\tau(z,\tau) - u_{zz}(z,\tau) = 0$. Namely, the solutions of (23) have the form

$$w(x,t) = u(z,\tau)e^{\frac{x^2}{2}+(1-c)t}$$

with $z = xe^{2t}$ and $\tau = \frac{1}{4}\left(e^{4t} - 1\right)$, where $u(z, \tau)$ are solutions of the heat equation. Now, choosing

$$u(z, \tau) = \operatorname{erfc}\left(\frac{z}{2\sqrt{\tau}}\right) = \operatorname{erfc}\left(\frac{xe^{2t}}{\sqrt{\left(e^{4t} - 1\right)}}\right),$$

we obtain that

$$w(x, t) = e^{\frac{x^2}{2} + (1-c)t} \operatorname{erfc}\left(\frac{xe^{2t}}{\sqrt{\left(e^{4t} - 1\right)}}\right)$$

is a solution of (23). We have then

$$w(0, t) = e^{(1-c)t}, \quad w_x(0, t) = -\frac{2}{\sqrt{\pi}} \frac{e^{(3-c)t}}{\sqrt{\left(e^{4t} - 1\right)}}$$

and

$$w(1, t) = e^{(1-c)t + 1/2} \operatorname{erfc}\left(\frac{e^{2t}}{\sqrt{\left(e^{4t} - 1\right)}}\right).$$

It is not difficult to see that

$$A(\lambda) = \mathcal{L}\left[w(0, t)\right](\lambda) = \mathcal{L}\left[e^{(1-c)t}\right](\lambda) = \frac{1}{\lambda + c - 1},$$

$$B(\lambda) = \mathcal{L}\left[w_x(0, t)\right](\lambda) = -\frac{1}{2} \frac{\Gamma\left(\frac{\lambda + c - 1}{4}\right)}{\Gamma\left(\frac{\lambda + c + 1}{4}\right)}.$$

The Laplace transform $G(\lambda)$ of $w(1, t)$ was computed numerically with the aid of the Matlab routine '*integral*'. The values of the functions $A(\lambda)$, $B(\lambda)$ and $G(\lambda)$ were obtained at 101 negative values of λ distributed uniformly from $\lambda = -(3\pi)^2$ to $\lambda = -0.01$ and at 101 positive values, distributed uniformly from $\lambda = 0.0001$ to $\lambda = \pi^2$. The numerical solution of the inverse problem was performed for $c = 16 + \pi i$. The value of N in (18) was chosen as $N = 4$. Thus, $\omega(L)$, $q^-(L)$, $q^+(L)$ and the sets $\{\varphi_n(L)\}_{n=1}^4$, $\{\sigma_n(L)\}_{n=1}^4$ were computed. It is worth mentioning that at this stage the absolute error of the computed $\omega(L)$ was $5 \cdot 10^{-7}$, while the absolute error of $q(0)$ and $q(L)$, obtained by (19), resulted in $6.6 \cdot 10^{-6}$ and $2.1 \cdot 10^{-4}$, respectively. With the coefficients $\omega(L)$, $q^-(L)$, $q^+(L)$ and $\{\varphi_n(L), \sigma_n(L)\}_{n=1}^4$ computed in the first step, the values (20) were computed at 1001 points γ_k logarithmically equally spaced

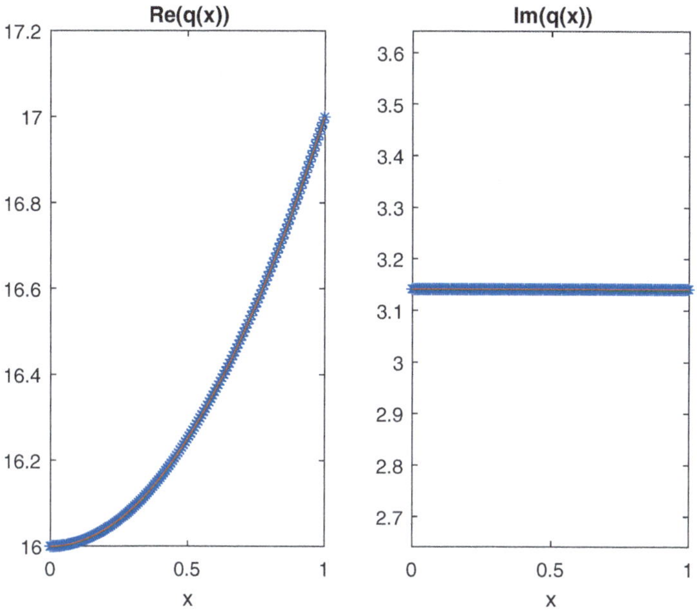

Fig. 2 Potential from Example 2, with $c = 16 + \pi i$, recovered from the values of the functions $A(\lambda)$, $B(\lambda)$ and $G(\lambda)$ that were obtained at 202 values of λ distributed in the interval $\left[-(3\pi)^2, \pi^2\right]$. The maximum absolute error resulted in $2.1 \cdot 10^{-4}$ (at $x = 1$)

on the segment [0.1, 1300]. These values were used to write the main system (21) with $N = 6$. System (21) was solved at 151 points x_j, uniformly distributed on the segment [0, 1], with obvious simplifications at the endpoints of the segment. Figure 2 presents the recovered potential. The maximum absolute error resulted in $2.1 \cdot 10^{-4}$ (at $x = 1$).

Funding Information CONAHCYT, Mexico, grant "Ciencia de Frontera" FORDECYT - PRONACES/61517/2020.

Data Availability The data that support the findings of this study are available upon reasonable request.

Conflict of Interest This work does not have any conflict of interest.

References

1. Abramovitz, M., Stegun, I.A.: Handbook of Mathematical Functions. Dover, New York (1972)
2. Anikonov, Yu.E.: Formulas in Inverse and Ill-posed Problems. De Gruyter, Berlin (1997)
3. Attas, H.A., Boumenir, A.: Reconstruction of a heat equation from one point observations. Commun. Appl. Math. Comput. **4**, 1280–1292 (2022)

4. Avdonin, S.A., Kravchenko, V.V.: Method for solving inverse spectral problems on quantum star graphs. J. Inverse Ill-posed Probl. **31**, 31–42 (2023)
5. Avdonin, S.A., Khmelnytskaya, K.V., Kravchenko, V.V.: Recovery of a potential on a quantum star graph from Weyl's matrix. Inverse Probl. Imaging **18**(1), 311–325 (2024)
6. Avdonin, S.A., Khmelnytskaya, K.V., Kravchenko, V.V.: Reconstruction techniques for quantum trees. Math. Methods Appl. Sci. **47**, 7182–7197 (2024)
7. Boumenir, A., Tuan, V.K.: An inverse problem for the heat equation. Proc. Amer. Math. Soc. **138**, 3911–3921 (2010)
8. Boumenir, A., Tuan, V.K.: Recovery of the heat coefficient by two measurements. Inverse Probl. Imaging **5**(4), 775–791 (2011)
9. Boumenir, A., Tuan, V.K.: Recovery of the heat equation from a single boundary measurement. Appl. Anal. **97**(10), 1667–1676 (2018)
10. Boumenir, A., Tuan, V.K.: One point recovery of a parabolic equation. J. Math. Anal. Appl. **463**(1), 161–166 (2018)
11. Cao, K., Lesnic, D.: Reconstruction of the perfusion coefficient from temperature measurements using the conjugate gradient method. Int. J. Comput. Math. **95**(4), 797–814 (2018)
12. Çetinkaya, F.A., Khmelnytskaya, K.V., Kravchenko, V.V.: Neumann series of Bessel functions for inverse coefficient problems. Math. Methods Appl. Sci. **47**(16), 12373–12387 (2024). https://doi.org/10.1002/mma.9855
13. Cordaro, P.D., Kawano, A.: A uniqueness result for the recovery of a coefficient of the heat conduction equation. Inverse Probl. **23**(3), 1069–1085 (2007)
14. Danilaev, P.G.: Coefficient Inverse Problems for Parabolic Type Equations and their Application. De Gruyter, Berlin (2001)
15. Deng, Z.C., Yu, J.N., Yang, L.: Optimization method for an evolutional type inverse heat conduction problem. J. Phys. A: Math. Theor. **41**, 035201 (2008)
16. Ignatiev, M., Yurko, V.: Numerical methods for solving inverse Sturm-Liouville problems. Results Math. **52**, 63–74 (2008)
17. Kabanikhin, S.I.: Inverse and Ill-posed Problems: Theory and Applications. De Gruyter, Berlin (2012)
18. Karapetyants, A.N., Kravchenko, V.V.: Methods of Mathematical Physics: Classical and Modern. Birkhäuser, Cham (2022)
19. Kirsch, A.: An Introduction to the Mathematical Theory of Inverse Problems. Applied Mathematical Sciences, vol. 120. Springer, New York (1996)
20. Kravchenko, V.V.: Direct and Inverse Sturm-Liouville Problems: A Method of Solution. Birkhäuser, Cham (2020)
21. Kravchenko, V.V.: Spectrum completion and inverse Sturm-Liouville problems. Math. Methods Appl. Sci. **46**, 5821–5835 (2023)
22. Kravchenko, V.V.: Reconstruction techniques for complex potentials. J. Math. Phys. **65**(3), 033501 (2024)
23. Kravchenko, V.V., Torba, S.M.: Asymptotics with respect to the spectral parameter and Neumann series of Bessel functions for solutions of the one-dimensional Schrödinger equation. J. Math. Phys. **58**, 122107 (2017)
24. Kravchenko, V.V., Navarro, L.J., Torba, S.M.: Representation of solutions to the one-dimensional Schrödinger equation in terms of Neumann series of Bessel functions. Appl. Math. Comput. **314**, 173–192 (2017)
25. Kravchenko, V.V., Khmelnytskaya, K.V., Çetinkaya, F.A.: Recovery of inhomogeneity from output boundary data. Mathematics **10**, 4349 (2022). https://doi.org/10.3390/math10224349
26. Pierce, A.: Unique determination of eigenvalues and coefficients in a parabolic problem. SIAM J. Control Optim. **17**, 494–499 (1979)
27. Polyanin, A.D.: Handbook of Linear Partial Differential Equations for Engineers and Scientists. Chapman & Hall/CRC, Boca Raton (2002)
28. Prilepko, A.I., Orlovsky, D.G., Vasin, I.A.: Methods for Solving Inverse Problems in Mathematical Physics. Chapman & Hall, CRC Pure and Applied by Global Express, Boca Raton (2000)

29. Ramm, A.G.: Inverse Problems: Mathematical and Analytical Techniques with Applications to Engineering. Springer, Boston (2005)
30. Romanov, V.G.: Inverse Problems of Mathematical Physics. VNU Science Press, Utrecht (1987)
31. Rundell, W.: An inverse problem for a parabolic partial differential equation. Rocky Mountain J. Math. **13**, 679–688 (1983)
32. Rundell, W.: The determination of a parabolic equation from initial and final data. Proc. Amer. Math. Soc. **99**, 637–642 (1987)
33. Rundell, W., Sacks, P.E.: Reconstruction techniques for classical inverse Sturm–Liouville problems. Math. Comput. **58**, 161–183 (1992)

Characteristic Determinants for a Second Order Difference Equation on the Half-Line Arising in Hydrodynamics

Yuri Latushkin and Shibi Vasudevan

To Yuri Karlovich on his seventy fifth birthday

Abstract We study the point spectrum of a second order difference operator with complex potential on the half-line via Fredholm determinants of the corresponding Birman-Schwinger operator pencils, the Evans and the Jost functions. An application is given to instability of a generalization of the Kolmogorov flow for the Euler equation of ideal fluid on the two dimensional torus.

1 Introduction and Main Results

In this paper we continue the work began in [38] and study the eigenvalues of the following boundary value problem for the second order asymptotically autonomous difference equation on the half-line $\mathbb{Z}_+ = \{0, 1, \dots\}$,

$$z_{n-1} - z_{n+1} + (b_n c_n) z_n = \lambda z_n, \quad n \geq 0, \tag{1}$$

$$z_{-1} = 0, \tag{2}$$

where $\lambda \in \mathbb{C}$ is the spectral parameter,

$$(b_n)_{n\geq 0} \in \ell^2(\mathbb{Z}_+; \mathbb{C}) \text{ and } (c_n)_{n\geq 0} \in \ell^2(\mathbb{Z}_+; \mathbb{C}) \tag{3}$$

Y. Latushkin (✉)
University of Missouri, Columbia, MO, USA
e-mail: latushkiny@missouri.edu

S. Vasudevan
Krea University, Sri City, Andhra Pradesh, India
e-mail: shibi.vasudevan@krea.edu.in

A. Böttcher et al. (eds.), *Achievements and Challenges in the Field of Convolution Operators*, Operator Theory: Advances and Applications 306,
https://doi.org/10.1007/978-3-031-80486-1_12

are two given complex valued sequences that, in general, may depend on λ holomorphically. We are seeking to characterize the values of the spectral parameter such that (1) and (2) has a nontrivial solution $\mathbf{z} = (z_n)_{n\geq 0} \in \ell^2(\mathbb{Z}_+;\mathbb{C})$. This question is important in stability issues for special steady state solutions of the two-dimensional Euler equation, the so called *generalized Kolmogorov flow*, and our main application is a result on its instability. Specifically, in the current paper we define (and prove that they are being equal) four functions of the spectral parameter whose zeros are the eigenvalues of (1) and (2). We call the functions *characteristic determinants*. Our overall strategy is as in [38] where the full line case has been considered, but the treatment of the important half-line case is technically more challenging; in particular, in the current paper we have to prove from the outset for the half-line case analogs of some results obtained in [12] for the line case and essentially used in [38]. By reflection, analogous results hold for the equations on the negative half-line.

An important particular case of (1) and (2) is the following eigenvalue problem for a difference equation arising in stability analysis of the generalized Kolmogorov flow of the Euler equation of ideal fluids on 2D torus, as seen below and considered in [17] and [38],

$$z_{n-1} - z_{n+1} = \lambda z_n/\rho_n, \quad n \geq 0, \quad z_{-1} = 0, \tag{4}$$

where $(\rho_n)_{n\geq 0}$ is a given sequence satisfying the following conditions:

$$\rho_0 < 0,\ \rho_n \in (0, 1),\ n \geq 1,\ \text{and } \rho_n = 1 + O(1/n^2) \text{ as } |n| \to \infty. \tag{5}$$

The problem (4) is reduced to (1) and (2) by setting

$$b_n = -\lambda\sqrt{1-\rho_n}/\rho_n \text{ and } c_n = \sqrt{1-\rho_n}. \tag{6}$$

For (4) we use continued fractions to define yet another, fifth function, also proven to be equal to the previous four characteristic determinants, whose zeros are the eigenvalues. This result, in turn, yields instability of the generalized Kolmogorov flow.

It is convenient to re-write (1) and (2) as

$$\big(S - S^{-1} + \operatorname{diag}_{n\in\mathbb{Z}_+}\{b_n c_n\}\big)\mathbf{z} = \lambda\mathbf{z}, \tag{7}$$

where we denote by $S : (z_n)_{n\geq 0} \mapsto (0, z_0, z_1, \dots)$ the right shift operator on $\ell^2(\mathbb{Z}_+;\mathbb{C})$ and by $S^{-1} = S^* : \mathbf{z} \mapsto (z_1, z_2, \dots)$ the left shift operator such that $S^{-1}S = I_{\ell^2(\mathbb{Z}_+;\mathbb{C})}$, the identity operator, while $\operatorname{ran}(SS^{-1}) = (\operatorname{span}\{(1, 0, \dots)\})^{\perp}$. By the spectral mapping theorem $\operatorname{Spec}(S - S^{-1}) = [-2\mathrm{i}, 2\mathrm{i}]$ since $\operatorname{Spec}(S) = \{\lambda \in \mathbb{C} : |\lambda| \leq 1\}$, and so throughout the paper we assume that $\lambda \notin [-2\mathrm{i}, 2\mathrm{i}]$ thus looking for the *isolated eigenvalues* of (1) and (2).

Letting $y_n = \begin{bmatrix} z_n \\ z_{n-1} \end{bmatrix} \in \mathbb{C}^2$, the problem (1) and (2) is equivalent to the following boundary value problem for the first order (2×2)-system of difference equations,

$$y_{n+1} = A_n^{\times} y_n, \; n \geq 0, \tag{8}$$

$$y_0 \in \operatorname{ran}(Q_+). \tag{9}$$

Here and throughout the paper we use the following notations,

$$A_n^{\times} = A + B_n C_n := \begin{bmatrix} b_n c_n - \lambda & 1 \\ 1 & 0 \end{bmatrix} \in \mathbb{C}^{2\times 2}, \quad n \geq 0. \tag{10}$$

$$A = A(\lambda) = \begin{bmatrix} -\lambda & 1 \\ 1 & 0 \end{bmatrix}, \; Q_+ = \begin{bmatrix} 1 & 0 \\ 0 & 0 \end{bmatrix}, \; Q_- = \begin{bmatrix} 0 & 0 \\ 0 & 1 \end{bmatrix}, \; B_n = b_n Q_+, \; C_n = c_n Q_+. \tag{11}$$

The eigenvalues of the matrix $A(\lambda)$ solve the quadratic equation $\mu^2 + \lambda\mu - 1 = 0$, and so our standing assumption $\lambda \notin [-2\mathrm{i}, 2\mathrm{i}]$ is equivalent to the fact that the eigenvalues of $A(\lambda)$ are off the unit circle. We let $\mu_+ = \mu_+(\lambda)$ and $\mu_- = \mu_-(\lambda)$ denote the roots of the equation $\mu^2 + \lambda\mu - 1 = 0$ satisfying the inequalities

$$|\mu_+(\lambda)| < 1 < |\mu_-(\lambda)|$$

and denote by $P_\pm = P_\pm(\lambda)$ the spectral projections for $A(\lambda)$ in $\mathbb{C}^2$ such that $\operatorname{Spec}(A(\lambda)\big|_{\operatorname{ran} P_\pm}) = \{\mu_\pm(\lambda)\}$. Finally, we let $R_+ = R_+(\lambda)$ denote the projection in $\mathbb{C}^2$ onto $\operatorname{ran} P_+$ parallel to $\operatorname{ran} Q_+$, and set $R_- = I_{2\times 2} - R_+$.

The choice of the projection R_+ is important for the half-line case. Indeed, the constant coefficient difference equation $y_{n+1} = Ay_n$ has exponential dichotomy on $\mathbb{Z}_+$ with the dichotomy projection P_+ whose range is the uniquely determined subspace of the initial values of the bounded solutions to the equation. Unlike the full line case, the exponential dichotomy on $\mathbb{Z}_+$ is not unique, but the only requirement on the dichotomy projection is that its range must be equal to the subspace $\operatorname{ran} P_+$. The choice of R_+ whose kernel is $\operatorname{ran} Q_+$ as the dichotomy projection will allow us to satisfy the boundary condition in (2), see, e.g., formula (33) below.

We now proceed to define our first characteristic determinant associated with (1)–(2). We consider the following Birman-Schwinger-type pencil of operators acting in $\ell^2(\mathbb{Z}_+; \mathbb{C})$,

$$K_\lambda^+ = -\operatorname{diag}_{n\in\mathbb{Z}_+}\{c_n\}(S - S^{-1} - \lambda)^{-1} \operatorname{diag}_{n\in\mathbb{Z}_+}\{b_n\}, \tag{12}$$

analogous to K_λ studied in [38] for the full line. By (3), the operator K_λ^+ is of trace class and so we may define our first characteristic determinant $\det(I - K_\lambda^+)$.

Next, we re-write (8) as $\big(S^{-1} - \mathrm{diag}_{n\in\mathbb{Z}_+}\{A_n^\times\}\big)\mathbf{y} = 0$ for $\mathbf{y} = (y_n)_{n\in\mathbb{Z}_+}$ where we continue to denote by S^{-1} the shift acting in the space of vector valued sequences. We stress that the operator $S^{-1} - \mathrm{diag}_{n\in\mathbb{Z}_+}\{A_n^\times\}$ in (8) and (9) acts from the subspace

$$\ell^2_{\mathrm{ran}(Q_+)}(\mathbb{Z}_+;\mathbb{C}^2) := \{(y_n)_{n\geq 0} \in \ell^2(\mathbb{Z}_+;\mathbb{C}^2) : y_0 \in \mathrm{ran}(Q_+)\} \tag{13}$$

into $\ell^2(\mathbb{Z}_+;\mathbb{C}^2)$ and refer to [5] for results on equivalence of invertibility of the operator and dichotomy of the difference equation (8) on the half-line. We introduce the Birman-Schwinger-type pencil of operators acting in $\ell^2(\mathbb{Z}_+;\mathbb{C}^2)$,

$$\mathcal{T}_\lambda^+ = \mathrm{diag}_{n\in\mathbb{Z}_+}\{C_n\}\big(S^{-1} - \mathrm{diag}_{n\in\mathbb{Z}_+}\{A\}\big)^{-1}\mathrm{diag}_{n\in\mathbb{Z}_+}\{B_n\}, \tag{14}$$

analogous to $\mathcal{T}_\lambda$ studied in [38] for the full line case. The operator $\mathcal{T}_\lambda^+$ is of trace class by (3) and we define our second characteristic determinant $\det(I - \mathcal{T}_\lambda^+)$.

We now consider the (2×2)-*matrix valued Jost solution* $\mathbf{Y}^+ = \mathbf{Y}^+(\lambda) = (Y_n^+)_{n\geq 0}$, $Y_n^+ \in \mathbb{C}^{2\times 2}$, cf. [12, 25, 38], defined as the matrix valued functions whose columns are solution to the difference equation (8) (without the boundary condition) satisfying

$$\big\|(\mu_+)^{-n}\big(Y_n^+ - A^n R_+\big)\big\|_{\mathbb{C}^{2\times 2}} \to 0 \text{ as } n \to +\infty \text{ and } Y_0^+ = Y_0^+ R_+. \tag{15}$$

As we will prove below, this solution is unique. Also, $Y_n^+ = Y_n^+ R_+$ for all $n \geq 0$. Using the projection $R_- = R_-(\lambda)$ onto ran Q_+ parallel to ran $P_+(\lambda)$, analogously to $\mathcal{E}(\lambda)$ in [38] for the full line case, we introduce our third characteristic determinant, the *Evans function*, by the formula

$$\mathcal{E}^+(\lambda) = \det\big(Y_0^+(\lambda) + R_-(\lambda)\big). \tag{16}$$

We call a solution $\mathbf{z}^+ = (z_n^+)_{n\geq -1}$ of the second order difference equation (1) (without the boundary condition) the *Jost solution,* cf. [9], provided

$$(\mu_+)^{-n} z_n^+ - 1 \to 0 \text{ as } n \to +\infty, \tag{17}$$

and denote by $\mathbf{z}^{\mathrm{r}} = (z_n^{\mathrm{r}})_{n\geq -1}$ the *regular* solution of the boundary value problem (1) and (2) satisfying the additional boundary condition $z_0^{\mathrm{r}} = 1$ (which is of course unique but not necessarily bounded at infinity). As in the case of the discrete Schrödinger equations, the Jost solution $\mathbf{z}^+ = \mathbf{z}^+(\lambda)$ is unique. Introducing the notation $\mathcal{W}(\mathbf{u},\mathbf{v})_n = (-1)^n\big(u_{n-1}v_n - u_n v_{n-1}\big)$, $n \geq 0$, for the Wronskian of any two sequences $\mathbf{u} = (u_n)_{n\geq -1}$ and $\mathbf{v} = (v_n)_{n\geq -1}$, we note that the Wronskian of the solutions to (1) is n-independent, and that $\mathcal{W}(\mathbf{z}^+,\mathbf{z}^{\mathrm{r}})_0 = z_{-1}^+(\lambda)$. We define our fourth characteristic determinant, the *Jost function*, analogously to $\mathcal{F}$ from [38] for the full line case,

$$\mathcal{F}^+(\lambda) = \mu_+(\lambda)\mathcal{W}(\mathbf{z}^+(\lambda),\mathbf{z}^{\mathrm{r}}(\lambda))_0, \tag{18}$$

where the scaling factor $\mu_+(\lambda)$ is chosen such that $\mathcal{F}^+(\lambda) = 1$ provided $b_n c_n = 0$ for $n \geq 0$ when $z_n^+ = \mu_+^n$ for $n \geq -1$.

Our fifth characteristic determinant will be defined only for the difference equation (4) assuming (5) and uses continued fractions; we introduce the notation

$$g_+(\lambda) = \cfrac{1}{\frac{\lambda}{\rho_1} + \cfrac{1}{\frac{\lambda}{\rho_2} + \dots}}. \tag{19}$$

The continued fraction converges for $\mathrm{Re}(\lambda) > 0$ and $|\arg(\lambda)| \leq \pi/2 - \delta$ for any $\delta \in (0, \pi/2)$ by the classical Van Vleck Theorem [32, Theorem 4.29] and we may now define our fifth function of interest by the formula, cf. $\mathcal{G}$ from [38],

$$\mathcal{G}^+(\lambda) = \mu_+(\lambda)\, z_0^+(\lambda)\, \big(g_+(\lambda) + \lambda/\rho_0\big), \tag{20}$$

where $z_0^+(\lambda)$ is the 0-th entry of the Jost solution, $\mu_+(\lambda)$ is the eigenvalue of $A(\lambda)$ inside of the unit disk, and $g_+(\lambda)$ is defined in (19), and we assume that $\mathrm{Re}(\lambda) > 0$.

We are ready to formulate the main results of this paper.

Theorem 1 *Assume* $\lambda \notin [-2\mathrm{i}, 2\mathrm{i}]$ *and that the sequences in* (3) *depend on* λ *holomorphically. The functions introduced in* (12), (14), (16) *and* (18) *are holomorphic in* λ *and equal,*

$$\det(I - K_\lambda^+) = \det(I - \mathcal{T}_\lambda^+) = \mathcal{E}^+(\lambda) = \mathcal{F}^+(\lambda). \tag{21}$$

As a result, λ *is a simple discrete eigenvalue of* (1) *and* (2) *if and only if* λ *is a zero of each of the functions in* (21).

Corollary 1 *Assume* $\mathrm{Re}(\lambda) > 0$ *and* (5), *and consider the difference equation* (4). *Define* (b_n) *and* (c_n) *by* (6) *and use the sequences to construct all four characteristic determinants in* (21). *Then* $\mathcal{G}^+(\lambda)$ *defined in* (20) *via the continued fraction is equal to the functions in* (21).

We refer to [38] for a detailed discussion of the literature related to the results, and mention here only the following. Regarding the Birman-Schwinger operator pencils see [2, 8]. The papers most relevant to the current setup are [12, 25] and [36]. We are not aware of any literature on the Birman-Schwinger type pencils specific for the first order systems (1) on the half-line. For the full line case they have been studied in [12] which is a companion paper to [25] dealing with differential equations where, among many other things, the first equality in (21) was proved for the case of the Schrödinger differential operators. We are not aware of results of this sort for the difference equations of type (1) on the half-line. This equation is of course a particular case of eigenvalue problems for three diagonal (Jacobi) matrices, cf. [29, 31] and also [7, 18, 19, 30, 35, 45–47]. The dichotomies on the half-line were studied in great detail, see [4–6] and the references therein. Regarding the Evans function we refer to [34, 44] where one can find many other sources. The most relevant papers are again [12, 25], where, in particular, the second equality in (21) has been proved for quite general first order systems of difference and differential equations but on the full line, and so the results in the current paper for the half-line case seem to be new. The Evans function is a relatively new topic in the study of the 2D-Euler-related difference equations as in (4); we are aware of only [14, 38]. The Jost solutions are classical [9] and so is the equality of $\det(I - K_\lambda^+)$ and $\mathcal{F}^+$ in (21) for the Schrödinger case, see [33, 43] and generalizations in [24, 26]; however, it is quite possible that the use of the Jost solutions as well as the equality of $\mathcal{F}^+$ and $\mathcal{E}^+$ and $\det(I - K_\lambda^+)$ in the context of (1), see [38], appear to be new for the half-line case. Regarding continued fractions in this context see [17, 21, 32, 42, 48].

We now briefly discuss connections to the Euler equation again referring to [17, 38] for more details and concentrating on just one particularly important case of stability of the so-called unidirectional, or generalized Kolmogorov, flow for the Euler equations of ideal fluid on the two dimensional torus $\mathbb{T}^2$,

$$\partial_t \Omega + U \cdot \nabla\Omega = 0, \quad \operatorname{div} U = 0,\ \Omega = \operatorname{curl} U, \quad \mathbf{x} = (x_1, x_2) \in \mathbb{T}^2, \tag{22}$$

where the two-dimensional vector $U = U(\mathbf{x})$ is the velocity and the scalar Ω is the vorticity of the fluid. The unidirectional (or generalized Kolmogorov) flow is the steady state solution to the Euler equations on $\mathbb{T}^2$ of the form

$$\Omega^0(\mathbf{x}) = \alpha e^{i\mathbf{p}\cdot\mathbf{x}}/2 + \alpha e^{-i\mathbf{p}\cdot\mathbf{x}}/2 = \alpha\cos(\mathbf{p}\cdot\mathbf{x}), \tag{23}$$

where $\mathbf{p} \in \mathbb{Z}^2\backslash\{0\}$ is a given vector and $\alpha \in \mathbb{R}$. In particular, $\mathbf{p} = (m, 0) \in \mathbb{Z}^2$ for $m \in \mathbb{N}$ corresponds to the classical Kolmogorov flow. It is thus a classical problem to study (linear) stability of the flow given by (23). To this end, using Fourier series $\Omega(\mathbf{x}) = \sum_{\mathbf{k}\in\mathbb{Z}^2\backslash\{0\}} \omega_{\mathbf{k}} e^{i\mathbf{k}\cdot\mathbf{x}}$ for vorticity, we rewrite (22) as a system of nonlinear equations for $\omega_{\mathbf{k}}$, $\mathbf{k} \in \mathbb{Z}^2$ as in [17, 38–40]. Linearizing this system about the unidirectional flow, one obtains the following operator in $\ell^2(\mathbb{Z}^2; \mathbb{C})$,

$$L : (\omega_{\mathbf{k}})_{\mathbf{k}\in\mathbb{Z}^2} \mapsto \big(\alpha\beta(\mathbf{p}, \mathbf{k}-\mathbf{p})\omega_{\mathbf{k}-\mathbf{p}} - \alpha\beta(\mathbf{p}, \mathbf{k}+\mathbf{p})\omega_{\mathbf{k}+\mathbf{p}}\big)_{\mathbf{k}\in\mathbb{Z}^2}, \tag{24}$$

where the coefficients $\beta(\mathbf{p}, \mathbf{q})$ for $\mathbf{p} = (p_1, p_2)$, $\mathbf{q} = (q_1, q_2) \in \mathbb{Z}^2$ are defined as

$$\beta(\mathbf{p}, \mathbf{q}) = \frac{1}{2}\big(\|\mathbf{q}\|^{-2} - \|\mathbf{p}\|^{-2}\big)(\mathbf{p}\wedge\mathbf{q}), \quad \text{with } \mathbf{p}\wedge\mathbf{q} := \det\left[\begin{smallmatrix} p_1 & q_1 \\ p_2 & q_2 \end{smallmatrix}\right], \tag{25}$$

for $\mathbf{p} \neq 0, \mathbf{q} \neq 0$, and $\beta(\mathbf{p}, \mathbf{q}) = 0$ otherwise. The flow (23) is called (linearly) unstable if the operator L has nonimaginary spectrum. We refer to [37, 41] for results on stability and instability for the 2D Euler equations and related models.

To study the spectrum of L, (see the discussion in [17, pp. 2054–2057]; see also [38, 39]) we decompose this operator into a sum of operators $L_{\mathbf{q}}$, $\mathbf{q} \in \mathbb{Z}^2$, acting in the space $\ell^2(\mathbb{Z};\mathbb{C})$, by "slicing" the grid $\mathbb{Z}^2$ along lines parallel to $\mathbf{p}$ such that $\mathrm{Spec}(L) = \cup_{\mathbf{q}} \mathrm{Spec}(L_{\mathbf{q}})$, where

$$L_{\mathbf{q}} : (w_n) \mapsto \big(\alpha\beta(\mathbf{p}, \mathbf{q} + (n-1)\mathbf{p})w_{n-1} - \alpha\beta(\mathbf{p}, \mathbf{q} + (n+1)\mathbf{p})w_{n+1}\big),\ n \in \mathbb{Z}, \tag{26}$$

and for $\mathbf{k} = \mathbf{q} + n\mathbf{p}$ from (24) we denote $w_n = \omega_{\mathbf{q}+n\mathbf{p}}$. In other words, $L_{\mathbf{q}} = \alpha(S - S^*)\,\mathrm{diag}_{n\in\mathbb{Z}}\{\beta(\mathbf{p}, \mathbf{q} + n\mathbf{p})\}$. By (25), if $\mathbf{q}$ is parallel to $\mathbf{p}$ then $L_{\mathbf{q}} = 0$ and thus we assume throughout that $\mathbf{q}$ is not parallel to $\mathbf{p}$. Moreover, since $L_{\mathbf{q}}$ contains a scalar multiple $\alpha \in \mathbb{R}$, with no loss of generality we may rescale this operator or, equivalently, will assume the normalization condition $\alpha(\mathbf{q} \wedge \mathbf{p})\|\mathbf{p}\|^{-2}/2 = 1$.

The operators $L_{\mathbf{q}}$ are classified based on the location of the line $B_{\mathbf{q}} = \{\mathbf{q} + n\mathbf{p} : n \in \mathbb{Z}\}$ through the point $\mathbf{q} \in \mathbb{Z}^2$ relative to the disc of radius $\|\mathbf{p}\|$ centered at zero, see, again, the discussion in [17, pp. 2054–2057]; see also [38, 39]. Spectral properties of the operators drastically depend on the location. For instance, if none of the vectors $\mathbf{q}+n\mathbf{p}$, $n \in \mathbb{Z}$, is located inside the open disc then $L_{\mathbf{q}}$ has no unstable eigenvalues [39, 40]. In the current paper we consider only the case when $\mathbf{p}$ and $\mathbf{q}$ are such that

$$\|\mathbf{q}\| < \|\mathbf{p}\|,\ \|\mathbf{q} - \mathbf{p}\| = \|\mathbf{p}\| \text{ and } \|\mathbf{q} + n\mathbf{p}\| > \|\mathbf{p}\| \text{ for all } n \in \mathbb{Z} \setminus \{-1, 0\}; \tag{27}$$

a typical example of this is $\mathbf{p} = (3, 1)$, $\mathbf{q} = (2, -2)$. This assumption corresponds to the case I_- described in [17], and we refer to this paper for a discussion regarding other possible cases. The case I_+ when -1 in (27) is replaced by $+1$ can be treated similarly to I_- while the case I_0 when $\|\mathbf{q} + n\mathbf{p}\| > \|\mathbf{p}\|$ for all $n \neq 0$ has been considered in [38]. Geometrically, the case I_- and I_+ occur when one of the points in $B_{\mathbf{q}}$ is inside of the open disc of radius $\|\mathbf{p}\|$ and one of the points in $B_{\mathbf{q}}$ is on the boundary of the disk while in case I_0 one point of $B_{\mathbf{q}}$ is inside of the open disk while all others are outside of the closed disk. An interesting open question is to describe the spectrum of $L_{\mathbf{q}}$ in the case II when two points of $B_{\mathbf{q}}$ are located inside of the open disk of radius $\|\mathbf{p}\|$ centered at zero.

Assuming that we are in the case I_-, that is, that (27) holds, we introduce the sequence $\rho_n = 1 - \|\mathbf{p}\|^2\|\mathbf{q} + n\mathbf{p}\|^{-2}$, $n \in \mathbb{Z}$, such that the operator $L_{\mathbf{q}}$ in (26) reads $L_{\mathbf{q}} = (S - S^*)\,\mathrm{diag}_{n\in\mathbb{Z}}\{\rho_n\}$, see (25). The eigenvalue equation for $L_{\mathbf{q}}$ is

$$\rho_{n-1}w_{n-1} - \rho_{n+1}w_{n+1} = \lambda w_n,\ n \in \mathbb{Z}, \text{ with } (w_n)_{n\in\mathbb{Z}} \in \ell^2(\mathbb{Z};\mathbb{C}). \tag{28}$$

By (27) we have $\rho_{-1} = 0$, $\rho_0 < 0$, and $\rho_n \in (0, 1)$ for all $n \in \mathbb{Z} \setminus \{0, -1\}$. In particular, the sequence $(\rho_n)_{n\geq 0}$ satisfies (5). Now Theorem 1 and Corollary 1 can be applied because the solution of (28) must be supported on $\mathbb{Z}_+$ as seen in Lemma 2 given in the next section. Theorem 1 implies the following.

Corollary 2 *Assume that a given* $\mathbf{p} \in \mathbb{Z}^2$ *is such that there exists a* $\mathbf{q} \in \mathbb{Z}^2$ *satisfying* (27). *Then the eigenvalues of* $L_{\mathbf{q}}$ *with positive real parts are in one-to-one correspondence with zeros of each of the five functions in* (21) *and* (20). *Moreover,* *the operator $L_{\mathbf{q}}$ from* (26), *and thus L from* (24), *has a positive eigenvalue. As a result, the unidirectional flow* (23) *is linearly unstable.*

The instability of the unidirectional flow in the current setting has been established in [17, Theorem 2.9]. Nevertheless, the first assertion in Corollary 2 is an improvement of the part of [17, Theorem 2.9] where the correspondence was established between only the *positive* roots of the function $g_+(\lambda) + \lambda/\rho_0$ and the *positive* eigenvalues of $L_{\mathbf{q}}$ but under the additional assumption that the respective eigensequences satisfy some special property [17, Property 2.8]. By applying Theorem 1 in the proof of Corollary 2 we were able to show that this assumption is redundant.

We conclude this section with references on the literature on stability of unidirectional flows. This topic is quite classical and well-studied, and we refer to [1, 3, 20–22, 42]. The setup used herein was also used in many papers [15, 16, 39, 40, 48], but the closest to the current work is [17, 38]. We mention also [36] regarding connections to the Birman-Schwinger operators. Finally, connections between the Evans function and the linearization of the 2D Euler equation has been studied in a recent important paper [14].

2 Proofs

We begin with several general comments regarding the objects introduced in the previous section. For a start, it is sometimes convenient to diagonalize the matrix A from (11). To this end we introduce matrices

$$W = \begin{bmatrix} \mu_+ & \mu_- \\ 1 & 1 \end{bmatrix}, \quad W^{-1} = (\mu_+ - \mu_-)^{-1} \begin{bmatrix} 1 & -\mu_- \\ -1 & \mu_+ \end{bmatrix}, \quad \widetilde{A} = \operatorname{diag}\{\mu_+, \mu_-\}, \tag{29}$$

so that one has, cf. (11),

$$W^{-1}AW = \widetilde{A}, \quad W^{-1}P_\pm W = Q_\pm. \tag{30}$$

We will use below the explicit formulas

$$\begin{aligned} R_+ &= \begin{bmatrix} 0 & \mu_+ \\ 0 & 1 \end{bmatrix}, \; R_- = \begin{bmatrix} 1 & -\mu_+ \\ 0 & 0 \end{bmatrix}, \\ P_+ &= (\mu_+ - \mu_-)^{-1} \begin{bmatrix} \mu_+ & 1 \\ 1 & -\mu_- \end{bmatrix}, \; A^{-1} = \begin{bmatrix} 0 & 1 \\ 1 & \lambda \end{bmatrix}. \end{aligned} \tag{31}$$

The operator $S - S^{-1} - \lambda$ is invertible in $\ell^2(\mathbb{Z}_+; \mathbb{C})$ if and only if $\lambda \notin [-2\mathrm{i}, 2\mathrm{i}]$ and in this case the inverse operator $\left(S - S^{-1} - \lambda\right)^{-1}$ is given by the infinite matrix

$[a_{nk}]_{n,k=0}^{\infty}$ whose entries are defined by the formula

$$a_{nk} = (\mu_+ - \mu_-)^{-1} \times \begin{cases} (\mu_-)^{n-k}\big(1 - (\frac{\mu_+}{\mu_-})^{n+1}\big) & \text{for } n \leq k, \\ (\mu_+)^{n-k}\big(1 - (\frac{\mu_+}{\mu_-})^{k+1}\big) & \text{for } n > k. \end{cases} \tag{32}$$

One can check this either directly by multiplying the respective infinite matrices and using the equation $\mu^2 + \lambda\mu - 1 = 0$, or by using [27, Theorem 1.1] as follows: we factorize $S - S^{-1} - \lambda = \mu_-(I + \mu_+ S^{-1})(I + \mu_-^{-1} S)$ and use the Neumann series to obtain the inverse of each factor.

From now on we abbreviate $A = \operatorname{diag}_{n\in\mathbb{Z}_+}\{A\}$. The operator $S^{-1} - A$ acting from $\ell^2_{\operatorname{ran} Q_+}(\mathbb{Z}_+; \mathbb{C}^2)$ into $\ell^2(\mathbb{Z}_+; \mathbb{C}^2)$, cf. (13) and [5], is invertible if and only if $\lambda \notin [-2\mathrm{i}, 2\mathrm{i}]$ and in this case the inverse operator $\big(S^{-1} - A\big)^{-1}$ from $\ell^2(\mathbb{Z}_+; \mathbb{C}^2)$ onto $\ell^2_{\operatorname{ran} Q_+}(\mathbb{Z}_+; \mathbb{C}^2)$ for $\mathbf{u} = (u_n)_{n\geq 0} \in \ell^2(\mathbb{Z}_+; \mathbb{C}^2)$ is given by the formula

$$\Big(\big(S^{-1} - A\big)^{-1}\mathbf{u}\Big)_n = -\sum_{k=n}^{+\infty} A^n R_- A^{-(k+1)} u_k + \sum_{k=0}^{n-1} A^n R_+ A^{-(k+1)} u_k, \quad n \geq 0; \tag{33}$$

here and below we always set $\sum_{k=0}^{-1} = 0$, and so the RHS of (33) for $n = 0$ is a vector from $\operatorname{ran} R_- = \operatorname{ran} Q_+$ as required in (13). Formula (33) can be either taken from [5] or checked directly by multiplying $S^{-1} - A$ and the operator in (33) and taking into account that $\operatorname{ran} R_+ = \operatorname{ran} P_+$ is the set of the initial data for the *bounded* on $\mathbb{Z}_+$ solutions of the difference equation $y_{n+1} = Ay_n$, the projection P_+ is a dichotomy projection for this equation on $\mathbb{Z}_+$, and therefore the projection R_+ onto $\operatorname{ran} P_+$ parallel to $\operatorname{ran} Q_+$ is also a dichotomy projection since $\operatorname{ran} P_+ \oplus \operatorname{ran} Q_+ = \mathbb{C}^2$, cf. [4–6] and also [10, 11] or [13, Chapter 4] for discussions of dichotomies on the half-line.

Formula (33) shows that the operator $\mathcal{T}_\lambda^+$ in (14) is an operator with semi-separable kernel, that is, $\mathcal{T}_\lambda^+$ can be written as an infinite matrix $\mathcal{T}_\lambda^+ = [T_{nk}]_{n,k=0}^{+\infty}$ so that $(\mathcal{T}_\lambda^+\mathbf{u})_n = \sum_{k=0}^{+\infty} T_{nk}u_k$, where

$$T_{nk} = \begin{cases} -C_n A^n R_- A^{-(k+1)} B_k & \text{for } 0 \leq n \leq k, \\ -C_n A^n R_+ A^{-(k+1)} B_k & \text{for } 0 \leq k < n. \end{cases}$$

We refer to [28, Chapter IX] and [23] for a discussion of the operators with semi-separable kernels; the results therein are used below in the proof of the second equality in (21).

The matrix-valued Jost solution $\mathbf{Y}^+ = (Y_n^+)_{n\geq 0}$ of (8) (with no boundary condition) is obtained as a solution to the following Volterra equation,

$$Y_n^+ - A^n R_+ = -\sum_{k=n}^{+\infty} A^{n-(k+1)} B_k C_k Y_k^+, \tag{34}$$

defined first for $n \geq N$ with N large enough and then extended to all of $\mathbb{Z}_+$ as a solution to the difference equation (8) via $Y_n^+ := (A_n^\times)^{-1} Y_{n+1}^+$ for $n = 0, 1, \ldots, N-1$. The sequence $(Y_n^+)_{n\geq 0}$ thus defined will satisfy the Volterra equation (34) as the following inductive step shows: Suppose we know that (34) holds for $n = N$ and that $B_{N-1}C_{N-1}Y_{N-1}^+ = Y_N^+ - AY_{N-1}^+$. We then use this and (34) with $n = N$ in the following calculation yielding (34) for $n = N-1$,

$$A^{N-1}R_+ - \sum_{k=N-1}^{+\infty} A^{(N-1)-(k+1)} B_k C_k Y_k^+$$
$$= A^{-1}\big(A^N R_+ - \sum_{k=N}^{+\infty} A^{N-(k+1)} B_k C_k Y_k^+ - Y_N^+\big) + Y_{N-1}^+ = Y_{N-1}^+.$$

Lemma 1 *There is a large enough N such that Eq.* (34) *for $n \geq N$ has a unique solution thus yielding the matrix-valued Jost solution $\mathbf{Y}^+ = (Y_n^+)_{n\geq 0}$ of* (8) *satisfying* (15).

Proof We recall that $|\mu_+| < 1$ and introduce the space

$$\ell_N^\infty := \big\{\mathbf{u} = (u_n)_{n\geq N} : \|\mathbf{u}\|_{\ell_N^\infty} := \sup\big\{|\mu_+|^{-n}\|u_n\|_{\mathbb{C}^2} : n \geq N\big\} < \infty\big\}$$

of exponentially decaying at infinity $\mathbb{C}^2$-valued sequences on $[N, +\infty) \cap \mathbb{Z}_+$. Let T_N denote the operator on ℓ_N^∞ that appears in the RHS of (34),

$$(T_N\mathbf{u})_n = -\sum_{k=n}^{+\infty} A^{n-(k+1)} B_k C_k u_k, \quad \mathbf{u} = (u_n)_{n\geq N} \in \ell_N^\infty. \tag{35}$$

Denoting $q_N = \sum_{n=N}^{+\infty} \|B_k C_k\|_{\mathbb{C}^{2\times 2}}$, we have $q_N \to 0$ as $N \to \infty$ by (3) and (11). Using (30) we have[1] $\|A^{n-(k+1)}\|_{\mathbb{C}^{2\times 2}} \lesssim |\mu_+|^{n-(k+1)}$ for $k \geq n$. Thus, for $n \geq N$,

$$|\mu_+|^{-n}\|(T_N\mathbf{u})_n\|_{\mathbb{C}^2} \lesssim \sum_{k=n} |\mu_+|^{-n}|\mu_+|^{n-(k+1)}\|B_k C_k\|_{\mathbb{C}^{2\times 2}}\|u_k\|_{\mathbb{C}^2} \lesssim q_N\|\mathbf{u}\|_{\ell_N^\infty} \tag{36}$$

[1] We write $a \lesssim b$ if $a \leq cb$ for a constant c independent of any parameters contained in a and b.

and the norm of T_N in ℓ_N^∞ is dominated by q_N and therefore is less than, say, $1/2$ provided N is large enough. Then $(Y_n^+)_{n\geq N} = (I - T_N)^{-1}(A^n R_+)_{n\geq N}$ is the unique solution of (34) whose columns are in ℓ_N^∞ and so for $n \geq N$ we infer

$$\begin{aligned}|\mu_+|^{-n}\|Y_n^+ - A^n R_+\|_{\mathbb{C}^{2\times 2}} &\leq \|(Y_n^+)_{n\geq N} - (A^n R_+)_{n\geq N}\|_{\ell_N^\infty} = \|T_N (Y_n^+)_{n\geq N}\|_{\ell_N^\infty} \\ &\lesssim q_N \|(Y_n^+)_{n\geq N}\|_{\ell_N^\infty} = q_N \|(I - T_N)^{-1}(A^n R_+)_{n\geq N}\|_{\ell_N^\infty} \\ &\lesssim q_N \|(A^n R_+)_{n\geq N}\|_{\ell_N^\infty} \lesssim q_N\end{aligned}$$

because $\|(A^n R_+)_{n\geq N}\|_{\ell_N^\infty} \lesssim 1$ since ran $R_+ =$ ran P_+. This yields the first assertion in (15) while the second follows by multiplying (34) by R_+ from the right and using the uniqueness of the solution $(Y_n^+)_{n\geq N} = (Y_n^+ R_+)_{n\geq N}$. □

Remark 1 Assertion $Y_n^+ = Y_n^+ R_+$, cf. (15), and formula (29) for R_+ show that the first column of the matrix Y_n^+ is zero. Since $(Y_n^+)_{n\geq 0}$ solves the difference equation (8), we conclude that there is a sequence $(z_n)_{n\geq -1}$ that solves the difference equation (1) (with no boundary condition) such that

$$Y_n^+ = \begin{bmatrix} 0 & z_n \\ 0 & z_{n-1} \end{bmatrix}, \quad n \geq 0. \tag{37}$$

Next, we proceed to discuss the Jost solution $\mathbf{z}^+ = (z_n^+)_{n\geq -1}$ obtained as a solution to the following scalar Volterra equation,

$$z_n^+ - (\mu_+)^n = -(\mu_+ - \mu_-)^{-1} \sum_{k=n}^{+\infty} b_k c_k \big((\mu_+)^{n-k} - (\mu_-)^{n-k}\big) z_k^+, \tag{38}$$

at first for $n \geq N$ with N sufficiently large. A computation using $(\mu_+)^2 + \lambda\mu_+ - 1 = 0$ shows that z_n^+ from (38) satisfy (1), see [29] for a similar computation. Therefore, the solution $(z_n^+)_{n\geq N}$ to (38) can be propagated backward as solution to (1), and thus z_n^+ can be defined for all $n \geq 1$. We now record properties of the Jost solution.

Remark 2 The Jost solution $\mathbf{z}^+(\lambda)$ is unique and holomorphic in λ. This follows by a standard argument, cf. [12, 25] and the references therein, presented in [38, Remark 2.2] and based on passing in (38) to the new unknowns z_n^+/μ_+^n and proving that the RHS of the resulting equation is a strict contraction in the space of exponentially decaying at infinity sequences, see the analogous proof of Lemma 1. As in the lemma, this yields the property (17).

We stress that the Jost solution $\mathbf{z}^+ = (z_n^+(\lambda))_{n\geq -1}$ is defined for $n \geq -1$ as the solution of the difference equation (1), with no boundary condition (2). Theorem 1 shows, in particular, that $z_{-1}^+(\lambda) = 0$ if and only if λ is an eigenvalue (which is not surprising as in this case $\mathbf{z}^+(\lambda)$ is the exponentially decaying solution to (1) that also satisfies the boundary condition (2)).

Remark 3 Let us consider the difference equation (4) which is a particular case of (1) with b_n, c_n as in (6) and ρ_n satisfying (5). Then the proof of [38, Lemma 2.3] applies and shows that the Jost solution satisfies $z_n^+ \neq 0$ for all $n \geq 0$.

Proof of Theorem 1 The three equalities in (21) are proved as follows.

1. We prove $\det(I - K_\lambda^+) = \det(I - \mathcal{T}_\lambda^+)$. We claim that K_λ^+ from (12) is the $(1, 1)$-block of the operator $\mathcal{T}_\lambda^+$ from (14) in the decomposition

$$\ell^2(\mathbb{Z}_+; \mathbb{C}^2) = \operatorname{ran}(\operatorname{diag}\{Q_+\}) \oplus \operatorname{ran}(\operatorname{diag}\{Q_-\}),$$

with $\operatorname{diag} = \operatorname{diag}_{\mathbb{Z}_+}$, for the projections $Q_\pm$ from (11), that is, that

$$\operatorname{diag}\{C_n\}(S^{-1} - A)^{-1} \operatorname{diag}\{B_n\} = \begin{bmatrix} -\operatorname{diag}\{c_n\}(S - S^{-1} - \lambda)^{-1} \operatorname{diag}\{b_n\} & 0 \\ 0 & 0 \end{bmatrix}. \tag{39}$$

Clearly, this implies the required equality.

To begin the proof of the claim, we fix $\mathbf{u} = (u_n)_{n\geq 0} \in \ell^2(\mathbb{Z}_+; \mathbb{C}^2)$ and denote $v_n = \big(\operatorname{diag}_{n\in\mathbb{Z}_+}\{Q_+\}(S^{-1} - A) \operatorname{diag}_{n\in\mathbb{Z}_+}\{Q_+\}\mathbf{u}\big)_n$. Formulas (11) and (31) show that $AR_+ = \mu_+ R_+$ and $R_- A^{-1} = -\mu_+ R_-$, and thus (33) and (30) imply that v_n is equal to

$$-\sum_{k=n}^{+\infty} (-\mu_+)^{k+1} Q_+ W \widetilde{A}^n W^{-1} R_- Q_+ u_k + \sum_{k=0}^{n-1} \mu_+^n Q_+ R_+ W \widetilde{A}^{-(k+1)} W^{-1} Q_+ u_k. \tag{40}$$

We now use formulas (29) and (31) to compute the matrices $Q_+ W$, $W^{-1} R_- Q_+$, $Q_+ R_+ W$, $W^{-1} Q_+$. Plugging this into (40) and using that $\mu_\pm$ solve the equation $\mu^2 + \lambda\mu - 1 = 0$, after a tedious computation we conclude that the first component of the vector $v_n = Q_+ v_n \in \mathbb{C}^2$ is given by the formula

$$(\mu_+ - \mu_-)^{-1} \Big(\sum_{k=n}^{+\infty} (\mu_-)^{n-k} \big(1 - \big(\tfrac{\mu_+}{\mu_-}\big)^{n+1}\big) z_k + \sum_{k=0}^{n-1} (\mu_+)^{n-k} \big(1 - \big(\tfrac{\mu_+}{\mu_-}\big)^{k+1}\big) z_k \Big).$$

We now use (32) to recognize that the last expression is $((S - S^{-1} - \lambda)^{-1}\mathbf{z})_n$ where z_n denote the first component of the vector $u_n \in \mathbb{C}^2$, $n \geq 0$. Multiplying by c_n and b_n and recalling (11) yields the required claim (39).

2. We prove $\det(I - \mathcal{T}_\lambda^+) = \mathcal{E}^+(\lambda)$ in three steps. The first step is to show that it suffices to prove the equality only for finitely supported $(B_n)_{n\geq 0}$ and $(C_n)_{n\geq 0}$. This is fairly standard and follows, say, as the *claim* in the proof of [12, Theorem 4.6] or the proof of [23, Theorem 4.3]: Indeed, replace B_k, C_k in (35) by $\mathbf{1}_M(k)B_k$, $\mathbf{1}_M(k)C_k$ where $\mathbf{1}_M$ is the characteristic function of $[0, M]$ equals to

one on the segment and to zero outside. As in the proof of Lemma 1, we denote the respective operator by $T_N^{(M)}$ and the respective matrix-valued Jost solution by $\mathbf{Y}^{+,M}$. Choosing a large N and even larger M the proof of Lemma 1 yields $\|T_N - T_N^{(M)}\|_{\mathcal{B}(\ell_N^\infty)} \lesssim q_{M+1} \to 0$ as $M \to \infty$ for the operator norm on ℓ_N^∞. Then $\|Y_0^+ - Y_0^{+,M}\|_{\mathbb{C}^{2\times 2}} \to 0$ as in the proof of Lemma 1 and therefore $\mathcal{E}^{+,M}(\lambda) \to \mathcal{E}^+(\lambda)$ as $M \to \infty$ for the Evans function $\mathcal{E}^{+,M}$ obtained by replacing Y_0^+ in (16) by $Y_0^{+,M}$. Analogously, let $\mathcal{T}_\lambda^{+,M}$ denote the operator in (14) obtained by replacing B_k, C_k by $\mathbf{1}_M(k)B_k$, $\mathbf{1}_M(k)C_k$. Then (3) yields convergence of $\mathcal{T}_\lambda^{+,M}$ to $\mathcal{T}_\lambda^+$ in the trace class norm and so $\det(I - \mathcal{T}_\lambda^{+,M}) \to \det(I - \mathcal{T}_\lambda^+)$ as $M \to \infty$ completing the first step in the proof. From now on we thus assume that sequences $(B_n)_{n\geq 0}$, $(C_n)_{n\geq 0}$ are finitely supported.

The second step in the proof is a reduction of the infinite dimensional determinant $\det(I - \mathcal{T}_\lambda^+)$ to a finite dimensional one. It follows a well established path, see [23–25, 28]. Multiplying the operator in the RHS of (33) by $\operatorname{diag}_{n\in\mathbb{Z}_+}\{C_n\}$ and $\operatorname{diag}_{k\in\mathbb{Z}_+}\{B_k\}$ and adding and subtracting $\sum_{k=n}^{+\infty} C_n A^n R_- A^{-(k+1)} B_k u_k$ to the result we arrive at the identity

$$\mathcal{T}_\lambda^+ = H_+ + H_2 H_3, \text{ where } H_+ = H_0 + H_1,$$

and we introduce notations

$$H_0 = -\operatorname{diag}_{n\in\mathbb{Z}_+}\{C_n A^{-1} B_n\}, \quad (H_1\mathbf{u})_n = -\sum_{k=n+1}^{+\infty} C_n A^{n-(k+1)} B_k u_k,$$

$$(H_2 y)_n = C_n A^n R_+ y, \quad H_3\mathbf{u} = R_+ \sum_{k=0}^{+\infty} A^{-(k+1)} B_k u_k$$

for $y \in \mathbb{C}^2$ and $\mathbf{u} = (u_n)_{n\geq 0} \in \ell^2(\mathbb{Z}_+; \mathbb{C}^2)$. Here, the operators H_0 and H_1 in $\ell^2(\mathbb{Z}_+; \mathbb{C}^2)$ are of trace class by (3) while the operators H_2 and H_3 are of rank one as H_2 acts from $\operatorname{ran} R_+ = \operatorname{ran} P_+ \subset \mathbb{C}^2$ into $\ell^2(\mathbb{Z}_+; \mathbb{C}^2)$ while H_3 acts from $\ell^2(\mathbb{Z}_+; \mathbb{C}^2)$ into $\operatorname{ran} R_+ = \operatorname{ran} P_+ \subset \mathbb{C}^2$. The diagonal operator $I - H_0$ is invertible with $\det(I - H_0)$ being equal to 1 by (10) because

$$\prod_{n=0}^{+\infty} \det(I + C_n A^{-1} B_n) = \prod_{n=0}^{+\infty} \det(I + A^{-1} B_n C_n) = \prod_{n=0}^{+\infty} \det(A^{-1}) \det(A + B_n C_n).$$

The operator H_1 is block-lower-triangular, and thus $I - H_+$ is invertible with $\det(I - H_+) = 1$. Writing $I - \mathcal{T}_\lambda^+ = (I - H_+)\big(I - (I - H_+)^{-1} H_2 H_3\big)$ and changing the order of factors in the determinant, we arrive at the identity

$$\det(I_{\ell^2(\mathbb{Z}_+;\mathbb{C}^2)} - \mathcal{T}_\lambda^+) = \det_{\mathbb{C}^{2\times 2}}\big(I_{2\times 2} - H_3(I - H_+)^{-1} H_2\big) \tag{41}$$

that reduces the computation of the infinite dimensional determinant for $I - \mathcal{T}_\lambda^+$ to the finite dimensional determinant for the operator $H_3(I - H_+)^{-1}H_2$ acting in ran R_+ thus completing the second step in the proof.

The third step relates the RHS of (41) and the solution $(Y_n^+)_{n\geq 0}$ to the Volterra equation (34). Indeed, for any $y \in \mathbb{C}^2$ we let $u_n = C_n Y_n^+ y$ and $\mathbf{u} = (u_n)_{n\geq 0}$. Multiplying (34) from the left by C_n and applying the resulting matrices to y yields

$$u_n = C_n A^n R_+ y - \sum_{k=n}^{+\infty} C_n A^{n-(k+1)} B_k u_k = (H_2 y)_n + (H_+\mathbf{u})_n, \quad n \in \mathbb{Z}_+,$$

or $\mathbf{u} = (I - H_+)^{-1}H_2 y$. Multiplying by H_3 from the left, by the definition of H_3,

$$H_3(I - H_+)^{-1}H_2 y = H_3\mathbf{u} = R_+ \sum_{k=0}^{+\infty} A^{-(k+1)} B_k u_k.$$

Since y is arbitrary, the respective matrices are equal, that is,

$$H_3(I - H_+)^{-1}H_2 = R_+ \sum_{k=0}^{+\infty} A^{-(k+1)} B_k C_k Y_k^+.$$

Since $Y_n^+ = Y_n^+ R_+$ by (15), the last identity and formula (41) yield

$$\det(I - \mathcal{T}_\lambda^+) = \det\nolimits_{\mathbb{C}^{2\times 2}} \Big(I_{2\times 2} - R_+ \sum_{k=0}^{+\infty} A^{-(k+1)} B_k C_k Y_k^+ R_+\Big). \tag{42}$$

Writing matrices in the block form using the direct sum decomposition $\mathbb{C}^2 = \operatorname{ran} R_+ \oplus \operatorname{ran} R_-$ gives the equality of the determinants of the following two matrices,

$$I_{2\times 2} - \sum_{k=0}^{+\infty} A^{-(k+1)} B_k C_k Y_k^+ R_+ = \begin{bmatrix} R_+ - R_+ \sum_{k=0}^{+\infty} A^{-(k+1)} B_k C_k Y_k^+ R_+ & 0 \\ -R_- \sum_{k=0}^{+\infty} A^{-(k+1)} B_k C_k Y_k^+ R_+ & R_- \end{bmatrix},$$

$$I_{2\times 2} - R_+ \sum_{k=0}^{+\infty} A^{-(k+1)} B_k C_k Y_k^+ R_+$$

$$= \begin{bmatrix} R_+ - R_+ \sum_{k=0}^{+\infty} A^{-(k+1)} B_k C_k Y_k^+ R_+ & 0 \\ 0 & R_- \end{bmatrix}.$$

This, (34) with $n = 0$, and (42) show that $\det(I - \mathcal{T}_\lambda^+)$ is equal to the determinant of the matrix

$$I_{2\times 2} - \sum_{k=0}^{+\infty} A^{-(k+1)} B_k C_k Y_k^+ R_+ = R_+ - \sum_{k=0}^{+\infty} A^{-(k+1)} B_k C_k Y_k^+ + R_- = Y_0^+ + R_-,$$

completing the proof of the equality $\det(I - \mathcal{T}_\lambda^+) = \mathcal{E}^+(\lambda)$ by (16).

3. We prove $\mathcal{E}^+(\lambda) = \mathcal{F}^+(\lambda)$. Using the Jost solution $(z_n^+)_{n\geq -1}$ and the regular solution satisfying $z_0^{\mathrm{r}} = 1$, we recall that $\mathcal{W}(\mathbf{z}^+, \mathbf{z}^{\mathrm{r}})_0 = z_{-1}^+$ in (18). By Remark 1 the solution $(Y_n^+)_{n\geq 0}$ is of the form (37) with some $(z_n)_{n\geq -1}$. We claim that the solution $(z_n)_{n\geq -1}$ in formula (37) for Y_n^+ satisfies $z_n = \mu_+ z_n^+$ where $(z_n^+)_{n\geq -1}$ is the Jost solution from (38). Assuming the claim, we use formulas (16), (37), (31), and (18) to obtain the desired result,

$$\mathcal{E}^+(\lambda) = \det(Y_0^+ + R_-) = \det\begin{bmatrix} 1 & -\mu_+ + \mu_+ z_0^+ \\ 0 & \mu_+ z_{-1}^+ \end{bmatrix} = \mu_+(\lambda) z_{-1}^+(\lambda) = \mathcal{F}^+(\lambda).$$

It remains to prove the claim $z_n = \mu_+ z_n^+$ in formula (37). Using (29) we multiply Eq. (34) by W^{-1} from the left and W from the right passing to the new unknowns $\widetilde{Y}_n^+$ and using (29) and $(z_n)_{n\geq -1}$ from (37),

$$\widetilde{Y}_n^+ = W^{-1} Y_n^+ W = (\mu_+ - \mu_-)^{-1} \begin{bmatrix} z_n - \mu_- z_{n-1} & z_n - \mu_- z_{n-1} \\ -(z_n - \mu_+ z_{n-1}) & -(z_n - \mu_+ z_{n-1}) \end{bmatrix}.$$

Noting that $A^n R_+ = \mu_+^n R_+$ by (11) and (31), and explicitly computing all matrices that appear in the equation for $\widetilde{Y}_n^+$, we arrive at the following equations,

$$(\mu_+ - \mu_-)^{-1}\big(z_n - \mu_- z_{n-1}\big) = \mu_+^n - (\mu_+ - \mu_-)^{-1} \sum_{k=n}^{+\infty} b_k c_k \mu_+^{n-(k+1)} z_k,$$

$$(\mu_+ - \mu_-)^{-1}\big(-z_n + \mu_+ z_{n-1}\big) = -(\mu_+ - \mu_-)^{-1} \sum_{k=n}^{+\infty} b_k c_k \big(-\mu_-^{n-(k+1)}\big) z_k.$$

Multiplying the first equation by μ_+ and the second by μ_- and adding the results,

$$z_n = \mu_+^{n+1} - (\mu_+ - \mu_-)^{-1} \sum_{k=n}^{+\infty} b_k c_k \big(\mu_+^{n-k} - \mu_-^{n-k}\big) z_k.$$

Divided by μ_+, this is Eq. (38) for z_n/μ_+ and the uniqueness of the solution of the equation yields the claim thus completing the proof of (21) in the theorem.

4. To prove the last assertion in the theorem, we rely on the Birman-Schwinger principle saying that $\lambda \in \operatorname{Spec}\big(S - S^{-1} + \operatorname{diag}_{n\in\mathbb{Z}_+}\{b_n c_n\}\big)$ if and only if λ is a zero of $\det(I - K_\lambda^+)$ with the same multiplicity which follows from the fact that the operator $S - S^{-1} + \operatorname{diag}_{n\in\mathbb{Z}_+}\{b_n c_n\} - \lambda$ can be written as the product

$$\big(S - S^{-1} - \lambda\big)\big(I_{\ell^2(\mathbb{Z}_+;\mathbb{C})} - \big(S - S^{-1} - \lambda\big)^{-1}\operatorname{diag}_{n\in\mathbb{Z}_+}\{b_n\} \times \operatorname{diag}_{n\in\mathbb{Z}_+}\{c_n\}\big)$$

and the standard property $\det(I - K_1 \times K_2) = \det(I - K_2 \times K_1)$ of the operator determinants. That the eigenvalues λ are simple follows from the fact that $\mathbf{z}$ is an eigensequence if and only if $\mathbf{z}$ is proportional to $\mathbf{z}^+(\lambda)$ with λ satisfying $z_{-1}^+(\lambda) = 0$ (in other words, the dichotomy projection for (8) has rank one).

□

Proof of Corollary 1 We follow [38] and show that $\mathcal{F}^+(\lambda) = \mathcal{G}^+(\lambda)$ for $\mathcal{F}^+$ from (18) and $\mathcal{G}^+$ from (20). Here, we assume that $\operatorname{Re}(\lambda) > 0$. The function $\mathcal{F}^+$ is holomorphic in λ by Remark 2. The convergent continued fraction in (19) is holomorphic in λ by the classical Stieltjes-Vitali Theorem [32, Theorem 4.30]. Thus, it is enough to show the desired equality only for $\lambda > 0$ which we assume from now on. We recall that $z_n^+ \neq 0$ for $n \geq 0$ by Remark 3, introduce $v_n^\pm = z_{n-1}^\pm / z_n^\pm$ and re-write Eq. (4) for $\mathbf{z}^\pm$ as $v_n^+ = \frac{\lambda}{\rho_n} + \frac{1}{v_{n+1}^+}$ for $n \geq 0$. The last formula iterated forward produces the continued fraction (19). Because the continued fraction converges and $\lambda > 0$, an argument from [17, pp.2063] involving monotonicity of the sequences formed by the odd and even truncated continued fractions yields $v_0^+ = g_+(\lambda) + \lambda/\rho_0$. Therefore,

$$\mathcal{W}(\mathbf{z}^+, \mathbf{z}^{\mathrm{r}})_0 = z_{-1}^+ = z_0^+ v_0^+ = z_0^+ (g_+(\lambda) + \lambda/\rho_0),$$

and the desired equality of $\mathcal{F}^+$ and $\mathcal{G}^+$ follows by (18) and (20). □

Lemma 2 *Assume* (27) *and* $\lambda > 0$. *Then the solutions of* (28) *on* $\mathbb{Z}$ *are in one-to-one correspondence with the* $\ell^2(\mathbb{Z}_+;\mathbb{C})$*-solutions to the eigenvalue problem* (4) *given by* $z_n = \rho_n w_n$ *for* $n \geq -1$ *and* $w_n = 0$ *for* $n \leq -2$.

Proof Suppose that $(w_n)_{n\in\mathbb{Z}}$ solves (28) and let $z_n = \rho_n w_n$ for all $n \in \mathbb{Z}$. Then $z_{-1} = 0$ since $\rho_{-1} = 0$ and the sequence $(z_n)_{n\leq -1}$ solves the equation

$$z_{n-1} - z_{n+1} = \lambda z_n/\rho_n \text{ for } n \leq -2 \text{ and } z_{-1} = 0.$$

If $z_{-2} \neq 0$ then the sequence $(z_n)_{n\leq -2}$ due to $\lambda > 0$ exponentially grows as $n \to -\infty$ as shown in the proof of [38, Lemma 2.3] contradicting $(w_n)_{n\in\mathbb{Z}} \in \ell^2(\mathbb{Z};\mathbb{C})$ in (28). So one has $z_{-2} = 0$, $z_{-1} = 0$ and, as a result, $z_n = 0$ for all $n \leq -2$. This, in turn, implies that $w_n = 0$ for all $n \leq -2$.

Conversely, suppose that $(z_n)_{n\geq 0} \in \ell_2(\mathbb{Z}_+)$, $z_{-1} = 0$ solves (4) and let $w_n = z_n/\rho_n$ for $n \geq 0$, $w_{-1} = -z_0/\lambda$, and $w_n = 0$ for $n \leq -2$. Then $(w_n)_{n\in\mathbb{Z}}$ so defined solves (28). □

Proof of Corollary 2 We follow the proof of [38, Corollary 1.3]. Due to Lemma 2, Theorem 1 yields the first assertion in the corollary. To finish the proof it suffices to show that there is a positive root of the function $\mathcal{G}^+$ defined in (20), equivalently, that for some $\lambda > 0$ and g_+ defined in (19) one has $-\lambda/\rho_0 = g_+(\lambda)$. We recall that $\rho_0 < 0$ and $g_+(\lambda) > 0$ for $\lambda > 0$ because condition (5) holds. The proof is completed by using the relations $\lim_{\lambda\to 0^+} g_+(\lambda) = 1$ and $\lim_{\lambda\to +\infty} g_+(\lambda) = 0$ established in [17, Lemma 2.10(4)]. □

Acknowledgments We thank the (anonymous) referee for their comments which improved the exposition of the paper. Y.L. was supported by the NSF grant DMS-2106157, and would like to thank the Courant Institute of Mathematical Sciences at NYU and especially Prof. Lai-Sang Young for their hospitality.

References

1. Beck, M., Wayne, C.E.: Metastability and rapid convergence to quasi-stationary bar states for the two-dimensional Navier-Stokes equations. Proc. R. Soc. Edinb. Sect. A Math. **143**, 905–927 (2013)
2. Behrndt, J., ter Elst, A.F.M., Gesztesy, F.: The generalized Birman-Schwinger principle. Trans. Amer. Math. Soc. **375**, 799–845 (2022)
3. Belenkaya, L., Friedlander, S., Yudovich, V.: The unstable spectrum of oscillating shear flows. SIAM J. App. Math. **59**, 1701–1715 (1999)
4. Ben-Artzi, A., Gohberg, I.: Dichotomy of systems and invertibility of linear ordinary differential operators. Oper. Theory Adv. Appl. **56**, 90–119 (1992)
5. Ben-Artzi, A., Gohberg, I., Kaashoek, M.A.: Invertibility and dichotomy of singular difference equations. Oper. Theory Adv. Appl. **48**, 157–184 (1990)
6. Ben-Artzi, A., Gohberg, I., Kaashoek, M.A.: Invertibility and dichotomy of differential operators on a half line. J. Dynam. Differ. Equ. **5**, 1–36 (1993)
7. Bögli, S., Stampach, F.: On Lieb-Thirring inequalities for one-dimensional non-self-adjoint Jacobi and Schrödinger operators. J. Spectr. Theory **11**, 1391–1413 (2021)
8. Carey, A., Gesztesy, F., Levitina, G., Nichols, R., Sukochev, F., Zanin, D.: The Limiting Absorption Principle for Massless Dirac Operators, Properties of Spectral Shift Functions, and an Application to the Witten Index of Non-Fredholm Operators. Mem. European Math. Soc., vol. 4. European Math. Soc., Berlin (2023)
9. Chadan, K., Sabatier, P.C.: Inverse Problems in Quantum Scattering Theory, 2nd edn. Springer, New York (1989)
10. Chicone, C., Latushkin, Y.: Evolution Semigroups in Dynamical Systems and Differential Equations. Math. Surv. Monogr., vol. 70. AMS, Providence (1999)
11. Coppel, W.A.: Dichotomies in Stability Theory. Lect. Notes Math., vol. 629. Springer, New York (1978)
12. Cramer, D., Latushkin, Y.: Fredholm determinants and the Evans function for difference equations. In: Perspectives in Operator Theory. Banach Center Publ., vol. 75, pp. 111–135. Polish Acad. Sci. Inst. Math., Warsaw (2007)
13. Daleckij, J., Krein, M.: Stability of Differential Equations in Banach Space. Amer. Math. Soc., Providence, RI (1974)
14. Dullin, H., Marangell, R.: An Evans function for the linearised 2D Euler equations using Hill's determinant. Phys. D **457**, Paper No. 133954, 14 pp. (2024)
15. Dullin, H.R., Worthington, J.: Stability results for idealized shear flows on a rectangular periodic domain. J. Math. Fluid Mech. **20**, 473–484 (2018)
16. Dullin, H.R., Marangell, R., Worthington, J.: Instability of equilibria for the 2D Euler equations on the torus. SIAM J. Appl. Math. **76**, 1446–1470 (2016)

17. Dullin, H., Latushkin, Y., Marangell, R., Vasudevan, S., Worthington, J.: Instability of the unidirectional flows for the 2D α-Euler equations. Comm. Pure. Appl. Anal. **19**, 2051–2079 (2020)
18. Egorova, I., Golinskii, L.: On the location of the discrete spectrum for complex Jacobi matrices. Proc. Amer. Math. Soc. **133**, 3635–3641 (2005)
19. Frank, R.L., Laptev, A., Weidl, T.: Schrödinger Operators: Eigenvalues and Lieb-Thirring Inequalities. Cambridge University Press, Cambridge (2023)
20. Frenkel, A.L., Zhang, X.: Large-scale instability of generalized oscillating Kolmogorov flows. SIAM J. Appl. Math. **58**, 540–564 (1998)
21. Friedlander, S., Howard, L.: Instability in parallel flows revisited. Stud. Appl. Math. **101**, 1–21 (1998)
22. Friedlander, S., Vishik, M., Yudovich, V.: Unstable eigenvalues associated with inviscid fluid flows. J. Math. Fluid. Mech. **2**, 365–380 (2000)
23. Gesztesy, F., Makarov, K.: (Modified) Fredholm determinants for operators with matrix-valued semi-separable integral kernels revisited. Integr. Equ. Oper. Theory **48**, 561–602 (2004)
24. Gesztesy, F., Nichols, R.: A Jost-Pais-type reduction of (modified) Fredholm determinants for semi-separable operators in infinite dimensions. Oper. Theory Adv. Appl. **244**, 287–314 (2015)
25. Gesztesy, F., Latushkin, Y., Makarov, K.: Evans functions, Jost functions, and Fredholm determinants. Arch. Rational Mech. Anal. **186**, 361–421 (2007)
26. Gesztesy, F., Mitrea, M., Zinchenko, M.: Variations on a theme of Jost and Pais. J. Funct. Anal. **253**, 399–448 (2007)
27. Gohberg, I.C., Feldman, I.A.: Convolution Equations and Projection Methods for Their Solution. Transl. Math. Monogr., vol. 41. AMS, Providence, RI (1974)
28. Gohberg, I., Goldberg, S., Kaashoek, M. A.: Classes of Linear Operators. Vol. I. Birkhäuser Verlag, Basel (1990)
29. Golinskii, L.: Volterra-type discrete integral equations and spectra of non-self-adjoint Jacobi operators. Integr. Equ. Oper. Theory **93**, Paper No. 63 (2021)
30. Golinskii, L.: Perturbation determinants and discrete spectra of semi-infinite non-self-adjoint Jacobi operators. J. Spectr. Theory **12**, 835–856 (2022)
31. Janas, J., Naboko, S.: On the point spectrum of some Jacobi matrices. J. Oper. Theory **40**, 113–132 (1998)
32. Jones, W.B., Thron, W.J.: Continued Fractions: Analytic Theory and Applications. Cambridge University Press, Cambridge (1984)
33. Jost, R., Pais, A.: On the scattering of a particle by a static potential. Phys. Rev. **82**, 840–851 (1951)
34. Kapitula, T., Promislow, K.: Spectral and Dynamical Stability of Nonlinear Waves. Appl. Math. Sci., vol. 185. Springer, New York (2013)
35. Krejcirik, D., Laptev, A., Stampach, F.: Spectral enclosures and stability for non-self-adjoint discrete Schrödinger operators on the half-line. Bull. Lond. Math. Soc. **54**, 2379–2403 (2022)
36. Latushkin, Y., Vasudevan, S.: Eigenvalues of the linearized 2D Euler equations via Birman-Schwinger and Lin's operators. J. Math. Fluid. Mech. **20**, 1667–1680 (2018)
37. Latushkin, Y., Vasudevan, S.: Stability criteria for the 2D α-Euler equations. J. Math. Anal. Appl. **472**, 1631–1659 (2019)
38. Latushkin, Y., Vasudevan, S.: Fredholm determinants, continued fractions, Jost and Evans functions for a Jacobi matrix associated with the 2D-Euler equations. Oper. Theory Adv. Appl. **297**, 207–221 (2024)
39. Latushkin, Y., Li, Y.C., Stanislavova, M.: The spectrum of a linearized 2D Euler operator. Studies Appl. Math. **112**, 259–270 (2004)
40. Li, Y. (Charles): On 2D Euler equations. I. On the energy-Casimir stabilities and the spectra for linearized 2D Euler equations. J. Math. Phys. **41**, 728–758 (2000)
41. Lin, Z.: Some stability and instability criteria for ideal plane flows. Comm. Math. Phys. **246**, 87–112 (2004)

42. Meshalkin, L.D., Sinai, Ia.G.: Investigation of the stability of a stationary solution of a system of equations for the plane movement of an incompressible viscous liquid. J. Appl. Math. Mech. **25**, 1700–1705 (1961)
43. Newton, R.G.: Relation between the three-dimensional Fredholm determinant and the Jost function. J. Math. Phys. **13**, 880–883 (1972)
44. Sandstede, B.: Stability of travelling waves. In: Hasselblatt, B., Katok, A. (eds.) Handbook of Dynamical Systems, vol. 2, pp. 983–1055. North-Holland/Elsevier, Amsterdam (2002)
45. Stampach, F.: The characteristic function for complex doubly infinite Jacobi matrices. Integr. Equ. Oper. Theory **88** 501–534 (2017)
46. Stampach, F., Stovicek, P.: The characteristic function for Jacobi matrices with applications. Linear Algebra Appl. **438**, 4130–4155 (2013)
47. Stovicek, P.: On infinite Jacobi matrices with a trace class resolvent. J. Approx. Theory **249**, 105306 (2020)
48. Vasudevan, S.: Instability of unidirectional flows for the 2D Navier–Stokes equations and related α-models. J. Math. Fluid Mech. **23**, 1–29 (2021)

On a Singular Integral Operator with Two Shifts and Conjugation

Rui C. Marreiros

Dedicated to Professor Yuri I. Karlovich

Abstract On the Hilbert space $\widetilde{L}_2(\mathbb{T})$ the singular integral operator with two shifts and conjugation $K = P_+ + \left[aI + \left(\sum_{j=0}^{m} a_j U_\alpha^j \right) U_\beta C \right] P_-$ is considered, where $P_\pm$ are the Cauchy projectors, a, a_j, $j = \overline{0, m}$, are continuous functions on the unit circle $\mathbb{T}$, U_α and U_β are non-Carleman and Carleman shift operators, respectively, both preserving the orientation on $\mathbb{T}$, and C is the operator of complex conjugation. An estimate for the dimension of the kernel of the operator K is obtained. We also consider the operator $M = [aI + (a_0 I + a_1 U_\alpha) U_\gamma C] P_+ + P_-$, where U_γ is a Carleman shift operator changing the orientation on $\mathbb{T}$.

1 Introduction

Let $\mathbb{T}$ denote the unit circle in the complex plane, $\mathbb{T}_+$ and $\mathbb{T}_-$ denote the interior and the exterior (∞ included) of $\mathbb{T}$, respectively. On the Hilbert space $L_2(\mathbb{T})$ we consider the singular integral operator (SIO) with Cauchy kernel, defined almost everywhere on $\mathbb{T}$ by

$$(S\varphi)(t) = (\pi i)^{-1} \int_{\mathbb{T}} \varphi(\tau)(\tau - t)^{-1} d\tau,$$

R. C. Marreiros (✉)
Departamento de Matemática, Faculdade de Ciências e Tecnologia, Universidade do Algarve, Faro, Portugal
e-mail: rmarrei@ualg.pt

A. Böttcher et al. (eds.), *Achievements and Challenges in the Field of Convolution Operators*, Operator Theory: Advances and Applications 306,
https://doi.org/10.1007/978-3-031-80486-1_13

where the integral is understood in the sense of its principal value. The operator S is a bounded linear involutive operator ($S^2 = I$, where I is the identity operator on $L_2(\mathbb{T})$). Then it is possible to define in $L_2(\mathbb{T})$ a pair of complementary projection operators,

$$P_{\pm} = \frac{1}{2}(I \pm S),$$

and to decompose $L_2(\mathbb{T}) = L_2^+(\mathbb{T}) \oplus \overset{\circ}{L_2^-}(\mathbb{T})$, with $L_2^+(\mathbb{T}) = \operatorname{im} P_+$ and $\overset{\circ}{L_2^-}(\mathbb{T}) = \operatorname{im} P_-$. We also set $L_2^-(\mathbb{T}) = \overset{\circ}{L_2^-}(\mathbb{T}) \oplus \mathbb{C}$.

As usual, $L_\infty(\mathbb{T})$ denotes the space of all essentially bounded functions on $\mathbb{T}$. Let us introduce the concept of matrix function generalized factorization (see, for instance, [2] and [16]); we say that a matrix function $c \in L_\infty^{n\times n}(\mathbb{T})$ admits a right (left) generalized factorization in $L_2(\mathbb{T})$, if it can be represented as

$$c = c_-\Lambda c_+, \quad (c = c_+\Lambda c_-) \tag{1}$$

where

$$c_-^{\pm 1} \in \left[L_2^-(\mathbb{T})\right]^{n\times n}, \quad c_+^{\pm 1} \in \left[L_2^+(\mathbb{T})\right]^{n\times n}, \quad \Lambda(t) = \operatorname{diag}\{t^{\varkappa_j}\},$$

$\varkappa_j \in \mathbb{Z}$, $j = \overline{1,n}$, with $\varkappa_1 \geq \varkappa_2 \geq \cdots \geq \varkappa_n$, and $c_-P_+c_+I$ (resp., $c_+P_+c_-I$) represents a bounded linear operator in $L_2^n(\mathbb{T})$. The integers $\varkappa_j$ are uniquely determined by the matrix function c and are called its right (left) partial indices and their sum $\varkappa = \sum_{j=1}^n \varkappa_j$ is called the right (resp., left) total index.

Any non-singular continuous matrix function $c \in C^{n\times n}(\mathbb{T})$ admits a generalized factorization (1) in $L_2(\mathbb{T})$ (see, for instance, the above cited [2] and [16]); for our purposes, it will be assumed that

$$c_\pm^{\pm 1} \in C^{n\times n}(\mathbb{T}). \tag{2}$$

In this setting, left and right total indices of c coincide with the (only) left and right partial index of $\det c$ and are equal to its winding number.

Now let ω be a homeomorphism of $\mathbb{T}$ onto itself, which is differentiable on $\mathbb{T}$ and whose derivative does not vanish there. The function $\omega : \mathbb{T} \to \mathbb{T}$ is called a shift function or simply a shift on $\mathbb{T}$. By

$$\omega_k(t) \equiv \omega[\omega_{k-1}(t)], \quad \omega_1(t) \equiv \omega(t), \quad \omega_0(t) \equiv t, \quad t \in \mathbb{T},$$

we denote the k-th iteration of the shift, $k \geq 2$, $k \in \mathbb{N}$.

A shift ω is called a (generalized) Carleman shift of order $n \in \mathbb{N} \setminus \{1\}$ if $\omega_n(t) \equiv t$, but $\omega_k(t) \not\equiv t$ for $k = \overline{1, n-1}$. Otherwise, if ω is not a Carleman shift, it is called a non-Carleman shift. In what follows we will consider three different shifts, i.e., $\omega = \alpha, \beta, \gamma$: α is a non-Carleman shift, β and γ are Carleman shifts. Let us concretize:

(1) α is the linear fractional non-Carleman shift preserving the orientation on $\mathbb{T}$ with the only fixed point at 1,

$$\alpha(t) = \frac{\mu t + \nu}{\overline{\nu} t + \overline{\mu}}, \qquad t \in \mathbb{T}, \tag{3}$$

where $\mu, \nu \in \mathbb{C}$: $|\mu|^2 - |\nu|^2 = 1$, $|\mathrm{Re}\,\mu| = 1$, and $\mathrm{Im}\,\mu = -i\overline{\nu}$.[1]

The shift function α admits the factorization

$$\alpha(t) = \alpha_+(t) t \alpha_-(t),$$

where

$$\alpha_+(t) = \frac{1}{\overline{\nu} t + \overline{\mu}}, \qquad \alpha_-(t) = \frac{\mu t + \nu}{t} = \overline{(\alpha_+(t))}^{-1}.$$

We see that the functions $\alpha_\pm, \alpha_\pm^{-1}$ are analytic in $\mathbb{T}_\pm$ and continuous in the closure of $\mathbb{T}_\pm$, respectively.

(2) β is the linear fractional Carleman shift preserving the orientation on $\mathbb{T}$,

$$\beta(t) = \frac{t - \lambda}{\overline{\lambda} t - 1}, \qquad t \in \mathbb{T},$$

where $\lambda \in \mathbb{C} \setminus \mathbb{T} : |\lambda| < 1$.

The shift function β admits the factorization

$$\beta(t) = \beta_+(t) t \beta_-(t),$$

where

$$\beta_+(t) = \frac{\sqrt{1 - |\lambda|^2}}{\overline{\lambda} t - 1}, \qquad \beta_-(t) = \frac{t - \lambda}{t\sqrt{1 - |\lambda|^2}}.$$

---

[1] In general, the shift $\alpha(t)$ has two fixed points on $\mathbb{T}$, τ_1 and τ_2, given by the formula

$$\tau_{1,2} = \frac{\mu - \overline{\mu} \pm \sqrt{(\mu + \overline{\mu})^2 - 4}}{2\overline{\nu}},$$

where $\mu, \nu \in \mathbb{C}$: $|\mu|^2 - |\nu|^2 = 1$; obviously $\tau_1 \neq \tau_2$ if $|\mathrm{Re}\,\mu| \neq 1$. With $|\mathrm{Re}\,\mu| = 1$, and $\mathrm{Im}\,\mu = -i\overline{\nu}$, the shift $\alpha(t)$ has one fixed point at 1. To this case corresponds the shift on the real line $\alpha_r(t) = t + \sigma$, $t \in \mathring{\mathbb{R}} = \mathbb{R} \cup \{\infty\}$, σ is a fixed real number; the shift $\alpha_r(t)$ has the only fixed point at infinity.

(3) γ is the linear fractional Carleman shift changing the orientation on $\mathbb{T}$,

$$\gamma(t) = \frac{t-\lambda}{\overline{\lambda}t-1}, \quad t \in \mathbb{T},$$

where $\lambda \in \mathbb{C} \setminus \mathbb{T} : |\lambda| > 1$.

The shift function γ admits the factorization

$$\gamma(t) = \gamma_+(t)t^{-1}\gamma_-(t),$$

where

$$\gamma_+(t) = \frac{t-\lambda}{i\sqrt{|\lambda|^2-1}}, \quad \gamma_-(t) = \frac{it\sqrt{|\lambda|^2-1}}{\overline{\lambda}t-1}.$$

Let a, a_j, $j = \overline{0,m}$, be given continuous functions defined on $\mathbb{T}$. Let $\widetilde{L}_2(\mathbb{T})$ denote the Hilbert space $L_2(\mathbb{T})$ considered over the field of real numbers. On $\widetilde{L}_2(\mathbb{T})$, associated with the shifts α, β, and γ, we consider the shift operators U_α, U_β, and U_γ, defined by

$$(U_\alpha\varphi)(t) = \alpha_+(t)\varphi[\alpha(t)],$$
$$(U_\beta\varphi)(t) = -\beta_+(t)\varphi[\beta(t)],$$
$$(U_\gamma\varphi)(t) = \gamma_-(t)t^{-1}\varphi[\gamma(t)], \quad t \in \mathbb{T}.$$

The shift operators U_ω, $\omega = \alpha, \beta, \gamma$, satisfy the properties:

(i) U_ω is isometric, i.e., $\|U_\omega\varphi\|_{L_2} = \|\varphi\|_{L_2}$;
(ii) $U_\alpha S = SU_\alpha$, $U_\beta S = SU_\beta$, and $U_\gamma S = -SU_\gamma$.
We consider the bounded linear operator of complex conjugation C,

$$(C\varphi)(t) = t^{-1}\overline{\varphi(t)}.$$

The operators $P_\pm$, U_α, U_β, U_γ, and C, verify the properties

$$\begin{aligned} U_\alpha P_\pm = P_\pm U_\alpha, \quad U_\beta P_\pm = P_\pm U_\beta, \quad U_\gamma P_\pm = P_\mp U_\gamma, \\ U_\alpha C = CU_\alpha, \quad U_\beta C = -CU_\beta, \quad U_\gamma C = CU_\gamma, \\ CP_\pm = P_\mp C, \quad U_\alpha U_\beta = U_\beta U_\alpha, \quad U_\alpha U_\gamma = U_\gamma U_\alpha, \\ P_\pm^2 = P_\pm, \quad U_\beta^2 = I, \quad U_\gamma^2 = I, \quad C^2 = I. \end{aligned} \tag{4}$$

In this work we will study the singular integral operators (SIOs) with two shifts and conjugation defined on the unit circle,

$$K = P_+ + \left[aI + \left(\sum_{j=0}^{m} a_j U_\alpha^j \right) U_\beta C \right] P_-, \tag{5}$$

and

$$M = [aI + (a_0 I + a_1 U_\alpha) U_\gamma C] P_+ + P_-. \tag{6}$$

The classical theory of singular integral equations and boundary value problems for analytic functions began with B. Riemann (1857) and had proceeded with D. Hilbert (1904), that, in particular established the relation of the so-called Riemann-Hilbert problem with a Fredholm integral equation. Historically C. Haseman (1907) was the first to consider the boundary value problems with a shift. T. Carleman (1932) also studied a boundary value problems with a shift, that was later designated with his name. N. Vekua (1948) was the first to consider a singular integral equation with a general shift (in the sense, Carleman or non-Carleman shift) (see [19] (first edition in 1950)). In the early 1950s, I. Vekua showed how some mathematical physics problems lead to the solvability of boundary value problems with shift (see [20] (first edition in 1959)). The Fredholm theory of SIOs with Carleman shift was constructed in the 1960s and the 1970s of the twentieth century essentially in the work of D. Kvesevala, M. Krein, I. Gohberg, N. Krupnik, N. Karapetiants, S. Samko, and, most importantly, G. Litvinchuk (see [14]). For the case of non-Carleman shift, the Fredholm theory was completed in the 1980s mainly by Yu. Karlovich, V. Kravchenko and G. Litvinchuk (see [8]). The so-called solvability theory of SIO with shift (the calculation of the defect numbers, the construction of bases for the defect subspaces, and other spectral properties, of a given operator), is an ongoing work, both in the case of Carleman Shift (see [4, 6, 10, 11]), and in the case of non-Carleman shift (see [1, 7, 9, 17]). The solvability problem for SIO with non-Carleman shifts, according to G. Litvinchuk, is "... [a] new and very difficult question ..." (see [15, p. XVI]).

In [12] we studied a generalized Riemann boundary value problem with a non-Carleman shift and conjugation on the real line, through the study of the kernel of the operator $X = P_+ + [aI + (a_0 I + a_1 U_\alpha)C]P_-$ (with one shift, continuous coefficients, and conjugation), and in [13], the operator $Y_r = P_+ + (aI + AC)P_-$, where $A = \sum_{j=0}^{m} a_j U_\alpha^j$ (with iterations of one shift, continuous coefficients, and conjugation). In both cases, for the dimension of its kernel the following estimate was obtained:

$$\dim \ker Y_r \leq l(g) + \max(-\varkappa_1, 0) + \max(-\varkappa_2, 0).$$

We had noted that the influence of the coefficients $a_1, a_2, \ldots, a_m$ is restricted to the term $l(g)$ only; the terms $\varkappa_1$ and $\varkappa_2$ depend only on the coefficients a and a_0. In [3] we considered the operator Y_r on the unit circle, which we denote by Y,

$$Y = P_+ + (aI + AC)P_-, \quad A = \sum_{j=0}^{m} a_j U_\alpha^j, \tag{7}$$

and had obtained the estimate

$$\dim\ker Y \le l(f) + \max(-\varkappa_1, 0) + \max(-\varkappa_2, 0) + 1.$$

We note that, besides the terms present when we had considered Y on the real line (i.e., Y_r), there is an extra term in the right hand side. This term, 1, appeared as a consequence of the weight t^{-1} in the definition of the operator of complex conjugation, that we had treated separately.

In the present paper we consider the SIO (5) with two shifts and conjugation on the unit circle (Sect. 2). We show that estimate (26) holds. We also prove that similar estimate (27) holds for the operator (7), improving the previous result for this operator. Then we consider the operator (6). We obtain estimates (33) and (34), with some additional conditions on the respective coefficients a, a_0, a_1 (Sect. 3).

2 On the Dimension of the Kernel of the Operator *K*

In this section we present an estimate for the dimension of the kernel of the SIO (5) with two shifts and conjugation.

Proposition 1 *Let* $K_1 : \widetilde{L}_2^2(\mathbb{T}) \to \widetilde{L}_2^2(\mathbb{T})$ *be the SIO with shift*

$$K_1 = N_1 P_+ + N_2 P_-,$$

where N_1, N_2, *are the functional operators*

$$N_1 = \begin{pmatrix} I & -A \\ 0 & \delta\overline{a(\beta)}I \end{pmatrix}, \quad N_2 = \begin{pmatrix} aI & 0 \\ \delta\widetilde{A} & -I \end{pmatrix},$$

$$A = \sum_{j=0}^{m} a_j U_\alpha^j, \quad \widetilde{A} = \sum_{j=0}^{m} \overline{a_j(\beta)} U_\alpha^j, \quad \delta = \frac{\beta_+}{\beta};$$

then

$$\dim\ker K = \frac{1}{2}\dim\ker K_1. \tag{8}$$

Proof Making use of the properties (4), we obtain the following relation between the operators K and K_1, similar to the Gohberg-Krupnik matrix equality (see [5]),

$$N \operatorname{diag}\{K, \widetilde{K}\} N^{-1} = K_1,$$

where

$$\widetilde{K} = P_+ + (aI - AU_\beta C)P_-,$$

and N is the following invertible operator in $\widetilde{L}_2^2(\mathbb{T})$

$$N = \frac{1}{\sqrt{2}} \begin{pmatrix} I & I \\ U_\beta C & -U_\beta C \end{pmatrix}.$$

We then have

$$\dim \ker K + \dim \ker \widetilde{K} = \dim \ker K_1.$$

Since $(iI)^{-1} K (iI) = \widetilde{K}$, we have

$$\dim \ker K = \dim \ker \widetilde{K}.$$

Thus $2 \dim \ker K = \dim \ker K_1$, i.e., (8) holds. □

Assume that $a \in C(\mathbb{T})$, $a(t) \neq 0$, everywhere on $\mathbb{T}$. Then N_1, N_2 are invertible operators, so K_1 is a Fredholm operator (see [8]).

Let us consider the operator

$$K_2 = N_2^{-1} K_1, \tag{9}$$

where

$$N_2^{-1} = \begin{pmatrix} a^{-1}I & 0 \\ \delta \widetilde{A} a^{-1} I & -I \end{pmatrix}.$$

Taking into account Proposition 1 and (9) we have the following result.

Proposition 2 *Let $K_2 : \widetilde{L}_2^2(\mathbb{T}) \to \widetilde{L}_2^2(\mathbb{T})$ be the SIO with shift defined by*

$$K_2 = \sum_{j=0}^{2m} b_j U_\alpha^j P_+ + P_-,$$

where

$$b_j = \operatorname{diag}\{1, -\delta\} \widetilde{b}_j, \;\; j = \overline{0, 2m},$$

with

$$\widetilde{b}_0 = \begin{pmatrix} a^{-1} & -a^{-1}a_0 \\ -a^{-1}\overline{a_0(\beta)} & \overline{a(\beta)} + a^{-1}a_0\overline{a_0(\beta)} \end{pmatrix},$$

$$\widetilde{b}_1 = \begin{pmatrix} 0 & -a^{-1}a_1 \\ -\alpha_+ a^{-1}(\alpha)\overline{a_1(\beta)} & \varrho_1 \end{pmatrix},$$

$$\varrho_1 = a^{-1}\overline{a_0(\beta)}a_1 + \alpha_+ a^{-1}(\alpha)a_0(\alpha)\overline{a_1(\beta)},$$

$$\widetilde{b}_2 = \begin{pmatrix} 0 & -a^{-1}a_2 \\ -\alpha_+^2 a^{-1}(\alpha_2)\overline{a_2(\beta)} & \varrho_2 \end{pmatrix},$$

$$\varrho_2 = a^{-1}\overline{a_0(\beta)}a_2 + \alpha_+ a^{-1}(\alpha)\overline{a_1(\beta)}a_1(\alpha) + \alpha_+^2 a^{-1}(\alpha_2)\overline{a_2(\beta)}a_0(\alpha_2),$$

$$\dots,$$

$$\widetilde{b}_m = \begin{pmatrix} 0 & -a^{-1}a_m \\ -\alpha_+^m a^{-1}(\alpha_m)\overline{a_m(\beta)} & \varrho_m \end{pmatrix},$$

$$\varrho_m = a^{-1}\overline{a_0(\beta)}a_m + \alpha_+ a^{-1}(\alpha)\overline{a_1(\beta)}a_{m-1}(\alpha) + \dots + \alpha_+^m a^{-1}(\alpha_m)\overline{a_m(\beta)}a_0(\alpha_m),$$

$$\widetilde{b}_{m+1} = \begin{pmatrix} 0 & 0 \\ 0 & \varrho_{m+1} \end{pmatrix},$$

$$\begin{aligned} \varrho_{m+1} =& \alpha_+ a^{-1}(\alpha)\overline{a_1(\beta)}a_m(\alpha) + \alpha_+^2 a^{-1}(\alpha_2)\overline{a_2(\beta)}a_{m-1}(\alpha_2) \\ &+ \dots + \alpha_+^m a^{-1}(\alpha_m)\overline{a_m(\beta)}a_1(\alpha_m), \end{aligned}$$

$$\dots,$$

$$\widetilde{b}_{2m} = \begin{pmatrix} 0 & 0 \\ 0 & \alpha_+^m a^{-1}(\alpha_m)\overline{a_m(\beta)}a_m(\alpha_m) \end{pmatrix}.$$

Then

$$\dim\ker K = \frac{1}{2}\dim\ker K_2.$$

Let e_n denote the $(n \times n)$ identity matrix and, for simplicity, $e \equiv e_2$.

Proposition 3 *Let* $K_3 : \widetilde{L}_2^{4m}(\mathbb{T}) \to \widetilde{L}_2^{4m}(\mathbb{T})$ *be the SIO with shift*

$$K_3 = (c_0 I + c_1 U_\alpha)P_+ + P_-, \tag{10}$$

where c_0 and c_1 are the $(4m \times 4m)$ matrix functions

$$c_0 = \begin{pmatrix} b_0 & 0 \\ 0 & e_{4m-2} \end{pmatrix}, \quad c_1 = \begin{pmatrix} b_1 & b_2 & \cdots & b_{2m-1} & b_{2m} \\ -e & 0 & \cdots & 0 & 0 \\ 0 & -e & \ddots & \vdots & \vdots \\ \vdots & \vdots & \ddots & 0 & 0 \\ 0 & 0 & \cdots & -e & 0 \end{pmatrix}. \tag{11}$$

Then

$$\dim \ker K = \frac{1}{2} \dim \ker K_3.$$

Proof We take into account the result formulated in [13, Proposition 2.3], on the real line, considering the shift $\alpha_r(t) = t + \sigma$, $t \in \overset{\circ}{\mathbb{R}} = \mathbb{R} \cup \{\infty\}$, σ is a fixed real number (see the footnote in the Introduction). In a similar way, on the unit circle, we show that

$$\dim \ker K_3 = \dim \ker K_2.$$

With Proposition 2 the result follows. □

Now we analyze the matrix function b_0 defined in Proposition 2 in more detail,

$$b_0 = \begin{pmatrix} a^{-1} & -a^{-1}a_0 \\ \delta a^{-1}\overline{a_0(\beta)} & -\delta\left(\overline{a(\beta)} + a^{-1}a_0\overline{a_0(\beta)}\right) \end{pmatrix}. \tag{12}$$

Note that $\det b_0(t) \neq 0$ for all $t \in \mathbb{T}$. So the non-singular continuous matrix function b_0 admits a right generalized factorization (1) in $L_2(\mathbb{T})$, and (2) will be assumed,

$$b_0 = b_-\Lambda b_+. \tag{13}$$

Proposition 4 *Let $a \in C(\mathbb{T})$, $a(t) \neq 0$, everywhere on $\mathbb{T}$, and c_1 be the matrix function defined by* (11). *Let b_0 be the matrix function defined by* (12), (13) *be a right generalized factorization of b_0 in $L_2(\mathbb{T})$, and $\varkappa_1$, $\varkappa_2$ its right partial indices. Then*

$$\dim \ker K \leq \frac{1}{2}(\dim \ker K_4 - 2\varkappa_1^- - 2\varkappa_2^-),$$

where $K_4 : \widetilde{L}_2^{4m}(\mathbb{T}) \to \widetilde{L}_2^{4m}(\mathbb{T})$ is the SIO with shift

$$K_4 = (I + fU_\alpha)P_+ + P_-, \tag{14}$$

f is the $(4m \times 4m)$ *matrix function*

$$f = \operatorname{diag}(\Lambda_-^{-1} b_-^{-1}, e_{4m-2}) c_1 \operatorname{diag}(b_+^{-1}(\alpha)\Lambda_+^{-1}(\alpha), e_{4m-2}), \tag{15}$$

with

$$\Lambda_\pm : \Lambda = \Lambda_- \Lambda_+, \quad \Lambda_\pm = \operatorname{diag}(t^{\varkappa_1^\pm}, t^{\varkappa_2^\pm}),$$

and

$$\varkappa_j^\pm : \varkappa_j = \varkappa_j^+ + \varkappa_j^-, \quad \varkappa_j^\pm = \frac{1}{2}(\varkappa_j \pm |\varkappa_j|), \quad j = 1, 2.$$

Proof The operator K_3 defined by (10) admits the factorization

$$K_3 = \operatorname{diag}\{b_-, e_{4m-2}\} \widetilde{K}_3 [\operatorname{diag}\{b_+, e_{4m-2}\} P_+ + \operatorname{diag}\{b_-^{-1}, e_{4m-2}\} P_-], \tag{16}$$

where

$$\widetilde{K}_3 = \left[\operatorname{diag}\{\Lambda, e_{4m-2}\} I + \widetilde{f} U_\alpha\right] P_+ + P_-,$$

with

$$\widetilde{f} = \operatorname{diag}\{b_-^{-1}, e_{4m-2}\} c_1 \operatorname{diag}\{b_+^{-1}(\alpha), e_{4m-2}\}.$$

The first and the third operators in left member of (16) are invertible, therefore

$$\dim \ker K_3 = \dim \ker \widetilde{K}_3. \tag{17}$$

Now we consider the left invertible operators

$$K_- = P_+ + \operatorname{diag}\{\Lambda_-, e_{4m-2}\} P_-, \quad K_+ = \operatorname{diag}\{\Lambda_+, e_{4m-2}\} P_+ + P_-,$$

and the operator

$$\widetilde{K}_4 = \left[\operatorname{diag}\{\Lambda_+, e_{4m-2}\} I + \operatorname{diag}\{\Lambda_-^{-1}, e_{4m-2}\} \widetilde{f} U_\alpha\right] P_+ + P_-.$$

The following equalities hold

$$\widetilde{K}_3 K_- = \operatorname{diag}\{\Lambda_-, e_{4m-2}\} \widetilde{K}_4, \tag{18}$$

$$\widetilde{K}_4 = K_4 K_+, \tag{19}$$

where K_4 is the operator defined by (14).

It follows from (18) that

$$\dim \ker \widetilde{K}_3 \le \dim \ker \widetilde{K}_4 + \dim \operatorname{coker} K_-, \tag{20}$$

and from (19)

$$\dim \ker \widetilde{K}_4 \le \dim \ker K_4. \tag{21}$$

It is known that (see [18])[2]

$$\dim \operatorname{coker} K_- = -2\varkappa_1^- - 2\varkappa_2^-. \tag{22}$$

Putting together (17), (20), (21), and (22) we obtain

$$\dim \ker K_3 \le \dim \ker K_4 - 2\varkappa_1^- - 2\varkappa_2^-.$$

It is now left to apply Proposition 3. □

Thus, it remains to estimate $\dim \ker K_4$. As usual, let $\sigma(\xi)$ and $\|\xi\|_2$ denote the spectrum and the spectral norm of a matrix $\xi \in \mathbb{C}^{n\times n}$, respectively. We will make use of some results from [9]; recall that 1 is the fixed point of the shift α defined by (3).

Lemma 1 ([9]) *For every continuous matrix function $\zeta \in C^{n\times n}(\mathbb{T})$ such that*

$$\sigma[\zeta(1)] \subset \mathbb{T}_+,$$

there exists a polynomial matrix s satisfying the conditions

$$\max_{t\in\mathbb{T}} \left\| s(t)\zeta(t)s^{-1}(\alpha(t)) \right\|_2 < 1$$

and

$$P_+ s^{\pm 1} P_+ = s^{\pm 1} P_+.$$

Let R_ζ denote the set of all such polynomial matrices s,

$$l_1(s) = \sum_{i=1}^{n} \max_{j=\overline{1,n}} l_{i,j},$$

[2] We have "2" in the left hand side of the equality because the operator acts in the space $\widetilde{L}_2^{4m}(\mathbb{T})$.

where $l_{i,j}$ is the degree of the element $s_{i,j}(t)$of the polynomial matrix s and

$$l(\zeta) = \min_{s \in R_\zeta} \{l_1(s)\}. \tag{23}$$

Lemma 2 ([9]) *Let* $T_\zeta = (I - \zeta U_\alpha)P_+ + P_- : L_2^n(\mathbb{T}) \to L_2^n(\mathbb{T})$, *where the matrix function* ζ *satisfies the conditions of the Lemma 1, and let* $l(\zeta)$ *be the number defined by* (23) *for the matrix function* ζ. *Then the following estimate holds*

$$\dim \ker T_\zeta \leq l(\zeta).$$

Suppose now that a matrix $\eta \in C^{n \times n}(\mathbb{T})$ has the properties

$$\sigma[\eta(1)] \subset \mathbb{T}_-, \quad \det \eta(t) \neq 0, \quad \forall t \in \mathbb{T}. \tag{24}$$

The non-singular continuous matrix function η admits a right generalized factorization (1) in $L_2(\mathbb{T})$, and (2) will be assumed,

$$\eta = \eta_- \Lambda \eta_+. \tag{25}$$

Lemma 3 ([9]) *Let* $T_\eta = (I - \eta U_\alpha)P_+ + P_- : L_2^n(\mathbb{T}) \to L_2^n(\mathbb{T})$, *where the matrix function* η *satisfies the conditions* (24), (25), *and let* $\kappa_1, \kappa_2, \ldots, \kappa_n$ *be its right partial indices. Let* $\widetilde{\eta} = \eta_+ \eta^{-1} \eta_+^{-1}(\alpha_{-1})$, *and* $l(\widetilde{\eta})$ *be the number defined by* (23) *for the function* $\widetilde{\eta}$. *Then the following estimate holds*

$$\dim \ker T_\eta \leq l(\widetilde{\eta}) + \sum_{\kappa_j < 0} |\kappa_j|.$$

Proposition 5 *Let the conditions of Proposition 4 be satisfied and let* K_4 *be the SIO defined by* (14). *Then*

$$\dim \ker K_4 \leq 2l(f),$$

where $l(f)$ *is the number defined by* (23) *for the matrix function* f *defined by* (15).

Proof Taking into account Lemmas 1 and 2, it suffices to show that

$$\sigma[f(1)] \subset \mathbb{T}_+.$$

From the factorization $b_0 = b_- \Lambda b_+$ of the matrix function b_0, we have

$$b_0(1) = b_-(1) b_+(1), \quad \text{so} \quad b_+^{-1}(1) = b_0^{-1}(1) b_-(1).$$

Now recalling (15), we can write

$$f(1) = \operatorname{diag}\{b_-^{-1}(1), e_{4m-2}\} c_1(1) \operatorname{diag}\{b_+^{-1}(1), e_{4m-2}\}$$

and so

$$f(1) = \operatorname{diag}\{b_-^{-1}(1), e_{4m-2}\} c_1(1) \operatorname{diag}\{b_0^{-1}(1), e_{4m-2}\} \operatorname{diag}\{b_-(1), e_{4m-2}\},$$

which means that the matrices $f(1)$ and $c_1(1) \operatorname{diag}\{b_0^{-1}(1), e_{4m-2}\}$ are similar.

From here, proceeding exactly as in the proof of [13, Proposition 2.6], we show that all the eigenvalues of the matrix

$$c_1(1) \operatorname{diag}\{b_0^{-1}(1), e_{4m-2}\} =$$

$$= \begin{pmatrix} b_1(1)b_0^{-1}(1) & b_2(1) & b_3(1) & \cdots & b_{2m-1}(1) & b_{2m}(1) \\ -b_0^{-1}(1) & 0 & 0 & \cdots & 0 & 0 \\ 0 & -e & 0 & \cdots & 0 & 0 \\ 0 & 0 & -e & \ddots & \vdots & \vdots \\ \vdots & & & \ddots & 0 & 0 \\ 0 & 0 & 0 & \cdots & -e & 0 \end{pmatrix},$$

are equal to 0. Thus $\sigma[f(1)] = \{0\}$. □

Finally, Propositions 4 and 5 allow us to establish our main result.

Theorem 1 *Let K be the SIO with two shifts and conjugation defined by* (5). *Let $a \in C(\mathbb{T})$, $a(t) \neq 0$, everywhere on $\mathbb{T}$, and $\varkappa_1$, $\varkappa_2$ be the right partial indices of the matrix function b_0 defined by* (12). *Let f be the matrix function defined by* (15), *and $l(f)$ be the number defined by* (23) *for the matrix function f. Then the following estimate holds*

$$\dim \ker K \leq l(f) + \max(-\varkappa_1, 0) + \max(-\varkappa_2, 0). \tag{26}$$

Remark 1 We note that all the results concerning the operator K, defined by (5), are valid for the operator Y, defined by (7), considering $\delta = t^{-1}$ in Proposition 1 and in the sequel. Therefore we can state the following result.

Corollary 1 *Let Y be the SIO with iterations of one shift and conjugation defined by* (7). *Let $a \in C(\mathbb{T})$, $a(t) \neq 0$, everywhere on $\mathbb{T}$, and $\varkappa_1$, $\varkappa_2$ be the right partial indices of the matrix function b_0 defined by* (12) *with $\delta = t^{-1}$. Let f be the matrix function defined by* (15), *and $l(f)$ be the number defined by* (23) *for the matrix function f. Then the following estimate holds*

$$\dim \ker Y \leq l(f) + \max(-\varkappa_1, 0) + \max(-\varkappa_2, 0). \tag{27}$$

3 On the Dimension of the Kernel of the Operator *M*

In this section we present two estimates for the dimension of the kernel of the SIO (6) with two shifts and conjugation.

Proposition 6 *Let* $M_1 : \widetilde{L}_2^2(\mathbb{T}) \to \widetilde{L}_2^2(\mathbb{T})$ *be the SIO with shift*

$$M_1 = (d_0 I + d_1 U_\alpha)P_+ + P_-,$$

where d_0 *and* d_1 *are the* (2×2) *matrix functions*

$$d_0 = \begin{pmatrix} a & a_0 \\ \overline{\rho\, a_0(\gamma)} & \overline{\rho\, a(\gamma)} \end{pmatrix}, \tag{28}$$

$$d_1 = \begin{pmatrix} 0 & a_1 \\ \overline{\rho\, a_1(\gamma)} & 0 \end{pmatrix}, \tag{29}$$

$$\rho = \frac{-\gamma_-(t)}{t\gamma(t)};$$

then

$$\dim\ker M = \frac{1}{2}\dim\ker M_1.$$

Proof Making use of the properties (4), we obtain the following relation between the operators M and M_1,

$$Z\,\mathrm{diag}\{M, \widetilde{M}\}Z^{-1} = M_1,$$

where

$$\widetilde{M} = [aI - (a_0 I + a_1 U_\alpha)U_\gamma C]P_+ + P_+,$$

and Z is the following invertible operator in $\widetilde{L}_2^2(\mathbb{T})$

$$Z = \frac{1}{\sqrt{2}}\begin{pmatrix} I & I \\ U_\gamma C & -U_\gamma C \end{pmatrix};$$

and, analogously to the proof of Proposition 1, the result follows. □

Assume that $a(t) \neq a_0(t)$, everywhere on $\mathbb{T}$. Let us analyze the matrix function d_0 defined by (28); note that $\det d_0(t) \neq 0$ for all $t \in \mathbb{T}$. So the non-singular continuous matrix function d_0 admits a right generalized factorization (1) in $L_2(\mathbb{T})$, and (2) will be assumed,

$$d_0 = d_- \Lambda d_+. \tag{30}$$

Proposition 7 *Let $a(t) \neq a_0(t)$, everywhere on $\mathbb{T}$, and d_1 be the matrix function defined by* (29). *Let d_0 be the matrix function defined by* (28), (30) *be a right generalized factorization of d_0 in $L_2(\mathbb{T})$, and $\varkappa_1$, $\varkappa_2$ its right partial indices. Then*

$$\dim \ker M \leq \frac{1}{2}(\dim \ker M_2 - 2\varkappa_1^- - 2\varkappa_2^-),$$

where $M_2 : \widetilde{L}_2^2(\mathbb{T}) \to \widetilde{L}_2^2(\mathbb{T})$ is the SIO with shift

$$M_2 = (I + gU_\alpha)P_+ + P_-,$$

g is the (2×2) *matrix function*

$$g = \Lambda_-^{-1} d_-^{-1} d_1 d_+^{-1}(\alpha) \Lambda_+^{-1}(\alpha), \tag{31}$$

with

$$\Lambda_\pm : \Lambda = \Lambda_- \Lambda_+, \quad \Lambda_\pm = \operatorname{diag}(t^{\varkappa_1^\pm}, t^{\varkappa_2^\pm}),$$

and

$$\varkappa_j^\pm : \varkappa_j = \varkappa_j^+ + \varkappa_j^-, \quad \varkappa_j^\pm = \frac{1}{2}(\varkappa_j \pm |\varkappa_j|), \quad j = 1, 2.$$

Proof The proof is similar to the proof of Proposition 4, taking into account that the operator K_4 is defined on $\widetilde{L}_2^{4m}(\mathbb{T})$ and the operator M_2 is defined on $\widetilde{L}_2^2(\mathbb{T})$. □

It remains to estimate $\dim \ker M_2$. We point out that $\det g(t) \neq 0$ for all $t \in \mathbb{T}$, where g be the matrix function defined by (31). The non-singular continuous matrix function g admits a right generalized factorization (1) in $L_2(\mathbb{T})$, and (2) will be assumed,

$$g = g_- \Lambda g_+.$$

Now let us return to the matrix function d_0 defined by (28). From the factorization $d_0 = d_- \Lambda d_+$ of d_0, we have at 1,

$$d_0(1) = d_-(1) d_+(1), \quad \text{so} \quad d_+^{-1}(1) = d_0^{-1}(1) d_-(1).$$

Now recalling (31), we can write

$$g(1) = d_-^{-1}(1)d_1(1)d_+^{-1}(1)$$

and so

$$g(1) = d_-^{-1}(1)d_1(1)d_0^{-1}(1)d_-(1),$$

which means that the matrices $g(1)$ and $d_1(1)d_0^{-1}(1)$ are similar.

The characteristic polynomial of $d_1(1)d_0^{-1}(1)$ is, with $z \in \mathbb{C}$,

$$z^2 + \frac{\overline{a_0(\gamma)}a_1 + a_0\overline{a_1(\gamma)}}{\det d_0}z - \frac{a_1\overline{a_1(\gamma)}}{\det d_0}. \tag{32}$$

Taking into account Lemmas 1, 2, and 3, with Propositions 6 and 7, we can write the following result.

Theorem 2 *Let M be the SIO with two shifts and conjugation defined by* (6). *Let $a(t) \neq a_0(t)$, everywhere on $\mathbb{T}$, and $\varkappa_1$ and $\varkappa_2$ be the right partial indices of the matrix function d_0 defined by* (28). *Let κ_1 and κ_2 be the right partial indices of the matrix function g defined by* (31). *Then the following estimates hold.*

(i) Let the functions a, a_0, a_1 be such that all the roots of the polynomial (32) *are in $\mathbb{T}_+$; then*

$$\dim \ker M \leq l(g) + \max(-\varkappa_1, 0) + \max(-\varkappa_2, 0). \tag{33}$$

where $l(g)$ be the number defined by (23) *for the matrix function g.*

(ii) Let the functions a, a_0, a_1 be such that all the roots of the polynomial (32) *are in $\mathbb{T}_-$; then*

$$\dim \ker M \leq l(\widetilde{g}) + \max(-\kappa_1, 0) + \max(-\kappa_2, 0) + \max(-\varkappa_1, 0) + \max(-\varkappa_2, 0). \tag{34}$$

where $l(\widetilde{g})$ is the number defined by (23) *for the matrix function $\widetilde{g} = g_+g^{-1}g_+^{-1}(\alpha_{-1})$.*

Remark 2 Theorem 2 can be formulated "ipsis verbis" on the real line, considering the non-Carleman shift $\alpha_r(t) = t + \sigma$, the Carleman shift $\gamma_r(t) = -t + \varsigma$, $t \in \overset{\circ}{\mathbb{R}} = \mathbb{R} \cup \{\infty\}$, σ and ς are fixed real numbers, and the operator of complex conjugation C defined by $(C\varphi)(t) = \overline{\varphi(t)}$. The shift $\alpha_r(t)$ has the only fixed point at infinity (see the footnote in the Introduction).

References

1. Baturev, A.A., Kravchenko, V.G., Litvinchuk, G.S.: Approximate methods for singular integral equations with a non-Carleman shift. J. Integral Equ. Appl. **8**(1), 1–17 (1996)
2. Clancey, K.F., Gohberg, I.: Factorization of Matrix Functions and Singular Integral Operators. Oper. Theory Adv. Appl., vol. 3. Birkhäuser, Basel, MA (1981)
3. Conceição, A.C., Marreiros, R.C.: On the kernel of a singular integral operator with non-Carleman shift and conjugation. Oper. Matrices **9**(2), 433–456 (2015)
4. Ehrhardt, T.: Invertibility theory for Toeplitz plus Hankel operators and singular integral operators with flip. J. Funct. Anal. **208**(1), 64–106 (2004)
5. Gohberg, I., Krupnik, N.: One-dimensional singular integral operators with shift. In: Convolution Equations and Singular Integral Operators. Oper. Theory Adv. Appl., vol. 205, pp. 201–211. Birkhäuser, Basel (2010)
6. Karelin, A.A., Peres Lechuga, K., Tarasenko, A.A.: The Riemann problem and singular integral equations with coefficients generated by piecewise-constant functions. Differ. Equ. **44**(9), 1225–1235 (2008)
7. Karlovich, Yu.I., Kravchenko, V.G.: Singular integral equations with non-Carleman translations on an open contour. Differ. Equ. **17**, 1408–1417 (1982)
8. Kravchenko, V.G., Litvinchuk, G.S.: Introduction to the Theory of Singular Integral Operators with Shift. Mathematics and its Applications, vol. 289. Kluwer Academic Publishers Group, Dordrecht (1994)
9. Kravchenko, V.G., Marreiros, R.C.: On the dimension of the kernel of a singular integral operator with shift. In: Operator Theory, Operator Algebras and Applications. Oper. Theory Adv. Appl., vol. 242, pp. 197–220. Birkhäuser/Springer, Basel (2014)
10. Kravchenko, V.G., Lebre, A.B., Litvinchuk, G.S.: Spectrum problems for singular integral operators with Carleman shift. Math. Nachr. **226**, 129–151 (2001)
11. Kravchenko, V.G., Lebre, A.B., Rodríguez, J.S.: Factorization of singular integral operators with a Carleman shift and spectral problems. J. Integral Equ. Appl. **13**(4), 339–383 (2001)
12. Kravchenko, V.G., Marreiros, R.C., Rodriguez, J.C.: On an estimate for the number of solutions of the generalized Riemann boundary value problem with shift. In: Differential & Difference Equations and Applications, pp. 605–614. Hindawi Publ. Corp., New York (2006)
13. Kravchenko, V.G., Marreiros, R.C., Rodriguez, J.C.: An estimate for the number of solutions of a homogeneous generalized Riemann boundary value problem with shift. In: Recent Progress in Operator Theory and Its Applications. Oper. Theory Adv. Appl., vol. 220, pp. 163–177. Birkhäuser/Springer Basel AG, Basel (2012)
14. Litvinchuk, G.S.: Boundary Value Problems and Singular Integral Equations with Shift. Nauka, Moscow (1977), in Russian
15. Litvinchuk, G.S.: Solvability Theory of Boundary Value Problems and Singular Integral Equations with Shift. Mathematics and Its Applications, vol. 523. Kluwer Academic Publishers, Dordrecht (2000)
16. Litvinchuk, G.S., Spitkovskii, I.M.: Factorization of Measurable Matrix Functions. Oper. Theory Adv. Appl., vol. 25. Birkhäuser, Basel (1987)
17. Marreiros, R.C.: On the kernel of a singular integral operator with shift. Oper. Matrices **11**(4), 1119–1139 (2017)
18. Prössdorf, S.: Some Classes of Singular Equations. North-Holland Mathematical Library, vol. 17, North-Holland Publishing, Amsterdam (1978)
19. Vekua, N.P.: Systems of Singular Integral Equations and Certain Boundary Value Problems. Nauka, Moscow (1970), in Russian
20. Vekua, I.N.: Generalized Analytic Functions, 2nd edn. Nauka, Moscow (1988), in Russian

The manufacturer's authorised representative in the EU is Springer Nature Customer Service Centre GmbH, Europaplatz 3, 69115 Heidelberg, Germany. If you have any concerns regarding our products, please contact ProductSafety@springernature.com

Printed and bound by CPI Group (UK) Ltd, Croydon, CR0 4YY
01/10/2026
02225931-0003